Catastrophic Event Response Planning

Matthew Pope, CPP with James Biesterfeld

Cover photograph courtesy of AP/Wide World

Compilation copyright © 2005 by Corinthian Colleges, Inc.
All rights reserved.

This copyright covers material written expressly for this volume by the editor/s as well as the compilation itself. It does not cover the individual selections herein that first appeared elsewhere. Permission to reprint these has been obtained by Pearson Custom Publishing for this edition only. Further reproduction by any means, electronic or mechanical, including photocopying and recording, or by any information storage or retrieval system, must be arranged with the individual copyright holders noted.

While the information contained in this book has been compiled from sources believed to be reliable and correct at the time of original publication, the publisher does not make any warranty, express or implied, with respect to the use of any techniques, suggestions and ideas disclosed in this book. The publisher disclaims any and all liability for any damages of any kind or character, including without limitation any compensatory, incidental, direct or indirect, special, punitive, or consequential damages, loss of income or profit, loss of or damage to property or person, claims of third parties, or other losses of any kind or character arising out of or in connection with the use of this book or the accuracy of the information contained herein, even if the publisher has been advised of the possibility of such damages or losses.

All trademarks, service marks, registered trademarks, and registered service marks are the property of their respective owners and are used herein for identification purposes only.

Printed in the United States of America

13 14 V0CR 14 13 12

ISBN 0-536-95416-X

2005340040

EM/LD

Please visit our web site at *www.pearsoncustom.com*

PEARSON CUSTOM PUBLISHING
75 Arlington Street, Suite 300, Boston, MA 02116
A Pearson Education Company

Table of Contents

FRONTMATTER

Introduction vii

A Conversation with Jim Biesterfield x

ONE

Planning as a Front End Concept   1

Catastrophic Events: Understanding the Threats 3

Examining Recent Disasters 6

   UN International Strategy for Disaster Reduction—Fatality Statistics 6

Analyzing the Data 9

   2005 Federal Disaster Declarations 10

   Top Ten Natural Disasters (U.S.) 12

   UN Development Programme—Assessment of the
   December 2004 Tsunami 14

   Emergency Planning: Prevention 17

FEMA's Hazard Mitigation Grant Program 25

U.S. Department of Homeland Security—Ready.gov "Ready Business" 29

   Continuity of Operations Planning 29

   Emergency Planning for Employees 31

   Emergency Supplies 32

   Review Insurance Coverage 34

   Prepare for Utility Disruptions 34

Secure Facilities, Buildings and Plants 35
Secure Your Equipment 36
Assess Building Air Protection 37
Improve Cyber Security 38

TWO
Key Elements in Response Protocol 41

Key Elements 44
Case Study: The Space Shuttle Columbia Recovery Operation 47
Case Study: "Heroism and Horror" — An Excerpt from the
9/11 Commission 55
Federal Radiological Emergency Response Plan
(FRERP) — Operational Plan 97

THREE
Planning with Other Agencies 159

City of Los Angeles — Multi-Agency Coordination 161
Standardized Emergency Management System (SEMS) 171
    What is SEMS? 171
    Executive Overview of SEMS- Kern County (CA) 173
ICS Organization 177
Incident Command: Roles & Responsibilities 178
Examples of U.S. Government Agency Response Protocols &
Multi-Agency Cooperation 181
Case Study: U.S. Environmental Protection Agency:
Inside the Emergency Response Program 182
Case Study: Safety Management in Disaster and Terrorism Response 184
Case Study: National Response Plan (NRP) 189

FOUR
Lookout, Communications, Escape and Safety (LACES) 193

LCES 195
Lookouts, Communication, Escape Routes, Safety zones 197
LCES Checklist 202

FIVE
Logistics 213

Pre-designated Incident Facilities 222
Comprehensive Resource Management 223
Disaster Mortuary Requirements 224
Evidence Gathering: FBI "Disaster Squad" 225
Triage and Disaster Planning 228
Preparing for the Media Mega-Event 231
National Preparedness and a National Health Information Infrastructure . . .237
DMAT—National Disaster Medical System 241
Communications Challenges 242
Chicago Office of Emergency Management and Communications 243
Concepts of EMS Communications 244
Logistical Planning—Lessons Learned from 9/11 .245
Community "Arks" 252
Mobile Operations Capability Guide for Emergency Managers
and Planners 263

SIX
First Responder Responsibilities 265

Job Description: Certified First Responder 267
First Responder Responsibilities 270
Roles of Emergency Services at Inter Agency Scenes 271

SEVEN
Emergency Operations Planning (EOP) vs. Incident Management System 275

Nation Incident Management System (NIMS) 277
Emergency Operations Planning 278

EIGHT
Mass Casualty/Fatality Planning 329

Mass Casualty Incidents 331
NTSB and DMORT—Key Partners in Transportation Disaster Response 340
DMORT: Triage Protocol 341

NINE

Federal Response Planning (FRP)   353

National Response Plan (NRP) 355
Concepts of Operation 356

TEN

On Scene Complication Planning   361

What Can Go Wrong—Will! 363
The Psychology of Command 364
Legal Considerations in Command and Control 377
Incident Risk Management 386
Case Study: Valuejet, May 11, 1996 401
Ethical Issues and Lessons from the SARS Outbreak 408

APPENDIX A
Northern Virginia Mass Casualty Incident Plan   415

APPENDIX B
FEMA: Roles and Responsibilities in a Terrorist Incident   467

APPENDIX C
Disaster Preparedness Guide for Incidents Involving Chemical, Biological, Radiological and Environmental Agents   537

APPENDIX D
Catastrophic Event—Equipment Requirements   557

APPENDIX E
Homeland Security Program Development Team   573

ENDNOTES
Endnotes   587

Courtesy of Corbis Images.

Introduction: Catastrophic Events

Catastrophe: (ke-'tas-tre-(,) fe) *n*.1: a great disaster or misfortune; 2; an utter failure.

The word "catastrophic" implies a devastating loss, critical disruption or total system break down.

For instance in 1996, Trans World Airlines Flight 800 from New York to Paris exploded eleven minutes into its flight, resulting in the total destruction of the 747-200 aircraft and the loss of all 230 lives aboard. The accident investigation by the National Transportation Safety Board concluded that the explosion was most likely caused by an exposed wire in the mostly empty centerline fuel tank that introduced a spark in the fuel-vapor-rich environment. The tank exploded causing a *catastrophic* failure of the aircraft. The use of the word "catastrophic" in this case is not merely used for dramatic effect, it holds investigative significance as the word denotes the complete loss of the 747 aircraft and everyone on board.

"Catastrophic" can be used to describe everything from 1999's "Melissa" computer virus that spread to hundreds of networks and completely shut down email on over 100,000 computer systems, to the Indian Ocean tsunami of December 26, 2004 that killed hundreds of thousands of people and cut a swath of devastation across multiple nations.

A catastrophic event—regardless of whether it is mechanical, technological, biological, physical, chemical, or meteorological—is generally characterized by an utter failure or total disruption. While, for instance, the phrase "catastrophic failure" could be applied to the complete crash of a personal computer's operating system after being infected with a virus, for the purposes of this textbook, we will be primarily framing the phenomenon of "catastrophe" through the lens of a large-scale destructive event—be it man-made or natural—which has the capacity to impact and destroy the lives, property, and livelihood of many.

Catastrophic Events in the New Millennium

The world grows closer together through the phenomena of globalization, the proliferation of information technology, and advances in rapid transport; and the population of the world continues to grow exponentially topping six billion near the turn of the twentieth century; sadly amidst all this progress, the capacity for catastrophic events increases as well. The more interconnected we are, the more accessible people, information and institutions become; and the more densely concentrated our community development is, the greater the likelihood that things can and will go wrong. And, when they do, tens, hundreds, thousands or even millions will be impacted. The goal and obligation of anyone involved in the study and application of domestic, civil, electronic, and national security then is to understand the principles of planning and effective response to catastrophic events.

Looking at the problem one way, the possibilities for catastrophic events are more varied and prolific today than in any time in the past. We have seen passenger airliners turned into improvised explosive devices; 100-story-plus skyscrapers built far above the reach of firefighting equipment; a global, unrestricted electronic infrastructure that is vulnerable to the transmission of malicious software code; an efficient intercontinental transportation system that is capable of spreading new strains of virus and infection across the world; nuclear, chemical, biological and high-explosive materials at risk of falling into the hands of terrorists; and densely packed urban and coastal centers threatened by all manner of manmade and natural disasters. These things all represent a modern world that is frighteningly at risk.

By the same token, however, our world has also developed and implemented important solutions toward becoming safer. Better seismic and fire prevention design in modern architecture have dramatically reduced losses in many cities from fire and earthquake; improved medical technology and processes have eradicated (to date) or significantly scaled back diseases such as polio and smallpox; new purification and sanitation methods have proved useful in much of the developed world in vastly curtailing the transmission of waterborne diseases; and security and life-safety systems can provide reasonable controls and preventive measures to the transport and use of some of the most dangerous substances on earth.

In any case, the best standard of catastrophic event management still remains that of preparation, planning, and prevention. By its very nature a catastrophe represents great loss. Thus, even the most impressive response efforts often are overshadowed by the shocking losses stemming from the initial event. Consider the events of September 11, 2001: Although almost 25,000 people managed to get out of the twin towers before they collapsed, due in part to prior training in building-evacuation procedures, it is still the loss of 2,800 first responders and victims trapped above the impact points that haunt our collective memory.

Therefore, it is imperative that we study, learn from and master that art and science of catastrophic event management—particularly in the disciplines of preparation and prevention. We must ready ourselves as both professionals and as citizens to anticipate, mitigate, and eliminate opportunities for catastrophic loss. Our technology, our processes, our education and training, and our imagination must exceed the destructive capabilities of calamity, carelessness, and chaos. This is the contribution we make to a world that continues to grow more interconnected, more complex, and in some very real ways more vulnerable every day.

A Conversation with Jim Biesterfield . . .

When it comes to predicting and preventing catastrophic events, security consultant, educator, and retired intelligence officer Jim Biesterfield has "been there and done that." Having worked in some of the most far-flung corners of the globe, tracking and interdicting some of the most dangerous personalities on Earth, Jim has become an expert in pre-incident planning and preventative analysis.

Today Jim shares with us his vast experience and knowledge with public safety and private security agencies in order to help counter-terrorism, law enforcement, and emergency management professionals prepare for and, ideally, prevent acts of catastrophic terrorism. The following is a conversation with Jim where he shares his insight and expertise in counter-terrorism and pre-incident planning.

POPE: *What are the steps/factors of a good pre-event analysis?*

BIESTERFELD: It depends on what you are analyzing. If this is an analysis of the current plan, policies, or procedures, there would be several factors in the analysis. First, we would want to determine how old the original plan is. Typically, anything older than five (5) years should be analyzed for content, changes in the climate (political, social, etc), changes in technology, and so forth. Second, once the plan has been reviewed in-house by the various agencies involved in the plan, it should be reviewed by an objective entity—inside or, preferably, outside of government. Government insiders all seem to have personal or agency agendas that can adversely impact the plan. An outside corporate entity has no such agenda to get in the way of an objective critique.

If the analysis is directed towards a catastrophic event, especially man-made, the analysis should begin with prioritization of a target list. Again, care must be given to avoid any political or agency agenda (normally involving competition for budget money). Research must be conducted to identify any priorities for potential terrorist groups (group manuals, documents, web sites, etc) to aid in prioritization and identification of potential targets.

Once this has been accomplished, a comprehensive Vulnerability Analysis (VA) of each target should be conducted. This may seem like a monumental task given that there are probably a couple of thousand potential targets in every state. However, law enforcement agencies can conduct some of these VAs, while private companies can conduct others. There is even the potential (given the ever-present political agenda) that the targets themselves might be able to conduct such a VA on itself. In this event, the VA should be reviewed by another entity to ensure that nothing was overlooked or over-stated.

The next step involves the identification of known terrorist groups operating in a particular area. It should be noted that—with the exception of Al Qaeda and the Muslim Brotherhood—most terrorist groups tend to be local or regional in scope. Thus, efforts to identify such groups' activities is somewhat simplified.

In conjunction with all of this, an evaluation of the readiness of first responders, planners, and emergency managers must be conducted. This is actually a process that should never end. Such evaluations should be conducted at least semi-annually—quarterly would be better.

This is the "nutshell" of a pre-event analysis. Within each segment, there are many and varied tasks to be performed and I will leave it to you to determine how you might want to "flesh out" your process.

POPE: *What are some pre-event indicators?*

BIESTERFELD: Right up my alley!! This is, to my mind, THE most important aspect of the terrorist timeline. I am sure that you have seen some of the surveillance tapes captured from Al Qaeda that depict buildings in New York, Washington D.C., Las Vegas, and other locales (I have some of these videos that I show in my classes). My question has always been: "Why aren't we arresting these guys while they are shooting film?" Isn't conspiracy a crime in and of itself?

The terrorist timeline comes in three phases:

- The Pre-Event (Planning) Phase
- The Event
- The Post Event Phase

While efforts must be waged within all three phases, I feel that the pre-event phase is the area that has received the least effort.

Building Blocks to an Attack

Group Membership	Fundraise	Select (acquire) Weapon
Select Target	Select Date	Conduct Reconnaissance
Move Weapon to Location	Terrorist Egress	Activate Weapon
Media Attention	Terrorist Claim of Responsibility	Undermine the Government

This Matrix was designed to identify the actions taken by terrorist groups from beginning to end, so to speak. (Sources include: FBI, CIA, NYPD, CHP, DoJ.)

In response, the following Matrix demonstrates government response to these actions:

Government Response

Identify Group	Remove Financial Support	Monitor Weapons/Materials
Threat/Bulnerability Assessment	Awareness of Important Dates	Counter Surveillance
Target Hardening	Awareness of Suspicious Behavior	Response Activities to Mitigate
Media Awareness of Terrorist Problem	Law Enforcement Investigation	Government Maintain Strong Appearance

Most importantly, the pre-event indicators are most prevalent within the first eight categories:

1. Group Identification
2. Removal of Financial Support
3. Monitor Weapons
4. Threat/Vulnerability Assessments
5. Awareness of Important Dates
6. Counter-Surveillance
7. Target Hardening
8. Awareness of Suspicious Behavior

These individual indicators can be further broken down, but basically, we must be more aware of their Intelligence Collection against targets in the U.S. and elsewhere and how Intelligence tradecraft bears on their efforts.

POPE: *What would you say are some real-world examples of either good or poor pre-event analysis?*

BIESTERFELD: Perhaps the most glaring example of poor analysis was the September 11th attacks. There were many indicators dating back to 1995 regarding such an attack and while I do not think it was preventable, there were certainly things that might have been done to either forestall the effort or at least make it more difficult to execute. (See the *9/11 Commission Report* recommendations.) Even the FBI has agents who recognized some of the elements in advance and tried to warn their superiors—to no avail.

The Columbine High School attack was another situation where pre-event indicators were overlooked or ignored. I am convinced that this event was preventable.

On the more positive aspect, some law enforcement agencies have taken steps that have interdicted similar attempts to attack schools by having tip lines that allow students to provide information of pending attacks.

However, on the whole, pre-event analysis is still lacking, but advancements are being made in some of our larger cities.

POPE: *Terrorists often "case" and gather information on targets before an attack. What are some indicators of possible terrorist surveillance?*

BIESTERFELD: There are a lot of indicators, the most important of which is keeping alert and observant to surveillances. You CAN see them; and, terrorist groups spend a lot of time conducting surveillances against targets—look at Las Vegas [*referring to the vast amount of videotape surveillance of Las Vegas casinos, hotels, and attractions which has been captured from Al Qaeda operatives*].

POPE: *You actually met the world's most notorious terrorist: Osama bin Laden. What were your impressions of that meeting?*

BIESTERFELD: I would hardly call it a meeting, more of an introduction. This occurred in 1985 in Riyadh, Saudi Arabia. I was invited to a party with the Saudi's and he was one of a couple of hundred guests. I noticed him only because of his height (he is abnormally tall for a Saudi) and inquired as to who he was. My Saudi friend identified him as Osama bin Laden [*one of the sons of Mohammed bin Laden, a Saudi billionaire and construction magnate*] and I asked for an introduction. I was introduced and spent all of about 10 seconds speaking with him. It was evident from his demeanor that he had little interest in conversing. I found him to be somewhat arrogant and condescending. Had I spent more time with him, maybe I would have found a warm spot, but this was not to be.

ONE

PLANNING AS A FRONT-END CONCEPT

1

Overview

The primary key to ensuring survival, loss mitigation, minimal disruption, and rapid recovery following any catastrophic event is the amount and quality of effective planning invested before the event. Business continuity planning, disaster planning, contingency planning, risk assessment, and continuity of operations are all relatively fashionable terms for the continually emerging art and science of organizational and community-based emergency planning.

Response is a distinct concept from planning. While preparing for an effective emergency response program is an important *element* of a valid continuity-of-operations plan, it should be regarded as only one component—distinct from the other features of disaster planning—and a component that principally should be reserved as a last resort measure.

The goals of a disaster plan are, in order of descending importance: 1) prevention, 2) mitigation, and 3) reducing the recovery curve. Good catastrophic event planning is really an update of that old bit of folk wisdom: *"An ounce of prevention is worth a pound of cure."* In other words, no amount of response excellence can compete with a single crucial act of prevention. By the time heroic first responders are rushing to a scene in order to contain damage and save lives, the damage has already been done. While perhaps not as glamorous, an attentive security officer properly detecting, identifying, reporting, and thereby eliminating an incipient risk—such as an exposed bit of wiring or a suspicious person "casing" a facility—can take that step which will later prevent firefighters from having to rush into a burning building, or police tactical units negotiating a tense standoff with an armed hostage taker.

In this chapter we will begin our review of disaster planning principles and preventative measures. We will examine the value of a good disaster plan and the general components and considerations that must go in to it. Chapter One also will review the status, history, and data of catastrophic events and the different threat varieties that they represent.

Catastrophic Events: Understanding the Threats

So what exactly are catastrophic events? As discussed during the opening pages of this text, the word "catastrophe" is frequently applied to situations in which there is calamitous loss of life, property, and/or security—a total breakdown. Understanding this then, "What events are capable of producing such a breakdown?" The list is pretty much limited only by the imagination of the student. There are no hard-and-fast rules regarding what type of an event will produce a catastrophic loss and frequently such losses are generated by a confluence of events: a seemingly innocuous series of factors that come together at the wrong place at the wrong time. That being said however, it is important to understand that there are certain circumstances in which it is both reasonable and prudent to assume that the result could be a catastrophic loss.

The following data was compiled by the United Nations International Strategy for Disaster Reduction (UNISDR)—a global body dedicated to the study and categorization of international disaster data as a means of better understanding and combating disaster losses. There are two key features of the classification of the UNISDR data. The first feature is that the major global entities are divided into the Development Groups **OECD, CEE & CIS, Developing Countries**, and **Least Developed Countries**.

- **OECD** or Organization for Economic Cooperation and Development represents what is sometimes called "the first world," meaning that member nations tend to have stable governments and economies, experienced consistent economic growth for most of their recent history, a reasonably high standard of living, and most modern amenities in terms of lifestyle, commerce, education, healthcare, public utilities, infrastructure and rule-of-law. The OECD is mostly found in North America, Western Europe, Japan, and Australia.

- The **CEE & CIS** represent Central and Eastern European nations as well as the Commonwealth of Independent States (the former Soviet Union). These are countries that have been characterized by marked democratic and economic reforms over the past decade-and-a-half and while this has done much to modernize and liberalize these nations, they are still grappling with fundamental economic and political instabilities such as widespread organized crime, corruption, tenuous democratic systems of government, internal pressures from hard-lined ideologues, and economies struggling to fully convert from command to free-market systems.

- The UNISDR classifies **"Developing Countries"** as those nations that have begun reforms and momentum toward modernization and liberalization and are predicted to, in sum or in part, emerge as greater economic and political powers in the coming years. These are nations which are beginning to more fully embrace modern technology, open their markets to imported goods and services, take a more active role in international treaties, and reduce the presence of internal destabilizing forces such as crime, corruption, repressive government, terrorism, extremism, restrictive economic policy, and public mismanagement. Examples of Developing Countries include China, Brazil, Mexico, India, Libya, Egypt, Panama, South Africa, Vietnam, The Philippines, and Chile.

- Finally, the classification of **"Least Developed Countries"** is given to those places where reform seems the furthest off and considerable work will be required to create a civic and commercial infrastructure and a workable system of government deemed viable to sustain any type of substantive foreign investment, commercial markets, industrial base or democratic processes. These countries are often hobbled by corruption, tribal and civil warfare, lawlessness, inefficiency, extremism, poor and antiquated amenities and populations virtually devoid of a middle-class. The populace of these countries consist of a very few extremely wealthy and a great many very poor. Many nations in Sub-Saharan Africa, and Central and Southeast Asia are found in this category.

While these categories may seem initially esoteric for the purposes of our study, they in fact highlight some valuable distinctions. Why, for instance, in 1989 did a magnitude 7.1 earthquake in the San Francisco Bay Area result in 51 fatalities, while a similarly strong quake in Iran killed thousands? The answer is based on many factors: architecture, governance, previous experience, and money—but the fundamental answer is basic: preparation. The Bay Area was better prepared for such a seismic event. This is not meant to imply that the citizens and civic institutions in San Francisco are wiser than those in Iran; but rather it illustrates how a community fortunate enough to have a stronger local economy, more mature and efficient government systems, a relatively sophisticated infrastructure, and planning codes and law drawn from previous relevant experience will likely fare better than a community without such advantages.

This is also not to say that San Francisco was ideally prepared either. Fifty-one fatalities is still 51 too many. Property losses totaled close to $870 million; many still remember the dramatic images of collapses along the Bay Bridge and the Nimitz Freeway and the devastating fire that roared through the city's affluent Marina district. Fifteen years later, there are still structures in the Bay Area that need to be seismically retrofitted to new standards mandated after the 1989 Loma Prieta quake. Lack of funding from year to year, political pressures, and the difficulty of retrofitting a complex urban infrastructure produce some challenges unique to a city such as San Francisco and result in its own problems with readiness.

The implication is clear: The degree of sophistication and development is likely to play a significant role in the impact of a disaster event on a society. An earthquake, tsunami, or hurricane may wreak a great deal more havoc on people living in densely crowded, coastal shanty towns, mud huts, or ancient stone structures than on modern urban centers with robust architecture and effective building safety codes. By the same token, however, a 9/11 could only have taken place in a complex urban area such as New York and Washington, DC. While diseases such as dysentery and malaria still thrive in less developed nations due to a lack of modern hygiene and sanitation facilities; the outbreak scare of severe acute respiratory syndrome (SARS) in the Pacific Rim and North America last year was attributed largely to the extensive use of modern air travel as a means of rapidly spreading such ailments throughout the modern world.

Thus when you read and interpret the following UN data, pay particular attention to what types of events tend to produce the greatest losses in the different development categories. This leads to the second important distinction of data—the event classifications. For the purposes of organizing their data, the UNISDR identifies the following four major categories:

Hydrometeor refers to severe weather events based on water, precipitation, and oceanographic phenomena.

Hurricanes	Tsunamis	Floods
Droughts	Typhoons	Serious Storms

Geological involves solid earth events and landmass shifts:

Earthquakes	Landslides	Avalanches

Biological addresses the spread of various disease and germs that are highly infectious, produce particularly egregious symptoms and that can cause mass fatalities.

Epidemics	Pandemics	Outbreaks
Viruses	Infections	Bacterial Strains

Technological is the classification for various manmade events whether they are deliberate, accidental, or cumulative that result from years of recurrent human activity.

Industrial Accidents	Hazardous Materials Releases	Large Fires
Famine Related to Civil Strife	Refugee Crises	Terrorism and War

Examining Recent Disasters

Use the following data as a means of developing an understanding of what the risks are that are presented by various disasters and how different development groups are vulnerable to different phenomena.

United Nations International Strategy for Disaster Reduction-Fatality Statistics[1]

Average number of people killed per million inhabitants in major world aggregates
1994 - 2003

Development Group	hydrometeor	geological	biological	technological
OECD	4.465	2.316	0.023	0.910
CEE+CIS	1.085	0.499	0.156	1.374
Developing countries	3.719	1.567	0.534	1.502
Least developed countries	3.399	1.386	7.809	3.075

Average number of people killed per million inhabitants

By UN Regions 1994 - 2003

Region	hydrometeor	geological	biological	technological
Africa	1.661	0.354	7.436	3.654
Americas	7.613	0.410	0.103	1.318
Asia	2.696	2.412	0.322	1.275
Europe	5.904	0.310	0.054	1.097
Oceania	1.694	7.337	0.937	2.056

By countries in the major world aggregates (1994-2003)

Source of data: EM-DAT : The OFDA/CRED International Disaster Database. http://www.em-dat.net, UCL - Brussels, Belgium

Analyzing the Data

After inspecting the UNDISR reports, what patterns did you notice? Why are deaths from hydrometeor events more prominent in the Americas and Europe than they are in less developed parts of the world? Why are there more aggregate deaths per million in Africa due to biological causes?

While the answers are complex and typically based on diverse factors, we can begin to take into account different geographical and socio-economic factors that give us a better understanding. The Americas include North, Central, and South America, including island concentrations in the Caribbean, Pacific, and South Atlantic, the low-lying Latin American isthmus, developed and densely populated peninsular regions such as Florida, the Aleutians, California Baja, and the Yucatan, and thousands upon thousands of miles of coastline. In addition, factor in the great number of mountain regions in the Western hemisphere that tend to run rampant with rivers that flood during winter rains and spring thaws.

Developing nations in Africa, on the other hand, often suffer from inadequate infrastructure due to either a lack of an economy, or pervasive governmental corruption mismanaging revenues. The result is civic sanitation and hygiene systems that are inefficient and public health structures incapable of providing adequate treatment and care. Africa also has been ravaged for years by a growing human immunodeficiency virus (HIV) epidemic that is not commonly understood by many of the people who are most at risk for exposure.

Examining the impact and pervasiveness of catastrophic events with this type of global eye begins to put in focus the responsibility of the disaster planning professional. While no form of disaster preparation can ever be completely comprehensive, by knowing and understanding which events represent the greatest risk due to geography, sociology, meteorology, and history, the planning professional will be able to take an important step forward in projecting resource requirements, developing budgets and logistical schedules, implementing appropriate policies and procedures, and designing effective safeguards.

Now that we have looked at the issue of defining catastrophic events from a global perspective, let us focus our study more on the disaster preparation needs of a specific nation: in this case, the United States.

In the United States, the primary Federal entity in the United States responsible for tracking, studying, and preparing for disasters is known as FEMA (Federal Emergency Management Agency). Since 2003, FEMA has become part of the new Department of Homeland Security (DHS). DHS responsibilities include not only its most recognized role of counterterrorism, but also the mitigation of loss potential represented by natural or manmade community-wide disasters. FEMA is the arm of DHS with the greatest responsibility for disaster relief and recovery. Working in conjunction with the executive branch, FEMA often coordinates, manages, and supports disaster recovery efforts in areas affected by earthquakes, snowstorms, hurricanes, tornados, wildfires, and floods. FEMA is also instrumental in advising the president on the status of a community affected by a disaster and helps the executive branch determine if an area should be declared a "Federal Disaster Area," meaning the local community's resources will be augmented by those of the United States government, and that U.S. tax dollars can be allocated to help local recovery and relief efforts. To illustrate how governments (e.g., the U.S.) plan for and respond to community-level catastrophes, the following study from FEMA has been provided.

2005 Federal Disaster Declarations[2]

Major Disaster Declarations

Number	Date	State	Title
1582	02/18	American Samoa	Tropical Cyclone Olaf, including High Winds, High Surf, and Heavy Rainfall
1581	02/17	Arizona	Severe Storms and Flooding
1580	02/15	Ohio	Severe Winter Storms, Flooding and Mudslides
1579	02/08	Kansas	Severe Winter Storms, Heavy Rains, and Flooding
1578	02/08	Kentucky	Severe Winter Storm and Record Snow
1577	02/04	California	Severe Storms, Flooding, Debris Flows, and Mudslides
1576	02/01	Utah	Severe Storms and Flooding
1575	02/01	Hawaii	Severe Storms and Flash Flooding
1574	02/01	West Virginia	Severe Storms, Flooding, and Landslides
1573	01/21	Indiana	Severe Winter Storms and Flooding

Number	Date	State	Title
3204	02/23	Nevada	Snow
3203	02/17	Rhode Island	Snow
3202	02/17	Nevada	Snow
3201	02/17	Massachusetts	Snow
3200	02/17	Connecticut	Snow
3199	02/01	Illinois	Snow
3198	01/11	Ohio	Snow
3197	01/11	Indiana	Snow

This information captures data from a relatively short period of time: Just January and February of 2005. Clearly there is a broad geographical area represented by the data—everywhere from American Samoa in the central Pacific Ocean to the midwest "Heartland" of Indiana and Illinois to the far northeast of New England. Considering how diverse these geographies are, and how much they vary in terms of climate and topography, one would wonder how this data would provide any useful information beyond simply cataloging events occurring in various places around the United States and its territories.

Imagine for a moment that you are a planner for FEMA, and that in your job description is the responsibility for deciding how to organize and arrange funds, logistics, and personnel for a coming year's relief or recovery operations.

- What, if any consistent themes, would you be able to extrapolate from this data?
- How would it be able to help you in your planning?
- What similarities do you see in the occurrence of catastrophic event by region?
    - Notice for example that in the Western region of the country, California, Arizona, and Utah, "severe storms and flooding" are prevalent concerns.
    - In February, three New England states qualified as disaster areas due to extremely heavy snowfall.
    - Tropical areas in the U.S—namely Hawaii and Samoa—were hit by heavy storm activity resulting in flooding and damage.
    - The Midwest was pounded by severe storms and flooding and two neighboring states, Ohio and West Virginia, were both affected by mud/landslides.

Suppose that FEMA knows that typical flood operations require X millions of dollars for local community relief and recovery, require the expertise of outside agencies such as the American Red Cross, and on average will require X number of Federal personnel to be on the ground for a set number of months. This preceding data will likely prove useful in determining, for example, how many and what type of resources will be required in the first two months of the year in order to support multiple flood sites in the Pacific, West, and Midwest. It is this type of data gathering and analysis that forms the earliest element of community-wide disaster planning. By tracking and understanding the likely needs and requirements of multiple regions, FEMA is better prepared to deploy its resources sensibly and in such a way as to maximize the efficacy of recovery and relief operations.

What is the Cost?

Catastrophic events cost a lot of money, which is one of the reasons why they are "catastrophic." This may strike some as a distasteful point to dwell on since these events typically result in mass fatalities and casualties, but the economics of catastrophes cannot be overlooked by the planning professional. Volunteers can only volunteer for so long; logistics and resources only can be donated to a finite degree. Ultimately, in a disaster event, professional rescuers must get paid, relief supplies must be procured, and fuel for helicopters, cargo planes, cranes, backhoes, and trailer trucks must be continually purchased.

Moreover, after the initial relief and recovery operations have ended, reconstruction efforts will begin and those often run in to the millions, if not billions, of dollars as armies of construction and engineering crews move in to clear out wreckage and begin re-establishing some semblance of infrastructure and sense of civic order.

Insurance payments in catastrophic events are usually on an order of magnitude that far exceeds normal expectations. A catastrophe will often result in the destruction of multiple facilities and resources, and many might be underwritten by the same insurance concerns. An insurance company may find itself having to make numerous payouts for total losses at the same time; to be sure, in the face of such payouts insurance companies will often search every legal resource in order to explore any means of avoiding or mitigating loss expenditures.

Lastly, there are the longer-term costs resulting from loss of livelihood:

- The air travel industry has languished under financial difficulties in the years following 9/11. Of the two airlines principally affected by that day's events, one had to file for Chapter 11 bankruptcy protection while the other has cut upwards of four billion dollars in operating costs in an effort to remain fiscally solvent. Lower Manhattan lost 30 percent of its usable office lease space on the morning of 9/11 and almost four years later still has not recouped most of it.

- After the 2004 Tsunami, the tourism industry that is the mainstay of Phuket and the Thai coast saw steep declines in terms of vacation bookings, hotel occupancy, and dollars being spent on local shops and restaurants. While all across the Indian Ocean hundreds of fishing villages lost the manpower, boats, and dock facilities that made up the core of their local economies.

- The U.S. hurricane season in 2004 resulted in four powerful hurricanes making landfall in Florida. The State lost billions in property, business, and local services. Interestingly, just prior to the start of hurricane season, a number of insurance companies that provided hurricane coverage to Florida residents issued notices to customers that their deductible payments were to be changed to a specific percentage of the total value of the homes. This had the effect of greatly increasing the amount of damage and recovery that homeowners had to pay out-of-pocket.

Top Ten Natural Disasters (U.S.)[3]

To give you an idea of what catastrophic events can cost, consider the following list from FEMA. This list reflects the top ten costliest natural disasters in U.S.

history in terms of the amount of FEMA dollars allocated to relief efforts. Do you notice a consistency among these disasters? What is the one thing they all have in common?

Ranked By FEMA Relief Costs

Event	Year	FEMA Funding*
Northridge Earthquake (CA)	1994	$6.967 billion
Hurricane Georges (AL, FL, LA, MS, PR, VI)	1998	$2.255 billion
Hurricane Andrew (FL, LA)	1992	$1.814 billion
Hurricane Hugo (NC, SC, PR, VI)	1989	$1.307 billion
Midwest Floods (IL, IA, KS, MN, MO, NE, ND, SD, WI)	1993	$1.140 billion
Tropical Storm Allison (FL, LA, MS, PA, TX)	2001	$1.375 billion
Hurricane Floyd (CT, DE, FL, ME, MD, NH, NJ, NY, NC, PA, SC, VT, VA)	1999	$1.054 billion
Loma Prieta Earthquake (CA)	1989	$865.8 million
Red River Valley Floods (MN, ND, SD)	1997	$741.2 million
Miami Floods (FL)	2000	$623.1 million

Amount obligated from the President's Disaster Relief Fund for FEMA's assistance programs, hazard mitigation grants, federal mission assignments, contractual services and administrative costs as of July 31, 2004. Figures do not include funding provided by other participating federal agencies, such as the disaster loan programs of the Small Business Administration and the Agriculture Department's Farm Service Agency.

Note: Funding amounts are stated in nominal dollars, unadjusted for inflation.

Of all of these disasters, the ten costliest in U.S. history to date have occurred within recent history and within twelve years of each other. U.S. insurance losses from disasters, according to FEMA, have increased exponentially every decade since the 1970s. This is largely due to increases in population and property values, and accessibility of new areas for development. Beach homes, hotels, and communities that now crowd the southeastern U.S. coast from the Gulf of Mexico, around the Florida Peninsula, up through the Carolinas, mid-Atlantic and even New England have become not only valuable pieces of real estate and property, but also prime targets for catastrophic loss during the annual hurricane and flood season from August through November. The State of California has exploded in population and development in the last 20 years and, as a result, also has become a much greater risk for losses of life, residences, and businesses from earthquake, wildfire, and mudslide

Despite the benefits realized, the rapid development of our modern world has increased not only the potential for disaster losses, but also the steep costs associated with relief, recovery, insurance payouts, and community and livelihood disruption.

UN Development Programme—Assessment of the December 2004 Tsunami[4]

Anticipating recovery needs, determining occurrence likelihood and projecting logistical, resource, manpower, and financial needs in advance of a catastrophic event are some of the essential elements of disaster planning. Another important step in the early stages of an event involves the initial assessment of damage and recovery needs. On a localized scale this process can be relatively simple. For a widespread catastrophic event, assessment will be far more complex and comprehensive. Such instances as total losses; community affect; impact on public health, livelihood, commerce, industry and infrastructure; and total recovery needs will be examined.

As a real-world illustration, the following case study has been provided. The subsequent pages are from the United Nation's Development Programme's initial assessment of conditions and recovery needs for the December 2004 South Asian tsunami disaster. Over and above being an excellent example of what is required for a catastrophic event assessment, the following study also reflects a solid examination of a catastrophic event's impact on a region.

Role of UNDP[5]

The focus of UNDP in the wake of the December 26 tsunami continues to be one of support to the Royal Thai Government and local Non-Governmental Organization in the longer-term recovery and rehabilitation efforts. UNDP's key areas of focus are: (i) community-based livelihood recovery, (ii) environmental rehabilitation and marine and coastal resource management, and (iii) promoting safe and sustainable shelter. In addition to action on the ground, UNDP is actively assisting in the coordination of international support to Thailand.

Summary Assessment

In Thailand, six southern provinces along the Andaman coastline have been severely affected by the December 26 tsunami: namely Phuket, Phang Nga, Krabi, Ranong, Trang and Satun.

The Department of Disaster Mitigation and Prevention of the Ministry of Interior recently released casualty and damage estimates from the tsunami. The casualty figures, as of February 15, stand at 5,395 dead, 8,457 injured and 3,001 missing. The severely affected areas cover 308 villages in 79 tambons (sub-districts) of 24 districts. In these areas, 12,068 households with a total of 54,672 people are considered to have been directly affected through loss of, or injury to, a family member. Phang-Nga, the most heavily impacted province, reported 61 villages in 15 tambons of 6 districts as severely affected. In Phang-Nga alone, over 4,000 households, or more than 19,000 people were directly affected. Over 3,600 houses were totally destroyed and almost 3,200 houses were partly damaged, with approximately 70% of the damage in Phang-Nga province.

Livelihoods Damaged

The disaster has seriously impeded the livelihoods of many local Thais, the majority making up populations of poorer fishing villages and local urban communities whose livelihood relies very much on tourism. The Royal Thai Government has moved quickly and effectively in providing immediate relief and temporary shelters for victims. Challenges in the next phase lie in management and coordination as well as moving towards recovery and rehabilitation with appropriate long-term planning towards sustainable livelihood.

Fisheries

It is estimated that over 120,000 people have been adversely affected by losses to the fishery sector. Nearly 500 fishing villages along the Andaman coast were seriously affected, nearly 30,000 households dependant on fisheries have lost their means of livelihood, over 4,500 fishing boats have been destroyed or damaged. The Department of Disaster Mitigation and Prevention reports that estimated damage to the Fishery sector is over 36 million dollars with half of the losses in Phang-Nga.

Livestock and Agriculture

The livestock and agricultural sectors are estimated to have sustained losses of 1.2 million and 6.2 million dollars respectively with over 90% of the losses in Phang-Nga.

Infrastructure

Damages to civil infrastructure, including roads, bridges and piers are estimated at 7.8 million dollars, with almost 70% of the damages in Phang-Nga.

Tourism

Tsunami related losses in the tourism sector have spread far beyond the directly affected coastal areas. There are reports that most of the hotels in the tsunami-hit provinces are managing to fill just 10% of their 35,000 rooms, while international passenger arrivals at Phuket international airport have reduced by 88% from the same period last year. The Tourism Ministry estimates that employment for some 200,000 workers is at risk. Another estimate by the Tourism Authority of Thailand suggests the affected provinces stand to lose more than one billion US dollars in tourism revenue, or five million tourists, in the first quarter of this year. It is further estimated that the three main tourist destinations of Phuket, Krabi and Phang-Nga would have generated over three billion U.S. dollars this year, had no disaster occurred. In Thailand, were tourism contributes 5-6% to total GDP, the estimated tsunami impact on the national economy translates into a drop in forecast GDP growth for 2005 from 6.3% to 5.6%. As one report clearly states, "People in the six Andaman sea provinces do not only need humanitarian aid but also jobs and opportunities to earn a living and rebuild their livelihoods."

Environment

At the same time, the tsunami had a very serious impact on the natural environment, with several marine and coastal national parks severely damaged, coral reefs destroyed by debris, and agricultural land affected by salt water intrusion. These environmental impacts are in turn having serious consequences for the tourism industry and people's livelihoods.

Participatory Recovery

Local governments in the affected areas are currently looking for a balance between public safety and rapid restoration of tourism activities. One proposal from the Krabi Government is for land on Phi Phi and other islands up to 15 metres above sea level to be labeled as vulnerable to tsunamis and therefore left as an open zone. According to the plan, which was submitted to the Interior Ministry recently, the risk areas will become open zones for recreational purposes while schools and hospitals must strictly be located on land higher then 15 metres above sea level. There have also been reports of illegal vendors protesting government bans on their activities in Phuket; however Phuket city issued an ordinance to keep the vendors off beaches after the tsunamis demolished unlicensed beachfront stalls. The city maintains that no vendors' stalls will be allowed back onto the beaches in the future. The rush to provide assistance to local communities has overlooked the need for a participatory approach towards disaster recovery. Some civil groups in Thailand have called on the government to adopt more community-oriented policies for assisting victims. A professor from Mahidol University is quoted as saying that "locals are getting hit by a second tidal wave, which we might call the "Tsunami of Mercy."

UNDP assessments

Joint Assessment Mission Jan. 4-8, 2005 – Livelihood Recovery and Environmental Rehabilitation

A joint UNDP, World Bank, FAO assessment of the mid- to long-term impact of the tsunami disaster and identification of areas of partnerships between Government agencies, local NGOs, and the UN Country Team in sustainable recovery of livelihoods among the local population in general, and fisheries in particular, as well as environment rehabilitation, sustainable coastal zone planning, and eco-tourism.

Joint Needs Assessment Mission Jan 10-13, 2005 – UN Country Team

A joint UNDP, UN-HABITAT, UNESCO, UNHCR, ILO, IOM and UNEP assessment mission on human settlements in Phuket and Phang Nga, the two most affected provinces. The assessment was conducted to determine needs of the government and communities, and to identify possible areas of collaboration between government agencies, civil society organizations and the UN Country Team for the recovery phase of the tsunami disaster.

Emergency Planning: Prevention

This chapter will conclude with some final examinations of preparation, and preventative and mitigation needs. So far we have examined and looked at case studies regarding the need for planning projection, recovery-cost estimation, and initial assessment. These are important steps but still are not as great as the value of creative and effective loss-mitigation solutions.

As a case study, the following examination from FEMA has been provided. This proposal studies the use of elevated homes in the coastal regions along the Gulf of Mexico as a means of mitigating losses resulting from hurricanes and flooding. This type of study employs creativity, technology, and forward-thinking in the search for solutions in preventing and/or minimizing loss potential from catastrophic events.

The first example is an overview of a pilot program from FEMA called Hazard Mitigation Grant Program (HMGP) which made funds available for innovative means of creating preventative strategies in reducing disaster losses. The study will discuss how a concept to raise several homes along the Gulf of Mexico on stilts helped prevent catastrophic home losses during a powerful hurricane.

CASE STUDY: FEMA'S HAZARD MITIGATION GRANT PROGRAM

MITIGATION

An INVESTMENT for the FUTURE[6]

Quantifying the Post - Disaster Benefits of Three Elevated Homes in Baldwin County, Alabama

Prepared by

FEMA Region IV Mitigation Division in cooperation with Alabama Emergency Management Agency and Baldwin County Emergency Management Agency

10 February 1999

Mitigation: An Investment for the Future

"The benefits of long-term hazard mitigation go beyond economics, as the reduction in vulnerability to disasters contributes to individual security, social stability and sustainable development. Nevertheless economic arguments built on a sound benefit-cost analysis are essential when one has to defend the use of scarce resources for investment in mitigation" [1]

Purpose

This report quantifies the benefits derived from elevating three homes with Hazard Mitigation Grant Program (HMGP) funds in Baldwin County, Alabama, a FEMA *Project Impact* community.

Location

Baldwin County is the largest county in Alabama with an area of 1,590 square miles. An estimated 120 miles of the county is bounded by Mobile Bay; another 60 miles faces the Gulf of Mexico. The county is particularly vulnerable to coastal storms and riverine flooding, primarily in the Fish River basin. Two of the homes are located in the floodplain of the Fish River and the third next to Mobile Bay.

Southern Alabama

Disaster Risk Reduction as a Development Strategy; Caribbean Disaster Mitigation Project; http://www.oas.org/EN/CDMP/document/lossredn.html.

History

The three residential properties experienced repetitive flood losses, most recently during Hurricane Danny in 1997. Financial assistance to elevate these buildings was provided from the HMGP after Hurricane Danny. Two of the sites experienced flooding during Hurricane Georges in September 1998 and the other came perilously close, but none of the three elevated buildings flooded.

HMGP Project Eligibility

FEMA and State Emergency Management Agencies use a software module to conduct a benefit cost analysis of mitigation projects, based on either riverine or coastal flooding parameters. The module projects savings derived from elevating properties above flood levels and is a key criterion for determining HMGP eligibility. The useful life of the project is designated and present values (PV) of the associated costs are calculated using an appropriate discount rate. Damages are calculated using a flood probability, an estimate of severity and vulnerability of a specific site.

Economic Monitoring

Direct Benefits   Until recently, it has not been common for sponsors to go back and monitor the economic return on projects, especially after subsequent floods. Opportunities presented by back-to-back floods of similar magnitude (both pre and post project) allow the actual or true benefits to be quantified on a one time basis. In this way one can validate the initial modules used to justify project funding and demonstrate to policy makers the return on investments, and in turn justify future mitigation expenditures.

Indirect Benefits   Indirect benefits, such as the operational cost savings from rescue and other emergency services, are beyond the scope of this paper, however, these savings are significant when examined from the perspective of limited emergency resources and the risk reduction to life and property. Other indirect benefits include reduced insurance premiums, fewer business interruptions and other benefits to the community at large. This report emphasizes that indirect benefits have real economic consequences to the property owner and the community. These benefits are dependent on less certain outcomes concerning health, safety, related economic issues and sociological impacts. To gain valuable insight to the indirect benefits, personal interviews with the homeowners were conducted in all three cases.

CASE 1.: SUMMERDALE, AL

History

A Baldwin County family owned a one-story home (Picture Set 1) on the Fish River. Its replacement value in 1997 was $90,002 and the contents $20,000. Flooding occurred four times between 1995 and 1997.

HMGP Project Eligibility

In 1997 a $26,585 HMGP project was approved to elevate the home four feet above the base flood (or 100-year flood) elevation (BFE). The maintenance costs over the 50-year life span were estimated to be $315 in today's dollars. Temporary housing costs based on 4 months were $2,896; making the total project cost $29,796. Present value of the benefits exceeded the project costs by a ratio of 1.14:1.

One Time Direct Savings – Post Hurricane Georges

In September 1998 what did this actually mean in dollars saved when Hurricane Georges caused the Fish River to flood two feet above the original floor elevation? The expected damages to just the residence based on the Riverine Flood Module were $34,000. Content damage and family displacement costs would have resulted in total damages of $70,191, far exceeding the original cost of the elevation project. Flood insurance premiums and coverage were adjusted to reflect the new configuration. The same premium now pays for a substantial increase in coverage.

SUMMERDALE RESIDENCE, Elevation Project 1997

Elevation Expenses

Project Cost	$26,585
PV of Maintenance Cost	315
Relocation/Displacement Cost	<u>2,896</u>
TOTAL	$29,796
Analyzed Benefit over 50 years*	
PV of Annualized Benefit	$33,983
Projected Net Benefit over Project Life	**$4,187**
Actual Benefit ** - Hurricane Georges	
Building(2000 ft^2) Damage Estimate ($65/ft^2)	$28,600
Contents ($85,000) Damage Estimate	28,050
Relocation/Displacement (126 days)	<u>3,541</u>
TOTAL	$60,191
Net Benefit Based on Hurricane Georges	**$30,395**

* Discount rate – 6%

** A two-foot flood of the structure & improved Building & Contents

Indirect Benefits – Post Hurricane Georges

Other benefits became apparent during interviews with homeowners after Hurricane Georges. During previous floods, disruptions to family life, social activities, job schedule and civic responsibilities were significant. Loss of productive work time after Hurricane Danny affected not only the homeowner but the community as well. The diversion of funds from their regular budget adversely impacted retirement plans and caused changes in other normal activities. None of these things happened after Georges because the house was high and dry and ready for occupancy as soon as the floodwaters allowed the families to return to their home.

During the recovery from previous events the family exposed themselves to potentially harmful elements of mold, pollution and other undesirable conditions. At the recommendation of the Health Department they and their helpers obtained tetanus inoculations. The stress they experienced had adverse effects on their wellbeing and caused considerable emotional distress. The disruption to their normal activities and the loss time experienced by the owners are unrecoverable. The intangible benefits from not repeating these events are significant but immeasurable.

Although they could not put a specific number to their total losses from the previous storms, the homeowners were adamant that the new enhanced feelings of security and the knowledge that their treasures would be safe was worth the cost. They also have successfully encouraged their neighbors to elevate rather than sell their property to the county in order to preserve a desirable community, maintain security, improve the tax base and consequently maintain quality services from the county....

CASE 2. :SILVERHILL, AL

History

A family of three lives approximately 800 feet from the Fish River. (Picture Set 2) Until Hurricane Danny, there had been no record of flooding. During Danny water rose so fast the family could not evacuate and was trapped. They were rescued by boat in heroic fashion by their son and neighbors. This devastating event gave the owners strong incentive to move or elevate. The building loss of nearly $55,000 was covered by insurance; however, the contents replacement value was over $58,000 of which only $20,000 was covered.

HMGP Project Eligibility

This one story residence had a floor elevation 6.0 feet below the BFE. The replacement value was $127,608 and the contents at $32,000. Because flooding from Danny exceeded the BFE by 2- 3 feet the structure was elevated four feet above the BFE. The elevation cost $44,150. Total project costs came to $48,537 including future maintenance and relocation costs. Based on a useful project life of 50 years, present value of the benefits exceeded the costs by a ratio of 2.46:1.

One Time Direct Savings – Simulated

Water from Hurricane Georges rose to within 200 feet of the building. Lower lying properties in the immediate vicinity experienced 2 – 4 feet of flooding. Since actual flooding of the property did not occur, the significant impact of the elevation is reflected in indirect benefits. However, a net benefit based on simulated flooding of two feet is summarized in the table below. Flood insurance premiums were reduced from $2,241 to $281 per year.

SILVERHILL RESIDENCE, Elevation Project 1998

Elevation Expenses	$44,150
Project Cost	315
PV of Maintenance Cost	4,072
Relocation/Displacement Cost	

TOTAL	$48,537
Analyzed Benefit over 50 years*	
PV of Annualized Benefit	$119,412
Projected Net Benefit over Project Life	**$70,875**
Benefit – Simulated Flood of Two Feet	
Building Damage Estimate ($78/ft^{2})	$28,074
Contents ($32,,000) Damage Estimate	10,560
Relocation/Displacement (126 days)	<u>4,776</u>
TOTAL	$43,410
Net Benefit Based on Simulated Flood	**(5,127)**

Discount rate – 6%

Indirect Benefits – Post Hurricane Georges

Both adults in this family teach in the public school system. After Danny, they were expected to report for work in two weeks. Their professional reference textbooks were a total loss. Without any clothes, a place to live, a vehicle and with a massive cleanup facing them, the family, with the help of volunteers and government, and tremendous inner resources, were able to start work on time. Housing was scarce and their first week was spent in a hotel....

CASE 3.:FAIRHOPE, AL

History

This house is nine miles south of Fairhope on the shore of Mobile Bay. The property is in a Velocity Flood Zone, with a BFE of 12 feet. The lowest floor was originally at 5.1 feet. The structure was damaged several times from wave action causing foundation erosion and failure. Most recently Hurricane Danny caused $6,000 in damage just to the foundation.

HMGP Project Eligibility

The pre-elevation replacement cost was $156,715. The content value was $47,015; however, the true content value at the time of the loss exceeded $100,000 due to special characteristics of the furnishings. Displacement cost was $1,085. The cost to elevate the house one-foot above BFE was $46,773. Based on a project life of 50 years the present value of benefits exceeded costs by a ratio of 1.63:1.

One Time Direct Savings – Post Hurricane Georges

Hurricane Georges created flood and wave conditions exceeding 9 feet in this area of Mobile Bay. With the original floor elevation of approximately 6 feet, substantial flood and erosion damage would have occurred to this dwelling. Because the original foundation was free standing piers resting on 2 inch thick concrete pads, it is very likely that the house would have shifted and caused significant damage. Using a depth-damage function based on 4 feet of flooding the predicted damage would have been $45,447.

Damage to furniture and appliances was conservatively estimated at $44,000. Costs for relocation/displacement were estimated at $6,932. Both the structure and contents of the elevated house escaped damage from Hurricane Georges. In this single event the cost of elevating was substantially exceeded by the damage that would have occurred.

FAIRHOPE RESIDENCE, Elevation Project 1998

Elevation Expenses

Project Cost	$46,773
PV of Maintenance Cost	315
Relocation/Displacement Cost	2,170
Additional Homeowner's Contribution	<u>6,450</u>
TOTAL	
$55,708	
Analyzed Benefit over 50 years*	
PV of Annualized Benefit	$90,725
Projected Net Benefit over Project Life	**$45,017**
Actual Benefit ** - Hurricane Georges	
Building Damage Estimate	$45,447
Contents ($100,000) Damage Estimate	44,000
Relocation/Displacement (126 days)	<u>6,932</u>
TOTAL	**$96,379**
Net Benefit Based on Hurricane Georges	**$40,671**

* Discount rate – 6%, ** Wave & Surge equated to a 4-foot flood of the structure & adjusted contents

Indirect Benefits – Post Hurricane Georges

The property owner was relieved from much of the stress and risk that comes from disaster preparation and the fear of the aftermath. In previous storms work time was lost and a great deal of personal effort was expended to put the household back to normal. The benefits from alleviating mental anguish and terrifying risks can not be underestimated.

Community Benefits

Indirect and intangible benefits have resulted from the three elevation projects that are not apparent in the casual analysis. Based on personal interviews with the residents the following list represents categories where savings have occurred or there has been a reduction of personal hardships.

Disaster housing	Use of emergency services	Other agencies support
Loss of job time	Disruption of education pursuits	Disruption of civic activities
Security risk	Disruption of social activities	Mental stress and anguish
Exposure to pollution	Exposure to unsafe environment	Diversion of funds from long range budget
Reduction in tax base	NFIP insurance claim	

Although we typically do not equate safety with monetary costs, reducing risk of injury or loss of life has positive economic effects. Emergency resources and medical care are frequently strained during disasters. Reducing the caseload for these vital services is fundamental to community recovery efforts. A provider's risk is also reduced if rescue teams do not have to be deployed.

There are few response agencies, and ultimately taxpayers, that do not gain some relief when a community fortifies against disasters. FEMA, the US Army Corp of Engineers, the Small Business Administration, the American Red Cross and the list goes on and on to include the several local and state agencies, must respond to emergencies. The benefit to any one agency because of one residential structure being elevated is small. If the value is summed over all the agencies that do not have to respond, the benefit/cost takes on a new dimension.

It must be emphasized that these three families would not have maintained the status quo of their structures in face of potential flooding. The monetary costs would be too high and the disruption and severe emotional impact would be too great to bear again. The trauma experienced by these families left an indelible mark on their lives. Without assistance to move out of harms way the distress would have been overwhelming. Notably, had the family in the third case initially known the ultimate out of pocket costs, they would not have undertaken the elevation project. When asked if they felt the same way after the elevation was complete they were emphatic that the exact opposite was true. The intangible and real benefits far outweighed the investment that they had made and they were extremely pleased with their decision.

Conclusions

The cost to elevate two of the repetitively flooded homes in Baldwin County was **$85,504.**

The direct damages avoided to buildings in these two instances on a one-time basis during Hurricane Georges was **$145,695**, a more than full recovery of the investment in four short years with a cost/benefit ratio of 1.7:1

Personal interviews with the three families identified a wide range of indirect benefits. While difficult to quantify, they are nevertheless real and significant.

Over the useful life (50 years) of the three elevation projects, benefits will continue to accrue as damage is avoided from future flooding.

Post disaster economic monitoring of mitigation projects is essential to demonstrate a quantifiable return on the initial investment and to provide communities justification to commit resources for mitigation measures.

FEMA's Hazard Mitigation Grant Program[7]

HMGP Frequently Asked Questions...

What is the Hazard Mitigation Grant Program?
Authorized under Section 404 of the Stafford Act, the Hazard Mitigation Grant Program (HMGP) administered by the Federal Emergency Management Agency (FEMA) provides grants to States and local governments to implement long-term hazard mitigation measures after a major disaster declaration. The purpose of the program is to reduce the loss of life and property due to natural disasters and to enable mitigation measures to be implemented during the immediate recovery from a disaster.

Who is eligible to apply?
Hazard Mitigation Grant Program funding is only available to applicants that reside within a Presidentially declared disaster area. Eligible applicants are

- State and local governments

- Indian tribes or other tribal organizations

- Certain non-profit organizations

Individual homeowners and businesses may not apply directly to the program; however a community may apply on their behalf.

What types of projects can be funded by the HMGP?
HMGP funds may be used to fund projects that will reduce or eliminate the losses from future disasters. Projects must provide a long-term solution to a problem, for example, elevation of a home to reduce the risk of flood damages as opposed to buying sandbags and pumps to fight the flood. In addition, a project's potential savings must be more than the cost of implementing the project. Funds may be used to protect either public or private property or to purchase property that has been subjected to, or is in danger of, repetitive damage. Examples of projects include, but are not limited to:

- Acquisition of real property for willing sellers and demolition or relocation of buildings to convert the property to open space use

- Retrofitting structures and facilities to minimize damages from high winds, earthquake, flood, wildfire, or other natural hazards

- Elevation of flood prone structures

- Development and initial implementation of vegetative management programs

- Minor flood control projects that do not duplicate the flood prevention activities of other Federal agencies

- Localized flood control projects, such as certain ring levees and floodwall systems, that are designed specifically to protect critical facilities

- Post-disaster building code related activities that support building code officials during the reconstruction process

How are potential projects selected and identified?
The State's administrative plan governs how projects are selected for funding. However, proposed projects must meet certain minimum criteria. These criteria are designed to ensure that the most cost-effective and appropriate projects are selected for funding. Both the law and the regulations require that the projects are part of an overall mitigation strategy for the disaster area.

The State prioritizes and selects project applications developed and submitted by local jurisdictions. The State forwards applications consistent with State mitigation planning objectives to FEMA for eligibility review. Funding for this grant program is limited and States and local communities must make difficult decisions as to the most effective use of grant funds....

What are the minimum project criteria?
There are five issues you must consider when determining the eligibility of a proposed project.

- Does your project conform to your State's Hazard Mitigation Plan?

- Does your project provide a beneficial impact on the disaster area—i.e. the State?

- Does your application meet the environmental requirements? ...

- Does your project solve a problem independently?

- Is your project cost-effective?

How much money is available in the HMGP?
The amount of funding available for the HMGP under a particular disaster dec-

laration is limited. The program may provide a State with up to 7.5 percent of the total disaster grants awarded by FEMA. States that meet higher mitigation planning criteria may qualify for 20 percent under the Disaster Mitigation Act of 2000.

FEMA can fund up to 75% of the eligible costs of each project. The State or grantee must provide a 25% match, which can be fashioned from a combination of cash and in-kind sources. Funding from other Federal sources cannot be used for the 25% share with one exception. Funding provided to States under the Community Development Block Grant program from the Department of Housing and Urban Development can be used to meet the non-federal share requirement.

How do I apply for the HMGP?
Following a disaster declaration, the State will advertise that HMGP funding is available to fund mitigation projects in the State. Those interested in applying to the HMGP should contact their local government to begin the application process. Local governments should contact their State Hazard Mitigation Officer.

What is the deadline for applying for HMGP funds?
Applications for mitigation projects are encouraged as soon as possible after the disaster occurs so that opportunities to do mitigation are not lost during reconstruction. The State will set a deadline for application submittal. You should contact your State Hazard Mitigation Officer for specific application dates.

How long will it take to get my project approved?
It is important for applicants to understand the approval process. Once eligible projects are selected by the State, they are forwarded to the FEMA Regional Office where they are reviewed to ensure compliance with Federal laws and regulations. One such law is the National Environmental Policy Act, passed by Congress in 1970, which requires FEMA to evaluate the potential environmental impacts of each proposed project. The time required for the environmental review depends on the complexity of the project.

Will I be forced to sell my home if my community is granted funding for an HMGP acquisition project?
Acquisition projects funded under the HMGP are voluntary and you are under no obligation to sell your home. Communities consider other options when preparing projects, but it may be determined by State and local officials that the most effective mitigation measure in a location is the acquisition of properties and the removal of residents and structures from the hazard area. Despite the effectiveness of property acquisitions, it may not make you or your family whole again. Acquisition projects are based on the principle of fair compensation for property. Property acquisitions present owners with an opportunity to recoup a large part of their investment in property that probably has lost some, if not most of its value due to damage. But, it will not compensate you or your family for your entire emotional and financial loss.

Why didn't I receive HMGP funds when some of my neighbors did?

The HMGP is administered by the State, which prioritizes and selects project applications developed and submitted by local jurisdictions. The State forwards applications consistent with State mitigation planning objectives to FEMA for eligibility review. Although individuals may not apply directly to the State for assistance, local governments may sponsor an application on their behalf. Funding for the grant program is limited and States and local communities must make difficult decisions as to the most effective use of available grant funds.

Will someone be able to rebuild and make a profit on the property I sell in an HMGP acquisition project?
Under the Stafford Act, any land purchased with HMGP funds must be restricted to open space, recreational, and wetlands management uses in perpetuity. Most often, a local government takes responsibility, but even if a State or Federal Agency takes ownership of the land, the deed restrictions still apply.

How can I get more information about the HMGP?
For more information on the Hazard Mitigation Grant Program, contact your State Hazard Mitigation Officer or the FEMA Mitigation Division in your Region.

HAZUS-MH Flood Loss Estimation Models[8]

The flood loss estimation methodology consists of two modules that carry out basic analytical processes: flood hazard analysis and flood loss estimation analysis. The hazard analysis module uses characteristics such as frequency, discharge, and ground elevation to estimate flood depth, flood elevation, and flow velocity. The loss estimation module calculates physical damage and economic loss from the results of the hazard analysis. The results are displayed in a series of reports and maps.

Users may perform three levels of analysis using the HAZUS-MH Flood Model. As noted below, the required input data and expertise vary according to the level of analysis:

Level 1
All of the information needed to produce a basic estimate of local flood losses are included as default data, based on national databases and nationally applicable methods.

Level 2
More detailed input data will be needed, including detailed information on local conditions. Modification of default databases will be required, along with the inclusion of local data and analyses.

Level 3
Detailed and site-specific input data are used to create state-of-the-art damage

estimates and situation assessment profiles. Level 3 is intended for the expert user.

The Flood Information Tool (FIT), released in 2002, is designed to process locally available flood information and convert it into data that can be used by the HAZUS-MN Flood Model. The FIT is a system of instructions, tutorials, and GIS analysis scripts. When provided with user-supplied inputs (e.g., ground elevations, flood elevations, and floodplain boundary information), the FIT calculates flood depth and elevation for riverine and coastal flood hazards. The FIT is intended to help users perform Level 2 or Level 3 flood hazard analyses. The user is allowed to input various combinations of data (i.e., default data provided with the model and community-specific data provided by the user) in order to customize the analysis.

Check here for additional information about the Flood Information Tool (FIT).

A HAZUS-MH Flood Model Patch, which works in any windows platform (NT,2000, XP), is provided on this website at www.fema.gov/hazus/flood-patch_build36.exe.

Another important example of mitigation and prevention requirements is the U.S. Department of Homeland Security's "www.ready.gov" initiative. While relatively simple and straightforward, this program seeks to provide businesses and citizens with simple suggestions for thinking about their own disaster readiness and mitigation needs.

U.S. Department of Homeland Security—www.ready.gov "Ready Business"

A successful continuity-of-operations plan is an integral component of catastrophic event readiness. A thorough assessment of an organization's contingency planning needs and recovery priorities affords key decision-makers with the capability to identify vulnerabilities and loss opportunities and to make essential, targeted preparations to minimize productivity disruptions, stem critical losses and expedite functional recovery.

To assist businesses with recovery planning and operations, the U.S. Department of Homeland Security has developed its www.ready.gov website which offers a series of useful recommendations for continuity of operations planning.

CONTINUITY OF OPERATIONS PLANNING[9]

How quickly your company can get back to business after a terrorist attack or tornado, fire or flood often depends on emergency planning done today. Start planning now to improve the likelihood that your company will survive and recover.

1. Carefully assess **how your company functions**, both internally and externally, to determine which staff, materials, procedures and equipment are absolutely necessary to keep the business operating.

    - Review your **business process flow chart** if one exists.

    - Identify **operations critical to survival** and recovery.

    - Include **emergency payroll, expedited financial decision-making and accounting systems** to track and document costs in the event of a disaster.

    - Establish procedures for **succession of management**. Include at least one person who is not at the company headquarters, if applicable.

2. Identify your **suppliers, shippers, resources and other businesses** you must interact with on a daily basis.

    - Develop **professional relationships** with **more than one** company to use in case your primary contractor cannot service your needs. A disaster that shuts down a key supplier can be devastating to your business.

    - **Create a contact list** for existing critical business contractors and others you plan to use in an emergency. Keep this list with other important documents on file, in your <u>**emergency supply kit**</u> and at an off-site location.

3. Plan what you will do if your **building, plant or store is not accessible**. This type of planning is often referred to as a continuity of operations plan, or COOP, and includes all facets of your business.

    - Consider if you can run the business from a different location or from your home.

    - Develop relationships with other companies to use their facilities in case a disaster makes your location unusable.

4. **Plan for payroll continuity**.

5. Decide **who should participate** in putting together your emergency plan.

    - Include co-workers from all levels in planning and as **active members** of the emergency management team.

    - Consider a **broad cross-section** of people from throughout your organization, but focus on those with expertise vital to daily **business functions**. These will likely include people with technical skills as well as managers and executives.

6. Define **crisis management procedures** and **individual responsibilities** in advance.

    - Make sure those involved know what they are supposed to do.

    - Train others in case you need back-up help.

7. Coordinate with others.

    - Meet with **other businesses in your building** or industrial complex.

    - Talk with first responders, emergency managers, community organizations and utility providers.

    - Plan with your suppliers, shippers and others you regularly do business with.

    - Share your plans and encourage other businesses to set in motion their own continuity planning and offer to help others.

8. **Review your emergency plans annually**. Just as your business changes over time, so do your preparedness needs. When you hire new employees or when there are changes in how your company functions, you should update your plans and inform your people.

EMERGENCY PLANNING FOR EMPLOYEES[10]

Your employees and co-workers are your business's most important and valuable asset. There are some procedures you can put in place before a disaster, but you should also learn about what people need to recover after a disaster. It is possible that your staff will need time to ensure the well-being of their family members, but getting back to work is important to the personal recovery of people who have experienced disasters. It is important to re-establish routines, when possible.

1. **Two-way communication is central** before, during and after a disaster.

    - Include emergency preparedness information in **newsletters, on company intranet, periodic employee emails** and other **internal communications** tools.

    - Consider setting up a telephone calling tree, a **password-protected page** on the company website, an email alert or a **call-in voice recording** to communicate with employees in an emergency.

- Designate an out-of-town phone number where employees can leave an "**I'm Okay**" message in a catastrophic disaster.

- Provide all co-workers with **wallet cards** detailing instructions on how to get company information in an emergency situation. Include telephone numbers or Internet passwords for easy reference.

- **Maintain** open communications where co-workers are free to bring questions and concerns to company leadership.

- Ensure you have established **staff members who are responsible for communicating** regularly to employees.

2. **Talk to co-workers with disabilities**. If you have employees with disabilities ask about **what assistance is needed**. People with disabilities typically know what assistance they will need in an emergency.

    - **Identify** co-workers in your organization with **special needs**.

    - Engage **people with disabilities in emergency planning**.

    - Ask about communications difficulties, physical limitations, equipment instructions and medication procedures.

    - Identify people willing to help co-workers with disabilities and be sure they are able to handle the job. This is particularly important if someone needs to be lifted or carried.

    - Plan **how you will alert people who cannot hear** an alarm or instructions.

3. **Frequently review and practice** what you intend to do during and after an emergency with **drills and exercises**.

EMERGENCY SUPPLIES[11]

When preparing for emergency situations, it's best to think first about the basics of survival: **fresh water, food, clean air and warmth**. Encourage everyone to have a **Portable Kit** customized to meet personal needs, such as essential medications.

1. **NOAA weather radio**

    - With tone-alert feature, if possible, that automatically alerts you when a **watch or warning** is issued in your area. Tone-alert is not available in some areas.

    - Include extra batteries.

- It is recommended that you have both a battery-powered commercial radio and a NOAA weather radio with an alert function. The NOAA weather radio can alert you to weather emergencies or announcements from the Department of Homeland Security. The commercial radio is a good source for news and information from local authorities.

2. Keep copies of **important records** such as site maps, building plans, insurance policies, employee contact and identification information, bank account records, supplier and shipping contact lists, computer backups, emergency or law enforcement contact information and other priority documents in a waterproof, fireproof portable container. Store a second set of records at an off-site location.

3. Talk to your co-workers about what **emergency supplies** the company can feasibly provide, if any, and which ones individuals should consider keeping on hand.

4. Recommended emergency supplies include the following:

    - **Water**, amounts for portable kits will vary. Individuals should determine what amount they are able to both store comfortably and to transport to other locations. If it is feasible, store one gallon of water per person per day, for drinking and sanitation

    - **Food**, at least a three-day supply of non-perishable food

    - **Battery-powered radio and extra batteries**

    - **Flashlight** and **extra batteries**

    - **First Aid kit**

    - **Whistle** to signal for help

    - **Dust or filter masks**, readily available in hardware stores, which are rated based on how small a particle they filter

    - **Moist towelettes** for sanitation

    - **Wrench** or **pliers** to turn off utilities

    - **Can opener** for food (if kit contains canned food)

    - **Plastic sheeting** and **duct tape** to "seal the room"

- **Garbage bags** and **plastic ties** for personal sanitation

REVIEW INSURANCE COVERAGE[12]

Inadequate insurance coverage can lead to major financial loss if your business is damaged, destroyed or simply interrupted for a period of time. Insurance policies vary, check with your agent or provider.

1. Meet with your insurance provider to **review current coverage** for such things as physical losses, flood coverage and business interruption.

2. Understand what it covers and what it does not.

3. Understand what your deductible is, if applicable.

4. Consider **how you will pay creditors and employees**.

5. Plan how you will provide for **your own income** if your business is interrupted.

6. Find out **what records** your insurance provider will want to see after an emergency and store them in a safe place.

PREPARE FOR UTILITY DISRUPTIONS[13]

Businesses are often dependent on electricity, gas, telecommunications, sewer and other utilities.

1. Plan ahead for **extended disruptions** during and after a disaster. Carefully examine which utilities are vital to your business's day-to-day operation. Speak with service providers about potential alternatives and identify back-up options.

2. Learn how and when to **turn off utilities**. If you turn the gas off, a professional must turn it back on. Do not attempt to turn the gas back on yourself.

3. Consider purchasing **portable generators** to power the vital aspects of your business in an emergency. Never use a generator inside as it may produce deadly carbon monoxide gas. It is a good idea to pre-wire the generator to the most important equipment. Periodically test the backup system's operability.

4. Decide **how you will communicate** with employees, customers, suppliers and others. Use cell phones, walkie-talkies, or other devices that do not rely on electricity as a back-up to your telecommunications system.

5. Plan a secondary means of **accessing the Internet** if it is vital to your company's day-to-day operations.

6. If **food storage** or refrigeration is an issue for your business, identify a vendor in advance that sells ice and dry ice in case you can't use refrigeration equipment.

SECURE FACILITIES, BUILDINGS AND PLANTS[14]

While there is no way to predict what will happen or what your business's circumstances will be, there are things you can do in advance to help protect your physical assets.

1. Install **fire extinguishers and smoke detectors** in appropriate places.

2. Locate and make available **building and site maps** with critical utility and emergency routes clearly marked.

   - Plan to provide a copy to fire fighters or other first responders in the event of a disaster.

   - Keep copies of these documents with your emergency plan and other important documents in your **emergency supply kit**.

3. Consider if you could benefit from **automatic fire sprinklers, alarm systems, closed circuit TV, access control, security guards** or other security systems.

4. **Secure ingress and egress.** Consider all the ways in which people, products, supplies and other things get into and leave your building or facility.

   - **Plan for mail safety.** The nation's battle against terrorism takes place on many fronts, including the mailrooms of U.S. companies. A properly informed and well-trained work force can overcome such threats.

   a. Teach employees to be able to **quickly identify suspect packages and letters**. Warning signs include:

   - Misspelled words

   - No return address

   - Excessive use of tape

   - Strange discoloration or odor

   b. The United States Postal Service suggests that **if a suspect letter or** package is identified:

- **Don't open, smell, touch or taste**
- Immediately **isolate** suspect packages and letters
- **Move out of the area** and don't let others in
- Quickly **wash with soap and water** and remove contaminated clothing
- **Contact** local law enforcement **authorities**

   c. Post emergency numbers for easy reference.

5. Identify what production machinery, computers, custom parts or other **essential equipment** is needed to keep the business open.

   - Plan how to replace or repair vital equipment if it is damaged or destroyed.
   - Identify more than one supplier who can replace or repair your equipment.

6. **Store extra supplies**, materials and equipment for use in an emergency.

7. Plan what you will do if your **building, plant or store is not usable**.

   - Consider if you can run the business from a different location or from your home.
   - Develop relationships with other companies to use their facilities in case a disaster makes your location unusable.

8. Identify and **comply with all local, state and federal codes** and other safety regulations that apply to your business.

9. Talk to your insurance provider about what impact any of these steps may have on your policy.

SECURE YOUR EQUIPMENT[15]

The force of some disasters can damage or destroy important equipment.

1. Conduct a **room-by-room walk-through** to determine what needs to be secured.

2. Attach **equipment and cabinets** to walls or other stable equipment.

3. Place **heavy or breakable objects on low shelves**.

4. **Move workstations** away from large windows, if possible.

5. **Elevate** equipment off the floor to avoid electrical hazards in the event of flooding.

ASSESS BUILDING AIR PROTECTION[16]

In some emergencies **microscopic particles** may be released into the air. For example, earthquakes often can release dust and debris into the air. A biological attack may release germs that can make you sick. And a dirty bomb can spread radioactive particles. Many of these things can only hurt you if they **get into your body**. A building can provide a **barrier** between contaminated air outside and people inside, but there are ways to **improve building air protection**.

Depending on the size of the building and the design and layout of the Heating, Ventilating and Air-Conditioning (HVAC) system, there may be simple steps building owners and managers can take to help protect people from some airborne threats. If you rent or lease your space, speak to the building owners and managers about HVAC maintenance. Ask if there are options for improving building air protection.

1. **Know the Heating, Ventilating and Air-Conditioning (HVAC) system**.

   - Building owners or managers, and employers should take a close look at the site's system and be sure it is **working properly and is well maintained**.

   - Be sure any security measures **do not adversely impact air quality or fire safety**.

2. Develop and practice **shut-down procedures** for the HVAC system.

3. **Secure outdoor air intakes**. HVAC systems can be an entry point and means of distributing biological, chemical and radiological threats.

   - **Limit access to air intake locations** to protect the people inside a building from airborne threats. Air intakes at or below ground level are most vulnerable because anyone can gain easy access.

   - Consider **relocating or extending** an exposed air intake, but do not permanently seal it.

4. **Determine if you can feasibly upgrade the building's filtration system**.

   - **Increasing filter efficiency** is one of the few things that can be done in advance to **consistently protect people** inside a building from biological and some other airborne threats.

- Carefully consider the **highest filtration efficiency** that will work with a building's HVAC system.

5. **HEPA (High Efficiency Particulate Arrester) Filter Fans**. These individual units have highly efficient filters that can capture very tiny particles, including many biological agents. Once trapped within a HEPA filter, contaminants cannot get into your body and make you sick. While these filters are excellent at filtering dander, dust, molds, smoke, many biological agents and other contaminants, they will not stop chemical gases.

IMPROVE CYBER SECURITY[17]

Protecting your data and information technology systems may require specialized expertise. Depending on the particular industry and the size and scope of the business, cyber security can be very complicated. However, even the smallest business can be better prepared.

Every computer can be **vulnerable to attack**. The consequences of such an attack can range from simple inconvenience to financial catastrophe. While a thief can only steal one car at a time, a single hacker can cause damage to a large number of computer networks and **can wreak havoc on both your business and the nation's critical infrastructure.**

Start with these simple steps:

1. **Use anti-virus software and keep it up-to-date**.

    - **Activate the software's auto-update** feature to ensure your cyber security is always up-to-date. Think of it as a regular flu shot for your computer to **stop viruses in their tracks!**

2. **Don't open email from unknown sources**.

    - Be suspicious of **unexpected emails that include attachments** whether they are from a known source or not.

    - When in doubt, **delete the file and the attachment**, and then **empty your computer's deleted items file**.

3. **Use hard-to-guess passwords**.

    - Passwords should have at least **8 characters with a mixture of uppercase and lowercase letters as well as numbers.**

- **Change** passwords frequently.

- **Do not give** your password to anyone.

4. **Protect your computer from Internet intruders by using firewalls.**

    - There are two forms of firewalls: **software firewalls** that run on your personal computer, and **hardware firewalls** that protect computer networks, or groups of computers.

    - Firewalls **keep out unwanted or dangerous traffic** while allowing acceptable data to reach your computer.

    - **Don't share access to your computers with strangers.**

    - Check your computer operating system to see if it allows others to access your hard-drive. Hard-drive access can open up your computer to infection.

    - Unless you really need the ability to share files, your best bet is to do away with it.

5. **Back up your computer data.** Many computer users have either already experienced the pain of losing valuable computer data or will at some point in the future. Back up your data regularly and consider keeping one version off-site.

6. **Regularly download security protection updates known as patches.** Patches are released by most major software companies to cover up security holes that may develop in their programs.

    - Regularly download and install the patches yourself, or check for automated patching features that do the work for you.

7. **Check your security on a regular basis.**

    - When you change your clocks for **Daylight Saving Time**, evaluate your computer security. The programs and operating system on your computer have security settings that you can adjust.

    - Do you have multiple door locks and a high-tech security system at your office? It could be that **tighter security for your computer system** is also what you need.

8. **Make sure your co-workers know what to do if your computer system becomes infected.**

- **Train employees** on how to update virus protection software, how to download security patches from software vendors, and how to create a proper password.

- **Designate a person** to contact for more information if there is a problem.

9. **Subscribe to** the Department of Homeland Security **National Cyber Alert System**, to receive free, timely alerts on new threats and learn how to better protect your area of cyberspace.

- US-CERT is a partnership between DHS and the public and private sectors. It was established to protect the Nation's Internet infrastructure through coordinated defense against and responses to cyber attacks.

Endnotes

[1] http://www.unisdr.org/disaster-statistics/impact-killed.htm

[2] http://www.fema.gov/news/disasters.fema

[3] http://www.fema.gov/library/df_8.shtm

[4] http://www.undp.or.th/tsunami/tsunami.htm

[5] Need to add source material reference here!

[6] http://www.fema.gov/doc/hazards/mit_baldwin.doc

[7] http://www.fema.gov/fima/hmgp/faqs.shtm

[8] http://www.fema.gov/hazus/fl_main.shtm

[9] http://www.ready.gov/business/st1-planning.html

[10] http://www.ready.gov/business/st1-empwellbeing.html

[11] http://www.ready.gov/business/st1-emersupply.html

[12] http://www.ready.gov/business/st3-reviewins.html

[13] http://www.ready.gov/business/st3-prepareutility.html

[14] http://www.ready.gov/business/st3-securefac.html

[15] http://www.ready.gov/business/st3-secureequip.html

[16] http://www.ready.gov/business/st3-assessair.html

[17] http://www.ready.gov/business/st3-improvecyber.html

TWO

KEY ELEMENTS IN RESPONSE PROTOCOL

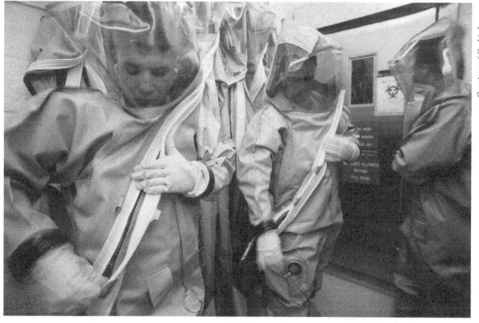

2

Overview

Realistically preventative measures cannot eliminate the possibility of every disaster. Should there be the occurrence of a catastrophic event, a response scenario will be generated which may require the allocation of many resources, assets, skills, and capabilities. Frequently the very nature of a catastrophic incident will demand a response effort that is comprised of multiple agencies and segments of the community. This requires a high level of planning, coordination, and control that we will begin to discuss in this chapter.

To define...

Before discussing the elements of a response protocol it may be useful to the student to understand some of the key differences in commonly used terms.

Response: The reaction triggered by an event or occurrence. In regards to emergency response, this term is generally applied to mean a community or organization's initial actions employed to mitigate, stabilize, and contain a disaster. Although simplistic and incomplete, the image that often characterizes this phase is one of "first responders" such as police, fire, and ambulance services racing to a scene.

Rescue: The actual effort of locating injured victims and initiating emergency treatment and care. In some circumstances this may also be known as a "search" phase.

Relief: Involves latter stage operations after the primary stages of response and rescue. Relief is typically thought of as such things as furnishing needed food, water, shelter and logistics, medicines, convalescence, communications with family, and psychological and spiritual counseling.

Recovery: Can be thought of as the "clean-up" and return to normalization stage. Recovery is when long term rebuilding and reconstruction begins. Recovery also has a sadder connotation as well. When a search operation moves from "rescue" to "recovery," it generally means that the assessment of professional rescuers is that they feel there is no longer much hope of rescuing a living victim, but are now more focused on recovering bodies.

Key Elements

The nature of a catastrophic event is that it is likely to overwhelm local resources. The capability of any one agency or community to sustain response, rescue, relief, and recovery operations on its own will be exceeded. Certain factors will have to be considered when preparing a large-scale response, rescue, relief, and recovery effort appropriate for a catastrophe level event. The following pages will begin an examination of these elements.

Multiple agencies

Catastrophic events typically trigger the need for various different responding agencies. Some commonly involved agencies and organizations are as follows:

- *Law Enforcement*: Such as local police, county sheriffs, state police, and highway patrols; law enforcement agencies may provide such services as scene security, assisting with evacuation, communications, tactical response, air support, search resources, explosives disposal, and traffic and crowd control.

- *Fire Department*: Fire suppression, emergency medical services, evacuation, hazardous material containment, communications, and urban search and rescue.

- *Emergency Medical Services*: Paramedics, emergency medical technicians, disaster medical teams, emergency room personnel—doctors, nurses, surgeons, physician's assistants; EMS can triage, evacuate, treat and mitigate injuries, diagnose symptoms, and identification of the possible presence of chemical, biological, or radiological contaminants.

- *Private and Facility Security*: Contract and private security functions employed by corporate, private, residential, and public facilities; assist with evacuation, communications, scene security, first-aid and medical assistance, and crowd and traffic control. In a community-wide catastrophic event, public safety agencies will be heavily taxed, if not completely overwhelmed. Corporate facilities, shopping centers, school and college campuses, and other private properties will most likely have to provide for their own emergency response, evacuation, first aid, shelter, logistics and communications without assistance from local public safety agencies, thus relying on their own internal, private security, and disaster relief functions.

- *Public Utilities*: Power, lights and electricity, civil engineering, public works, telecommunications, water treatment; utilities and public works personnel are as much a part of a community-wide disaster response and recovery effort as are fire, police, and other more commonly-identified emergency responders. Public utilities will often furnish the specialized training and equipment required to secure hazards such as downed power-lines, ruptured gas lines and broken water mains; flood control; hazardous materials containment and clean-up; building safety assessment; and removing debris, downed trees and vegetation

- *Construction, Engineering, and Technical Specialists*: Private construction and engineering firms may be called upon to play a key role in catastrophic event response. Iron and steel workers, welders, heavy equipment operators, truck drivers, crane operators may be needed to move heavy debris, clear paths blocked to rescuers; assist in search and rescue-particularly where there is widespread structural damage and collapse. Other tradespersons, such as electricians, plumbers, and carpenters may be needed to secure damaged systems and construct ramps, temporary roads, and structures for rescue crews. Finally, engineers and investigators from private contractors also may be required at the scene of a catastrophic event. In the case of a transportation accident for example, engineers involved in the design and manufacture of the vehicle involved may be called upon to consult, help determine root causes, and determine best response options.

- *Government Agencies and Non-Governmental Organizations (NGOs)*: Such as FEMA and National Guard; federal investigators from the FBI, BATFE, NTSB, DOT, EPA, and USCG, for example; international organizations from the UN or non-governmental disaster relief specialists such as International Red Cross/ Red Crescent; Salvation Army; Doctors without Borders, for example. These varying agencies all bring considerable talents and resources to disaster relief and recovery–everything from establishing temporary hospitals and shelters to distribution of relief supplies, logistics, immunizations, food, water treatment, and communications.

- *Convergent Volunteers*: Disasters, particularly community-level catastrophes can evoke a powerful outpouring from community members wishing to help, as well as activating skilled volunteers specifically trained and equipped for such response. Volunteer organizations may include corporate or community emergency response teams, local shortwave radio clubs, Civil Air Patrol, US Coast Guard Auxiliary, and Sheriff's Search and Rescue. Teams. A rush of volunteers can lend badly needed manpower and equipment to relief and recovery efforts, but also may sometimes present a challenge to on-scene commanders who may find a rush of spontaneous volunteers overwhelming.

Coordination

As we have determined, a catastrophic event will usually require a diverse number of responders, equipment, skills, and resources. Organizing, coordinating, and communicating these various elements is a substantial undertaking. In later chapters, we will begin examining the Standardized Emergency Management System (SEMS), which is a principal specifically designed to manage large-scale response and recovery operations. For this rudimentary discussion, however, we will just touch upon the topic of multi-agency and event coordination. The following are some key elements you should consider including in any large-scale catastrophic response plan.

- *Logistics*: What equipment is required? Logistical coordination may constitute a broad range of equipment ranging from blankets, medicines, communications equipment, potable water, and portable generators up to earth moving equipment, heavy-life cranes and helicopters. In a multi-agency response, coordinating the resources of multiple agencies under the aegis of a single unified command structure, and ensuring that resources are properly utilized and accounted becomes the primary logistical challenge.

- *Delegation and Decision Making*: Unified command concepts such as SEMS and ICS (Incident Command System) were born out of multi-agency catastrophic event responses in which unclear command or jurisdictional authority surfaced as a factor prohibiting an effective and coordinated response effort. The function of an incident commander (IC) and command staff is to establish a centralized response coordinator who will be recognized by all responders as the delegating and decision-making authority. Typically the selection of a "lead agency" for incident command will be predicated on factors such as jurisdiction, experience, expertise, or resources.

- *Communications*: Responders must be able to talk to one another; while this would seem to be a fairly obvious point, its importance cannot be understated. On 9/11 over 340 first responders were lost largely because radio equipment used by police, fire, and port authority personnel was not sufficiently interoperable. Critical observations from police helicopters were not able to be passed along to NYFD ICs; the command to evacuate the towers in advance of a collapse was not received by rescuers still in the buildings; even something as minor as the failure to correctly activate a high rise repeater station meant that NYFD chiefs were unable to talk to each other. As much as possible, communications systems need to be compatible, interoperable, and durable in order to meet the particular demands of a catastrophic event response.

- *Planning*: Any effective response protocol has to involve prior planning. First responders, community members, and local, state, and Federal authorities must conduct prior planning in order to determine and reinforce initial actions and response priorities. Building, hospital, and school evacuation drills; tabletop simulations; computer modeling; and large-scale disaster response exercises involving multiple agencies are valuable methods of identifying catastrophic event priorities. In a high rise building fire, for example, a building population that has been properly trained on, and able to self-initiate correct evacuation procedures is perhaps the best tool for minimizing injuries and fatalities in such an event. Prior planning will also cut down on confusion among responding agencies. If general ideas already have been established regarding response priorities and agency responsibilities, this can save valuable time and effort during the initial response and rescue phases of an event.

- *Public Information*: During the lessons on SEMS, we will examine the role of a public information officer (PIO) more thoroughly but it is important to note here that a key element of response protocol preparation is keeping the media, various authorities, and the general public adequately informed. Good control and coordination of public information helps reduce anxiety and panic, ensures that accurate information and instructions are presented to those who are potentially affected, and can even be useful in enlisting the public's help in avoiding certain areas, not tying up telecommunications networks, or enlisting volunteers. Effective communications with local, state, and Federal authorities can help facilitate the allocation of additional resources and good governmental cooperation.

- *Administration and Finance*: Lastly, a large-scale response effort is going to be an administrative and accounting nightmare if tight and effective controls are not employed early and implemented consistently throughout the event. Resources and logistics must be tracked and accounted for; usage and burn rates predicted and run to ensure that adequate supply chains are maintained—and the money supply cannot be allowed to run out! While not always the most exciting aspect of event response and recovery, effective administration and finance is integral to managing operations, ensuring adequate resource allocation, proper return of logistics to contributing agencies, accounting of finances, and the overall well being of ongoing recovery efforts. Post 9/11 many unfortunate, disappointing, and wasteful revelations came to light about relief funds mismanagement. This provides as excellent example of how significant the role of administration and finance is.

Now that we have spent some time focusing on the basic elements of a large-scale, coordinated catastrophic event response protocol, let's examine a case study of these principles in action. In early 2003, America and the world suffered the second catastrophic loss of a STS Space Shuttle orbiter. The recovery of this loss was a massive, coordinated multi-agency effort requiring a substantial investment of responders, resources, and finances.

Case Study: The Space Shuttle Columbia Recovery Operation[1]

On January 16, 2003, the Space Shuttle Columbia and her crew of seven launched on the 113th Shuttle flight from the Kennedy Space Center in Florida to begin a sixteen-day scientific research mission. On February 1, 2003, Columbia broke up upon re-entry over the western United States at an altitude of 200,000 feet and a speed of Mach 18. The break up resulted in a debris field that extended approximately 250 miles from near Dallas, Texas to Fort Polk, Louisiana. The hazardous nature of the shuttle material, the imperative to recover Columbia's crew, and the criticality of evidence to the investigation of the cause of this tragic accident prompted a substantial intergovernmental, interagency response.

Over the 100 days that followed, over 450 government agencies, private companies, and nonprofit organizations and 25,000 people would participate in what became the largest search and recovery effort in history.

On February 2nd, FEMA and NASA agreed that the mission of the response would include four primary goals: 1. Protect the public, 2. Recover the crew, 3.

Retrieve the evidence, and 4. Provide public assistance (i.e. financial assistance to local governments). To meet these goals, the operation proceeded in two discernable phases. The first phase began on February 1st, when the shuttle broke up, and continued until February 14th when the crew had been recovered.

During this phase a large number of local citizens and responders (many of whom were volunteers), the Texas Forest Service, the Texas department of Public Safety, the Texas Army National Guard, USFS, and many other agencies cooperated with NASA and the EPA to mitigate known hazards from material that had fallen in public areas and to search for the crew.

After the immediate public hazards were mitigated and the crew [was] recovered, attention turned to the retrieval of the evidence necessary to support the independent Columbia Accident Investigation Board's (CAIB's) determination of the cause of the accident. The second phase involved comprehensive searches on the ground, by air, and under water to find and retrieve shuttle material. This ground and air searches were directed by the NASA Mishap Investigation Board and conducted by the wild land fire service, which was activated under ESF-4 of the FRP. The Texas Forest Service fulfilled the role of state area command, and delegated authority to a total of twenty Type-1 and Type-2 IMT's that were ordered by the Texas Interagency Center (TICC) over the course of the 100-day incident to manage a grid search of a corridor four miles wide along the orbiter's flight path using hand crews provided from 44 states. Four base camps were established along the length of the search corridor, each run by an IMT and populated by 20-50 crews. In addition, a camp operated in Longview, TX to coordinate the mobilization and demobilization of the 597 crews that were deployed over the course of the incident. Thus, at any given time during the incident, there were five IMT's on the ground.

With the help of Urban Search and Rescue Teams, crews were trained to deploy and search with a 75 percent probability of detecting shuttle material. The IMT's worked closely with NASA engineers and EPA technicians--in many cases forming a unified command structure--to conduct the search, recover, and document the material that was discovered. The search conditions presented unique challenges and safety concerns for the fire service. For example, cold and wet weather, poison ivy, snakes, and thick brambles dictated logistical support requirements that were unusual for fire crews (such as more substantial shelter, chaps, and dry boots). In addition, the combination of contract crews and personnel from agencies outside the fire service that were trained to different standards, unevenly prepared for searching in difficult terrain, and housed at separate locations, required the IMT's and crew leadership to accommodate variable performance levels.

Ultimately, the ground search covered an area of over 680,000 acres, an area approximately equivalent in size to the state of Rhode Island. Over 84,000 pounds of material were recovered, about 38% of the landing-configured weight of the orbiter. The material recovered permitted reconstruction of key components of the shuttle. This, coupled with forensic evidence from the material, was central to the determination of the cause and nature of the break up. The search also produced particular items that were of critical importance to the investigation (such as the on-board Modular Auxiliary Data System recorder that contained important information from the sensors aboard the vehicle, and cabin video that was shot as the crew began reentry), as well as some of the research experiment payloads that produced useful scientific data.

IV. OPERATIONS AND MANAGEMENT FINDINGS

This section presents the findings of this study pertinent to the four research questions posed in Section II, above, and is organized around these questions. Much of what is presented here is substantiated with more detailed (and less redacted) data contained in Appendices 3 and 4.

Lessons from the shuttle Columbia recovery operation

The IMT's were asked to reflect on their experience during the shuttle Columbia recovery operation, and to identify factors that helped to make the operation was successful, as well as problems or challenges that should be addressed in future all-risk responses. Reported below are the consolidated and summarized points raised by the IMT's.

1. Mission focus.

By the time most of the IMT's began to be deployed (beginning at the end of the second week), the response itself and the relationships among the lead agencies were well organized. The IMT's operated in specific roles to support an unambiguous mission (the ground search for shuttle material) and to fulfill clear objectives. The teams praised the mission-focused nature of the effort. In particular, they noted that there was an unusual lack of jurisdictional tensions (so-called "turf battles"). Moreover, success was measured by what was accomplished: As one member explained, "It was about mission focus– identifying problems and solving them, not focusing on how the problem developed."

It should be noted that those who were "on the ground" at the beginning of the incident, and who comprised an ad hoc IMT, experienced a great deal of chaos and confusion during the first several days. This was partially attributed to a leadership vacuum that arose because FEMA was unaccustomed to its role as lead agency (rather than supporting a state government). This was also displayed in the mission assignments FEMA wrote, in which the IMT's reported that "the language was not clear and concise." Participants suggested that future incidents would succeed better if leaders were identified up front and agency roles and responsibilities were clarified early. Particularly in incidents that involve active threats to life and property, addressing leadership concerns quickly is vital to success.

2. Role clarity and empowerment.

Many members praised the fact that the teams and the agencies both seemed to understand their roles well, to be empowered to fulfill them, and to be open to working together. One mentioned that "Agencies were open to feedback and allowed the IMT's to have discretionary input." Several commented that one reason the search effort was so successful is that on this incident they were asked to use their full incident management capacity, rather than being limited to a supporting role, as they often are during non-wildfire incidents. Furthermore, they were allowed to employ their own incident command system, with which they are comfortable because it is "tried and true after 33 years."

One participant summarized, "Everyone had their role and was empowered to manage it, unlike working with FEMA in the past where there has been tug of war. This time we sensed no tug-of-war with FEMA. We credit the Texas Forest Service with helping this." In fact, most applauded the clear delegation of authority from the Texas Forest Service, which was a key step in defining roles. Others did, however, note the lack of a clear area command organization, since the Disaster Field Office seemed to fill this function, in addition to the roles of multi-agency group and unified incident command. Of this unusual arrangement, one member commented, "We adapted, but it was tough." In addition, it was noted that there is no national coordinating authority for all-risk, as there is for wildfires.

One characteristic of success the IMT's often mentioned is flexibility–the ability to adapt to unique circumstances and requirements. This requires sufficient discretion and independence, as was granted on the Columbia recovery. Some mentioned that this approach worked better than circumstances where they are required to conform to FEMA guidelines, described as "not functional." One participant explained, "FEMA fell into a support role and recognized the capabilities of the IMT's. Generally, this is the other way around. I found FEMA more accommodating this time." As another described it, "We were multi-tasked–appropriately handed the whole ball of wax to manage in the geographic area we were responsible for. We are trained and to multi-task; that's why we come with resources, logistics, etc., and FEMA does not understand that. FEMA has a tough time understanding what all the IMT can bring to the table." In this case, FEMA was flexible enough to allow the IMT's to apply the full range of their expertise.

3. *Continuity.*

In a long-duration incident, continuity of operations can be difficult to maintain, particularly as an incident progresses from its fast-paced and chaotic early stages, often dominated by a multitude of local agencies, to more established and predictable later stages. On the Columbia incident, the upper management was the same for the entire 100 days of the operation, which served an important stabilizing function as the incident objectives shifted. Some participants noted that, because of this dynamic climate, the IMT's need to be more in-tune with the initial phases of the incident and to monitor the effectiveness and appropriateness of the direction they are given as circumstances evolve. One IC explained, "A good IMT anticipates a few days ahead. This allows being proactive versus reactive, and provides feedback on performance." On a related point, several teams mentioned that the transitions from to worked well, and that the operation could still be running seamlessly. They credit their training in transitions, their established relationships, a clear transition process, and the uniform framework under which they operate for the smoothness with which fresh personnel could be brought in without any compromise to the operation.

4. *Communications.*

Several respondents recognized the vital role that good communications played in several dimensions of the Columbia incident. At a fundamental level, many agencies did not understand the structure, capabilities, and operation of the wildland fire service. The Texas Area Command is an organization established to: 1. Oversee the management of multiple incidents that are each being handled by an Incident Command System; or 2. Oversee the management of a very large incident that has multiple Incident Management Teams assigned to it.

Forest Service served as an intermediary to help the agencies understand one another and work together. At a command level, the teams mentioned how important it was that they were provided the information they needed to conduct their assigned missions. At an operational level, agencies successfully coordinated with each other using the daily planning framework ICS provides. Finally, at a physical level, interagency relationships benefited from co-location.

5. *Resources.*

One set of issues about which the teams made substantial comments was the availability and appropriateness of resources. Many complimented the capability of the Texas Interagency Center, which was able to interact effectively with the national resource distribution system. It was also pointed out that the IMT's had the support they needed to take good care of their crews and sustain their morale under working conditions to which they were not normally accustomed.

While the IMT system is adept at integrating and employing all types of resources, many of the teams attributed the constraints on resource availability to the fact that the wildland firefighting resource management system is not well-adapted to all-risk requirements. Several noted that the system is not prepared to handle requests 24 hours a day, 7 days a week, every day of the year. Many also commented about the level of support from their state and local agencies, reporting that agency administrators are reluctant to make personnel available for non-wildfire operations throughout the year. Some believe that their administrators do not feel a sense of urgency about non-wildfire incidents, and so did not want to release resources. Finally, some proposed that the length of commitment called for in the national mobilization guide (14 days) is not always suitable to meet all-risk requirements and could be extended to thirty days as some incidents might require.

In addition, the differences between purchasing and reimbursement authority for fire versus all-risk incidents caused considerable consternation for the teams, particularly with respect to the employment of state and private personnel. Teams also found it difficult to find the necessary people because name-requests or requests for personnel paid on a "portal to portal" basis were not honored initially, and because pay constraints with respect to both "base 8" and overtime kept local organizations from being willing to send personnel they would otherwise commit.

With respect to equipment, one proposed that a FEMA logistics liaison be assigned to the IMT's to provide clear purchasing guidance, information about what resources are available in the national warehouses, and better management of accountable property.

Early in the response, portal-to-portal name requests were not honored in an effort by managers at the DFO to control costs. Later in the incident, when critical personnel shortages arose, such requests were honored. Often IMT's prefer the flexibility of name requests and portal-to-portal compensation because these incentives make it easier for them to obtain the people they want. On the other hand, portal-to-portal compensation is relatively expensive, so fiscally responsible managers use this approach judiciously. Name requests can slow the ordering process because positions are not open to be filled by the first available resource in the system. When critical shortages arise, however, name requests can help to resolve them.

6. *Community relations.*

Teams were uniformly positive about their interactions with local communities during the Columbia incident. They were enthusiastic about the high level of local support and generosity. The teams were warmly welcomed by local residents, which was especially important since the search depended on access to a large amount of privately-owned land. At the same time, the teams worked hard to identify and address the needs and concerns of the communities where they were operating. Even absent a directive to do so, teams often developed a unified command approach with local public officials. As one member observed, "The combination of our team's sensitivity and the local support made the incident successful."

7. *Special concerns of all-risk incidents.*

The IMT's pointed out several issues raised by Columbia that apply to all-risk incidents more broadly. Many asserted that the national teams have the experience and expertise to manage all-risk responses, and that these assignments are a natural extension of their wildland firefighting incident management capacity. They recognize, however, that many all-risk incidents have characteristics that require technical expertise and resources that wildland IMT's do not habitually include. This suggests a need for all-risk incident analysis to characterize the complexity of each incident--and, moreover, to understand how multiple incidents interact with one another. Such analysis would permit the right teams to be assigned and appropriately tailored for the particular requirements of the circumstances they will face. One complication that is likely to arise in terrorist incidents is access to analysis, information, intelligence, which may be constrained by security concerns.

Another important concern is human resources management in environments where personnel from several agencies operating under different rules, regulations, and professional codes of conduct are convened to work together as teams. Several examples of the difficulties posed by such circumstances arose on the Columbia incident, where EPA contractors and contract crews were not bound by the same physical fitness and training requirements as the regular fire crews were. Additionally, contractors did not report to the same chain of command, and lived under a different set of rules governing acceptable personal behavior, such as with regard to drinking and meeting attendance. This problem was exacerbated by personnel shortages, which increased the number of contractors and contract crews involved.

The characteristics of successful incident management

The IMT's were asked to confer with their fellow members to provide a written list of the characteristics of successful incident management. The full, unedited list of the responses they provided is included in Appendix 4 [of the full report]. What follows is a synopsis of the most salient points they made as they expounded on their responses during the subsequent group discussion:

1. *Command and leadership*

The participants described several important characteristics of commanders and the command function. Three central points were reiterated by many of the IMT's:

First, leaders should have a problem-solving orientation, which requires them to be able to work independently to identify and react to problems by obtaining and deploying whatever tools are necessary. Successful problem-solving particularly involves the ability to anticipate requirements. As one person said, "plan three days ahead, but take the incident one day at a time."

Second, effective command demands the ability to be adaptable to rapidly changing conditions, and the flexibility to be responsive to unexpected events while maintaining focus on the primary mission. Often, bureaucratic procedures threaten the discretion necessary to be agile. Incident managers on the ground should be afforded adequate authority to solve the problems that would prevent the accomplishment of the mission.

Third, these qualities should be incorporated into a standard organizational model or protocol that sets the parameters for decision-making and operations. The wildland fire service uses its own well-developed and carefully communicated version of the Incident Command System, that specifies important dimensions of incident management, including the span of control, chain of command, authorities, functions, and planning processes. Having a common model like ICS promotes stability and consistency, so that the management can create order from the chaos brought by crisis events. The key is that the model needs to be shared by all agencies participating in the incident to reap its full benefits.

The IMT's also emphasized the value of trust, which allows the to function well internally, and allows it to interact with outside actors, such as the community, other agencies, and contractors. Building trust is a complex process that can be aided by several important actions: the display of professionalism; the exercise of social and political sensitivity to the needs, concerns, and circumstances of all involved in the incident; the use of continuous, honest feedback; and the demonstration that the safety and welfare of all personnel involved in the incident is of primary importance.

In the end, IMT's made the point that there is no substitute for qualified, experienced, competent, committed leaders willing to make tough choices and decisions. Developing leaders with appropriate knowledge, skills, and abilities requires a well-developed standardized professional training and mentorship program. Moreover, it requires a commitment to staff positions based on specific qualifications, not on rank. (Though rank and qualification may be correlated, they do not automatically correspond to one another.)

2. *Dynamics.*

Several participants pointed out that complex incidents demand that numerous personnel and a diverse array of competencies be brought together to mitigate the threats and hazards at hand. Facilitating collaboration among these parties and marshaling them to become a well-functioning, effective, compatible is at the core of successful incident management. Incident managers must therefore be skilled at developing positive interpersonal and interagency interactions, and at promoting a identity that subsumes individual egos.

As cohesiveness takes time to develop and incidents are characterized by urgency, incident management benefits greatly from the employment of pre-established teams with stable membership that habitually work together. Such teams have synergies that are hard to develop when teams are assembled hurriedly for a single incident or even only episodically. To reap these benefits, members must work together frequently and over a long period.

Regardless of the team's composition, however, all teams function best when they share a vision of what their mission and priorities are, how they will fulfill them, what role each member plays, and what responsibilities each member has. A shared vision emerges most powerfully when members view themselves as participating partners in developing it, and when they feel themselves to be accountable for upholding their responsibilities.

3. *Communications and interpersonal relationships.*

Leadership and effectiveness depend, in turn, on consistent, open, and clear, communication and with all involved. The IMT's defined communication broadly, to include timely transmission of accurate information, active listening, and providing and soliciting constructive criticism. They assume that communication must occur among all parties connected to an incident, including responders, public officials, citizens, and the media. They identified several tools that can enhance communications, as follows:

- Using a planning process that proceeds according to a set cycle.
- Conducting good incident briefings at regular intervals and careful briefings prior to and during transitions.

- Recognizing constraints on communications and working to overcome them.
- Establishing logistical support for communications up front.
- Making members reasonably accessible to others.

4. *Mission, objectives, and operations.*

The IMT's universally asserted that a shared, well-defined mission is at the core of successful incident management. The ability to specify the mission requires that the incident be well understood and that active jurisdictions and authorities be well-defined relative to it. If the incident managers discover areas where the mission, objectives, and priorities are unclear, unattainable, or participating agencies do not agree about them, they must work to generate clarity and explicit consensus. Moreover, they must be vigilant about maintaining clarity and consensus as the response evolves. To be successful in this, managers need to understand who their "customers" are and what the customers perceive their needs to be. This demands that managers be informed about and sensitive to the political, social, financial ramifications of the incident and the response.

In the wake of the terrorist attacks upon the U.S. on 9/11, a special commission was empanelled to conduct an exhaustive investigation of the circumstances which led to and allowed for the success of the attacks. Among the most heartbreaking and informative sections of the report is Chapter 9, "Heroism and Horror" which describes the multi-agency rescue efforts carried out that day in New York, Washington, DC, and Pennsylvania. The following study will illustrate the challenges of rescue efforts and coordination in a significant catastrophic event when time, magnitude, and a lack of sufficient prior planning are all factors working against the efforts of responders. This section should help illustrate for students the daunting challenge of coordinating a multi-agency response effort under the most stressful conditions and in the face of almost unimaginable horror.

Case Study: "Heroism and Horror"—An Excerpt from the 9/11 Commission[2]

HEROISM AND HORROR

9.1 PREPAREDNESS AS OF SEPTEMBER 11

Emergency response is a product of preparedness. On the morning of September 11, 2001, the last best hope for the community of people working in or visiting the World Trade Center rested not with national policymakers but with private firms and local public servants, especially the first responders: fire, police, emergency medical service, and building safety professionals.

Building Preparedness The World Trade Center. The World Trade Center (WTC) complex was built for the Port Authority of New York and New Jersey. Construction began in 1966, and tenants began to occupy its space in 1970. The Twin Towers came to occupy a unique and symbolic place in the culture of New York City and America.

The WTC actually consisted of seven buildings, including one hotel, spread across 16 acres of land. The buildings were connected by an underground mall (the concourse). The Twin Towers (1 WTC, or the North Tower, and 2 WTC, or the South Tower) were the signature structures, containing 10.4 million square feet of office space. Both towers had 110 stories, were about 1,350 feet high, and were square; each wall measured 208 feet in length. On any given workday, up to 50,000 office workers occupied the towers, and 40,000 people passed through the complex.[1]

Each tower contained three central stairwells, which ran essentially from top to bottom, and 99 elevators. Generally, elevators originating in the lobby ran to "sky lobbies" on higher floors, where additional elevators carried passengers to the tops of the buildings.[2]

Stairwells A and C ran from the 110th floor to the raised mezzanine level of the lobby. Stairwell B ran from the 107th floor to level B6, six floors below ground, and was accessible from the West Street lobby level, which was one floor below the mezzanine. All three stairwells ran essentially straight up and down, except for two deviations in stairwells A and C where the staircase jutted out toward the perimeter of the building. On the upper and lower boundaries of these deviations were transfer hallways contained within the stairwell proper. Each hallway contained smoke doors to prevent smoke from rising from lower to upper portions of the building; they were kept closed but not locked. Doors leading from tenant space into the stairwells were never kept locked; reentry from the stairwells was generally possible on at least every fourth floor.[3]

Doors leading to the roof were locked. There was no rooftop evacuation plan. The roofs of both the North Tower and the South Tower were sloped and cluttered surfaces with radiation hazards, making them impractical for helicopter landings and as staging areas for civilians. Although the South Tower roof had a helipad, it did not meet 1994 Federal Aviation Administration guidelines.[4]

The 1993 Terrorist Bombing of the WTC and the Port Authority's Response. Unlike most of America, New York City and specifically the World Trade Center had been the target of terrorist attacks before 9/11. At 12:18 P.M. on February 26, 1993, a 1,500-pound bomb stashed in a rental van was detonated on a parking garage ramp beneath the Twin Towers. The explosion killed six people, injured about 1,000 more, and exposed vulnerabilities in the World Trade Center's and the city's emergency preparedness.[5]

The towers lost power and communications capability. Generators had to be shut down to ensure safety, and elevators stopped. The public-address system and emergency lighting systems failed. The unlit stairwells filled with smoke and were so dark as to be impassable. Rescue efforts by the Fire Department of New York (FDNY) were hampered by the inability of its radios to function in buildings as large as the Twin Towers. The 911 emergency call system was overwhelmed. The general evacuation of the towers' occupants via the stairwells took more than four hours.[6]

Several small groups of people who were physically unable to descend the stairs were evacuated from the roof of the South Tower by New York Police Department (NYPD) helicopters. At least one person was lifted from the North Tower roof by the NYPD in a dangerous helicopter rappel operation- 15 hours after the bombing. General knowledge that these air rescues had occurred appears to have left a number of civilians who worked in the Twin Towers with the false impression that helicopter rescues were part of the WTC evacuation plan and that rescue from the roof was a viable, if not favored, option for those who worked on upper floors. Although they were considered after 1993, helicopter evacuations in fact were not incorporated into the WTC fire safety plan.[7]

To address the problems encountered during the response to the 1993 bombing, the Port Authority spent an initial $100 million to make physical, structural, and technological improvements to the WTC, as well as to enhance its fire safety plan and reorganize and bolster its fire safety and security staffs.[8]

Substantial enhancements were made to power sources and exits. Fluorescent signs and markings were added in and near stairwells. The Port Authority also installed a sophisticated computerized fire alarm system with redundant electronics and control panels, and state-of-the-art fire command stations were placed in the lobby of each tower.[9]

To manage fire emergency preparedness and operations, the Port Authority created the dedicated position of fire safety director. The director supervised a team of deputy fire safety directors, one of whom was on duty at the fire command station in the lobby of each tower at all times. He or she would be responsible for communicating with building occupants during an emergency.[10]

The Port Authority also sought to prepare civilians better for future emergencies. Deputy fire safety directors conducted fire drills at least twice a year, with advance notice to tenants. "Fire safety teams" were selected from among civilian employees on each floor and consisted of a fire warden, deputy fire wardens, and searchers. The standard procedure for fire drills was for fire wardens to lead co-workers in their respective areas to the center of the floor, where they would use the emergency intercom phone to obtain specific information on how to proceed. Some civilians have told us that their evacuation on September 11 was greatly aided by changes and training implemented by the Port Authority in response to the 1993 bombing.[11]

But during these drills, civilians were not directed into the stairwells, or provided with information about their configuration and about the existence of transfer hallways and smoke doors. Neither full nor partial evacuation drills were held. Moreover, participation in drills that were held varied greatly from tenant to tenant. In general, civilians were never told not to evacuate up. The standard fire drill announcement advised participants that in the event of an actual emergency, they would be directed to descend to at least three floors below the fire. Most civilians recall simply being taught to await the instructions that would be provided at the time of an emergency. Civilians were not informed that rooftop evacuations were not part of the evacuation plan, or that doors to the roof were kept locked. The Port Authority acknowledges that it had no protocol for rescuing people trapped above a fire in the towers.[12]

Six weeks before the September 11 attacks, control of the WTC was transferred by net lease to a private developer, Silverstein Properties. Select Port Authority employees were designated to assist with the transition. Others remained on-site but were no longer part of the official chain of command. However, on September 11, most Port Authority World Trade Department employees-including those not on the designated "transition team"- reported to their regular stations to provide assistance throughout the morn-ing. Although Silverstein Properties was in charge of the WTC on September 11, the WTC fire safety plan remained essentially the same.[13]

Preparedness of First Responders
On 9/11, the principal first responders were from the Fire Department of New York, the New York Police Department, the Port Authority Police Department (PAPD), and the Mayor's Office of Emergency Management (OEM).

Port Authority Police Department. On September 11, 2001, the Port Authority of New York and New Jersey Police Department consisted of 1,331 officers, many of whom were trained in fire suppression methods as well as in law enforcement. The PAPD was led by a superintendent. There was a separate PAPD command for each of the Port Authority's nine facilities, including the World Trade Center.[14]

Most Port Authority police commands used ultra-high-frequency radios. Although all the radios were capable of using more than one channel, most PAPD officers used one local channel. The local channels were low-wattage and worked only in the immediate vicinity of that command. The PAPD also had an agencywide channel, but not all commands could access it.[15]

As of September 11, the Port Authority lacked any standard operating procedures to govern how officers from multiple commands would respond to and then be staged and utilized at a major incident at the WTC. In particular, there were no standard operating procedures covering how different commands should communicate via radio during such an incident.

The New York Police Department. The 40,000-officer NYPD was headed by a police commissioner, whose duties were not primarily operational but who retained operational authority. Much of the NYPD's operational activities were run by the chief of department. In the event of a major emergency, a leading role would be played by the Special Operations Division. This division included the Aviation Unit, which provided helicopters for surveys and rescues, and the Emergency Service Unit (ESU), which carried out specialized rescue missions. The NYPD had specific and detailed standard operating procedures for the dispatch of officers to an incident, depending on the incident's magnitude.[16]

The NYPD precincts were divided into 35 different radio zones, with a central radio dispatcher assigned to each. In addition, there were several radio channels for citywide operations. Officers had portable radios with 20 or more available channels, so that the user could respond outside his or her precinct. ESU teams also had these channels but at an operation would use a separate point-to-point channel (which was not monitored by a dispatcher).[17]

The NYPD also supervised the city's 911 emergency call system. Its approximately 1,200 operators, radio dispatchers, and supervisors were civilian employees of the NYPD. They were trained in the rudiments of emergency response. When a 911 call concerned a fire, it was transferred to FDNY dispatch.[18]

The Fire Department of New York. The 11,000-member FDNY was headed by a fire commissioner who, unlike the police commissioner, lacked operational authority. Operations were headed by the chief of department- the sole five-star chief.[19]

The FDNY was organized in nine separate geographic divisions. Each division was further divided into between four to seven battalions. Each battalion contained typically between three and four engine companies and two to four ladder companies. In total, the FDNY had 205 engine companies and 133 ladder companies. On-duty ladder companies consisted of a captain or lieutenant and five firefighters; on-duty engine companies consisted of a captain or lieutenant and normally four firefighters. Ladder companies' primary function was to conduct rescues; engine companies focused on extinguishing fires.[20]

The FDNY's Specialized Operations Command (SOC) contained a limited number of units that were of particular importance in responding to a terrorist attack or other major incident. The department's five rescue companies and seven squad companies performed specialized and highly risky rescue operations.[21]

The logistics of fire operations were directed by Fire Dispatch Operations Division, which had a center in each of the five boroughs. All 911 calls concerning fire emergencies were transferred to FDNY dispatch.[22]

As of September 11, FDNY companies and chiefs responding to a fire used analog, point-to-point radios that had six normal operating channels. Typically, the companies would operate on the same tactical channel, which chiefs on the scene would monitor and use to communicate with the firefighters. Chiefs at a fire operation also would use a separate command channel. Because these point-to-point radios had weak signal strength, communications on them could be heard only by other FDNY personnel in the immediate vicinity. In addition, the FDNY had a dispatch frequency for each of the five boroughs; these were not point-to-point channels and could be monitored from around the city.[23]

The FDNY's radios performed poorly during the 1993 WTC bombing for two reasons. First, the radios signals often did not succeed in penetrating the numerous steel and concrete floors that separated companies attempting to communicate; and second, so many different companies were attempting to use the same point-to-point channel that communications became unintelligible.[24]

The Port Authority installed, at its own expense, a repeater system in 1994 to greatly enhance FDNY radio communications in the difficult high-rise environment of the Twin Towers. The Port Authority recommended leaving the repeater system on at all times. The FDNY requested, however, that the repeater be turned on only when it was actually needed because the channel could cause interference with other FDNY operations in Lower Manhattan. The repeater system was installed at the Port Authority police desk in 5 WTC, to be activated by members of the Port Authority police when the FDNY units responding to the WTC complex so requested. However, in the spring of 2000 the FDNY asked that an activation console for the repeater system be placed instead in the lobby fire safety desk of each of the towers, making FDNY personnel entirely responsible for its activation. The Port Authority complied.[25]

Between 1998 and 2000, fewer people died from fires in New York City than in any three-year period since accurate measurements began in 1946. Fire-fighter deaths-a total of 22 during the 1990s-compared favorably with the most tranquil periods in the department's history.[26]

Office of Emergency Management and Interagency Preparedness. In 1996, Mayor Rudolph Giuliani created the Mayor's Office of Emergency Management, which had three basic functions. First, OEM's Watch Command was to monitor the city's key communications channels-including radio frequencies of FDNY dispatch and the NYPD-and other data. A second purpose of the OEM was to improve New York City's response to major incidents, including terrorist attacks, by planning and conducting exercises and drills that would involve multiple city agencies, particularly the NYPD and FDNY. Third, the OEM would play a crucial role in managing the city's overall response to an incident. After OEM's Emergency Operations Center was activated, designated liaisons from relevant agencies, as well as the mayor and his or her senior staff, would respond there. In addition, an OEM field responder would be sent to the scene to ensure that the response was coordinated.[27]

The OEM's headquarters was located at 7 WTC. Some questioned locating it both so close to a previous terrorist target and on the 23rd floor of a building (difficult to access should elevators become inoperable). There was no backup site.[28]

In July 2001, Mayor Giuliani updated a directive titled "Direction and Control of Emergencies in the City of New York." Its purpose was to eliminate "potential conflict among responding agencies which may have areas of overlapping expertise and responsibility." The directive sought to accomplish this objective by designating, for different types of emergencies, an appropriate agency as "Incident Commander." This Incident Commander would be "responsible for the management of the City's response to the emergency," while the OEM was "designated the 'On Scene Interagency Coordinator.'"[29]

Nevertheless, the FDNY and NYPD each considered itself operationally autonomous. As of September 11, they were not prepared to comprehensively coordinate their efforts in responding to a major incident. The OEM had not overcome this problem.

9.2 SEPTEMBER 11, 2001

As we turn to the events of September 11, we are mindful of the unfair perspective afforded by hindsight. Nevertheless, we will try to describe what happened in the following 102 minutes:

- the 17 minutes from the crash of the hijacked American Airlines Flight 11 into 1 World Trade Center (the North Tower) at 8:46 until the South Tower was hit

- the 56 minutes from the crash of the hijacked United Airlines Flight 175 into 2 World Trade Center (the South Tower) at 9:03 until the collapse of the South Tower

- the 29 minutes from the collapse of the South Tower at 9:59 until the collapse of the North Tower at 10:28

From 8:46 until 9:03 A.M. At 8:46:40, the hijacked American Airlines Flight 11 flew into the upper portion of the North Tower, cutting through floors 93 to 99. Evidence suggests that all three of the building's stairwells became impassable from the 92nd floor up. Hundreds of civilians were killed instantly by the impact. Hundreds more remained alive but trapped.[30]

Civilians, Fire Safety Personnel, and 911 Calls North Tower. A jet fuel fireball erupted upon impact and shot down at least one bank of elevators. The fireball exploded onto numerous lower floors, including the 77th and 22nd; the West Street lobby level; and the B4 level, four stories below ground. The burning jet fuel immediately created thick, black smoke that enveloped the upper floors and roof of the North Tower. The roof of the South Tower was also engulfed in smoke because of prevailing light winds from the northwest.[31]

Within minutes, New York City's 911 system was flooded with eyewitness accounts of the event. Most callers correctly identified the target of the attack. Some identified the plane as a commercial airliner.[32]

The first response came from private firms and individuals-the people and companies in the building. Everything that would happen to them during the next few minutes would turn on their circumstances and their preparedness, assisted by building personnel on-site.

Hundreds of civilians trapped on or above the 92nd floor gathered in large and small groups, primarily between the 103rd and 106th floors. A large group was reported on the 92nd floor, technically below the impact but unable to descend. Civilians were also trapped in elevators. Other civilians below the impact zone-mostly on floors in the 70s and 80s, but also on at least the 47th and 22nd floors-were either trapped or waiting for assistance.[33]

It is unclear when the first full building evacuation order was attempted over the public-address system. The deputy fire safety director in the lobby, while immediately aware that a major incident had occurred, did not know for approximately ten minutes that a commercial jet had directly hit the building. Following protocol, he initially gave announcements to those floors that had generated computerized alarms, advising those tenants to descend to points of safety-at least two floors below the smoke or fire-and to wait there for further instructions. The deputy fire safety director has told us that he began instructing a full evacuation within about ten minutes of the explosion. But the first FDNY chiefs to arrive in the lobby were advised by the Port Authority fire safety director-who had reported to the lobby although he was no longer the designated fire safety director-that the full building evacuation announcement had been made within one minute of the building being hit.[34]

Because of damage to building systems caused by the impact of the plane, public-address announcements were not heard in many locations. For the same reason, many civilians may have been unable to use the emergency intercom phones, as they had been advised to do in fire drills. Many called 911.[35]

The 911 system was not equipped to handle the enormous volume of calls it received. Some callers were unable to connect with 911 operators, receiving an "all circuits busy" message. Standard operating procedure was for calls relating to fire emergencies to be transferred from 911 operators to FDNY dispatch operators in the appropriate borough (in this case, Manhattan).Transfers were often plagued by delays and were in some cases unsuccessful. Many calls were also prematurely disconnected.[36]

The 911 operators and FDNY dispatchers had no information about either the location or the magnitude of the impact zone and were therefore unable to provide information as fundamental as whether callers were above or below the fire. Because the operators were not informed of NYPD Aviation's determination of the impossibility of rooftop rescues from the Twin Towers on that day, they could not knowledgeably answer when callers asked whether to go up or down. In most instances, therefore, the operators and the FDNY dispatchers relied on standard operating procedures for high-rise fires-that civilians should stay low, remain where they are, and wait for emergency personnel to reach them. This advice was given to callers from the North Tower for locations both above and below the impact zone. Fire chiefs told us that the evacuation of tens of thousands of people from skyscrapers can create many new problems, especially for individuals who are disabled or in poor health. Many of the injuries after the 1993 bombing occurred during the evacuation.[37]

Although the guidance to stay in place may seem understandable in cases of conventional high-rise fires, FDNY chiefs in the North Tower lobby determined at once that all building occupants should attempt to evacuate immediately. By 8:57, FDNY chiefs had instructed the PAPD and building personnel to evacuate the South Tower as well, because of the magnitude of the damage caused by the first plane's impact.[38]

These critical decisions were not conveyed to 911 operators or to FDNY dispatchers. Departing from protocol, a number of operators told callers that they could break windows, and several operators advised callers to evacuate if they could.[39] Civilians who called the Port Authority police desk located at 5 WTC were advised to leave if they could.[40]

Most civilians who were not obstructed from proceeding began evacuating without waiting for instructions over the intercom system. Some remained to wait for help, as advised by 911 operators. Others simply continued to work or delayed to collect personal items, but in many cases were urged to leave by others. Some Port Authority civilian employees remained on various upper floors to help civilians who were trapped and to assist in the evacuation.[41]

While evacuating, some civilians had trouble reaching the exits because of damage caused by the impact. Some were confused by deviations in the increasingly crowded stairwells, and impeded by doors that appeared to be locked but actually were jammed by debris or shifting that resulted from the impact of the plane. Despite these obstacles, the evacuation was relatively calm and orderly.[42]

Within ten minutes of impact, smoke was beginning to rise to the upper floors in debilitating volumes and isolated fires were reported, although there were some pockets of refuge. Faced with insufferable heat, smoke, and fire, and with no prospect for relief, some jumped or fell from the building.[43]

South Tower. Many civilians in the South Tower were initially unaware of what had happened in the other tower. Some believed an incident had occurred in their building; others were aware that a major explosion had occurred on the upper floors of the North Tower. Many people decided to leave, and some were advised to do so by fire wardens. In addition, Morgan Stanley, which occupied more than 20 floors of the South Tower, evacuated its employees by the decision of company security officials.[44]

Consistent with protocol, at 8:49 the deputy fire safety director in the South Tower told his counterpart in the North Tower that he would wait to hear from "the boss from the Fire Department or somebody" before ordering an evacuation.[45] At about this time, an announcement over the public-address system in the South Tower stated that the incident had occurred in the other building and advised tenants, generally, that their building was safe and that they should remain on or return to their offices or floors. A statement from the deputy fire safety director informing tenants that the incident had occurred in the other building was consistent with protocol; the expanded advice did not correspond to any existing written protocol, and did not reflect any instruction known to have been given to the deputy fire safety director that day. We do not know the reason for the announcement, as both the deputy fire safety director believed to have made it and the director of fire safety for the WTC complex perished in the South Tower's collapse. Clearly, however, the prospect of another plane hitting the second building was beyond the contemplation of anyone giving advice. According to one of the first fire chiefs to arrive, such a scenario was unimaginable, "beyond our consciousness." As a result of the announcement, many civilians remained on their floors. Others reversed their evacuation and went back up.[46]

Similar advice was given in person by security officials in both the ground-floor lobby-where a group of 20 that had descended by the elevators was personally instructed to go back upstairs-and in the upper sky lobby, where many waited for express elevators to take them down. Security officials who gave this advice were not part of the fire safety staff.[47]

Several South Tower occupants called the Port Authority police desk in 5 WTC. Some were advised to stand by for further instructions; others were strongly advised to leave.[48]

It is not known whether the order by the FDNY to evacuate the South Tower was received by the deputy fire safety director making announcements there. However, at approximately 9:02-less than a minute before the building was hit-an instruction over the South Tower's public-address system advised civilians, generally, that they could begin an orderly evacuation if conditions warranted. Like the earlier advice to remain in place, it did not correspond to any prewritten emergency instruction.[49]

FDNY Initial Response Mobilization. The FDNY response began within five seconds of the crash. By 9:00, many senior FDNY leaders, including 7 of the 11 most highly ranked chiefs in the department, as well as the Commissioner and many of his deputies and assistants, had begun responding from headquarters in Brooklyn. While en route over the Brooklyn Bridge, the Chief of Department and the Chief of Operations had a clear view of the situation on the upper floors of the North Tower. They determined that because of the fire's magnitude and location near the top of the building, their mission would be primarily one of rescue. They called for a fifth alarm, which would bring additional engine and ladder companies, as well as for two more elite rescue units. The Chief of Department arrived at about 9:00; general FDNY Incident Command was transferred to his location on the West Side Highway. In all, 22 of the 32 senior chiefs and commissioners arrived at the WTC before 10:00.[50]

As of 9:00, the units that were dispatched (including senior chiefs responding to headquarters) included approximately 235 firefighters. These units consisted of 21 engine companies, nine ladder companies, four of the department's elite rescue teams, the department's single Hazmat team, two of the city's elite squad companies, and support staff. In addition, at 8:53 nine Brooklyn units were staged on the Brooklyn side of the Brooklyn-Battery Tunnel to await possible dispatch orders.[51]

Operations. A battalion chief and two ladder and two engine companies arrived at the North Tower at approximately 8:52. As they entered the lobby, they encountered badly burned civilians who had been caught in the path of the fireball. Floor-to-ceiling windows in the northwest corner of the West Street level of the lobby had been blown out; some large marble tiles had been dislodged from the walls; one entire elevator bank was destroyed by the fireball. Lights were functioning, however, and the air was clear of smoke.[52]

As the highest-ranking officer on the scene, the battalion chief initially was the FDNY incident commander. Minutes later, the on-duty division chief for Lower Manhattan arrived and took over. Both chiefs immediately began speaking with the former fire safety director and other building personnel to learn whether building systems were working. They were advised that all 99 elevators in the North Tower appeared to be out, and there were no assurances that sprinklers or standpipes were working on upper floors. Chiefs also spoke with Port Authority police personnel and an OEM representative.[53]

After conferring with the chiefs in the lobby, one engine and one ladder company began climbing stairwell C at about 8:57, with the goal of approaching the impact zone as scouting units and reporting back to the chiefs in the lobby. The radio channel they used was tactical 1. Following FDNY high-rise fire protocols, other units did not begin climbing immediately, as the chiefs worked to formulate a plan before sending them up. Units began mobilizing in the lobby, lining up and awaiting their marching orders.[54]

Also by approximately 8:57, FDNY chiefs had asked both building personnel and a Port Authority police officer to evacuate the South Tower, because in their judgment the impact of the plane into the North Tower made the entire complex unsafe-not because of concerns about a possible second plane.[55]

The FDNY chiefs in the increasingly crowded North Tower lobby were confronting critical choices with little to no information. They had ordered units up the stairs to report back on conditions, but did not know what the impact floors were; they did not know if any stairwells into the impact zone were clear; and they did not know whether water for firefighting would be available on the upper floors. They also did not know what the fire and impact zone looked like from the outside.[56]

They did know that the explosion had been large enough to send down a fireball that blew out elevators and windows in the lobby and that conditions were so dire that some civilians on upper floors were jumping or falling from the building. They also knew from building personnel that some civilians were trapped in elevators and on specific floors. According to Division Chief for Lower Manhattan Peter Hayden, "We had a very strong sense we would lose firefighters and that we were in deep trouble, but we had estimates of 25,000 to 50,000 civilians, and we had to try to rescue them."[57]

The chiefs concluded that this would be a rescue operation, not a firefighting operation. One of the chiefs present explained:

We realized that, because of the impact of the plane, that there was some structural damage to the building, and most likely that the fire suppression systems within the building were probably damaged and possibly inoperable....We knew that at the height of the day there were as many as 50,000 people in this building. We had a large volume of fire on the upper floors. Each floor was approximately an acre in size. Several floors of fire would have been beyond the fire-extinguishing capability of the forces that we had on hand. So we determined, very early on, that this was going to be strictly a rescue mission. We were going to vacate the building, get everybody out, and then we were going to get out.[58]

The specifics of the mission were harder to determine, as they had almost no information about the situation 80 or more stories above them. They also received advice from senior FDNY chiefs that while the building might eventually suffer a partial collapse on upper floors, such structural failure was not imminent. No one anticipated the possibility of a total collapse.[59]

Emergency medical services (EMS) personnel were directed to one of four triage areas being set up around the perimeter of the WTC. Some entered the lobby to respond to specific casualty reports. In addition, many ambulance paramedics from private hospitals were rushing to the WTC complex.[60]

NYPD Initial Response
Numerous NYPD officers saw the plane strike the North Tower and immediately reported it to NYPD communications dispatchers.[61]

At 8:58, while en route, the NYPD Chief of Department raised the NYPD's mobilization to level 4, thereby sending to the WTC approximately 22 lieutenants, 100 sergeants, and 800 police officers from all over the city. The Chief of Department arrived at Church and Vesey at 9:00.[62]

At 9:01, the NYPD patrol mobilization point was moved to West and Vesey in order to handle the greater number of patrol officers dispatched in the higher-level mobilization. These officers would be stationed around the perimeter of the complex to direct the evacuation of civilians. Many were diverted on the way to the scene by intervening emergencies related to the attack.[63]

At 8:50, the Aviation Unit of the NYPD dispatched two helicopters to the WTC to report on conditions and assess the feasibility of a rooftop landing or of special rescue operations. En route, the two helicopters communicated with air traffic controllers at the area's three major airports and informed them of the commercial airplane crash at the World Trade Center. The air traffic controllers had been unaware of the incident.[64]

At 8:56, an NYPD ESU team asked to be picked up at the Wall Street heliport to initiate rooftop rescues. At 8:58, however, after assessing the North Tower roof, a helicopter pilot advised the ESU team that they could not land on the roof, because "it is too engulfed in flames and heavy smoke condition."[65]

By 9:00, a third NYPD helicopter was responding to the WTC complex. NYPD helicopters and ESU officers remained on the scene throughout the morning, prepared to commence rescue operations on the roof if conditions improved. Both FDNY and NYPD protocols called for FDNY personnel to be placed in NYPD helicopters in the event of an attempted rooftop rescue at a high-rise fire. No FDNY personnel were placed in NYPD helicopters on September 11.[66]

The 911 operators and FDNY dispatchers were not advised that rooftop rescues were not being undertaken. They thus were not able to communicate this fact to callers, some of whom spoke of attempting to climb to the roof.[67]

Two on-duty NYPD officers were on the 20th floor of the North Tower at 8:46.They climbed to the 29th floor, urging civilians to evacuate, but did not locate a group of civilians trapped on the 22nd floor.[68]

Just before 9:00, an ESU team began to walk from Church and Vesey to the North Tower lobby, with the goal of climbing toward and setting up a triage center on the upper floors for the severely injured. A second ESU team would follow them to assist in removing those individuals.[69]

Numerous officers responded in order to help injured civilians and to urge those who could walk to vacate the area immediately. Putting themselves in danger of falling debris, several officers entered the plaza and successfully rescued at least one injured, nonambulatory civilian, and attempted to rescue others.[70]

Also by about 9:00, transit officers began shutting down subway stations in the vicinity of the World Trade Center and evacuating civilians from those stations.[71]

Around the city, the NYPD cleared major thoroughfares for emergency vehicles to access the WTC. The NYPD and PAPD coordinated the closing of bridges and tunnels into Manhattan.[72]

PAPD Initial Response
The Port Authority's on-site commanding police officer was standing in the concourse when a fireball erupted out of elevator shafts and exploded onto the mall concourse, causing him to dive for cover. The on-duty sergeant initially instructed the officers in the WTC Command to meet at the police desk in 5 WTC. Soon thereafter, he instructed officers arriving from outside commands to meet him at the fire safety desk in the North Tower lobby. A few of these officers from outside commands were given WTC Command radios.[73]

One Port Authority police officer at the WTC immediately began climbing stairwell C in the North Tower.[74] Other officers began performing rescue and evacuation operations on the ground floors and in the PATH (Port Authority Trans-Hudson) station below the WTC complex.

Within minutes of impact, Port Authority police officers from the PATH, bridges, tunnels, and airport commands began responding to the WTC. The PAPD lacked written standard operating procedures for personnel responding from outside commands to the WTC during a major incident. In addition, officers from some PAPD commands lacked interoperable radio frequencies. As a result, there was no comprehensive coordination of PAPD's overall response.[75]

At 9:00, the PAPD commanding officer of the WTC ordered an evacuation of all civilians in the World Trade Center complex, because of the magnitude of the calamity in the North Tower. This order was given over WTC police radio channel W, which could not be heard by the deputy fire safety director in the South Tower.[76]

Also at 9:00, the PAPD Superintendent and Chief of Department arrived separately and made their way to the North Tower.[77]

OEM Initial Response
By 8:48, officials in OEM headquarters on the 23rd floor of 7 WTC-just to the

north of the North Tower-began to activate the Emergency Operations Center by calling such agencies as the FDNY, NYPD, Department of Health, and the Greater Hospital Association and instructing them to send their designated representatives to the OEM. In addition, the Federal Emergency Management Agency (FEMA) was called and asked to send at least five federal Urban Search and Rescue Teams (such teams are located throughout the United States). At approximately 8:50, a senior representative from the OEM arrived in the lobby of the North Tower and began to act as the OEM field responder to the incident. He soon was joined by several other OEM officials, including the OEM Director.[78]

Summary
In the 17-minute period between 8:46 and 9:03 A.M. on September 11, New York City and the Port Authority of New York and New Jersey had mobilized the largest rescue operation in the city's history. Well over a thousand first responders had been deployed, an evacuation had begun, and the critical decision that the fire could not be fought had been made.

Then the second plane hit.

From 9:03 until 9:59 A.M. At 9:03:11, the hijacked United Airlines Flight 175 hit 2 WTC (the South Tower) from the south, crashing through the 77th to 85th floors. What had been the largest and most complicated rescue operation in city history instantly doubled in magnitude. The plane banked as it hit the building, leaving portions of the building undamaged on impact floors. As a consequence-and in contrast to the situation in the North Tower-one of the stairwells (A) initially remained passable from at least the 91st floor down, and likely from top to bottom.[79]

Civilians, Fire Safety Personnel, and 911 Calls South Tower. At the lower end of the impact, the 78th-floor sky lobby, hundreds had been waiting to evacuate when the plane hit. Many had attempted but failed to squeeze into packed express elevators. Upon impact, many were killed or severely injured; others were relatively unharmed. We know of at least one civilian who seized the initiative and shouted that anyone who could walk should walk to the stairs, and anyone who could help should help others in need of assistance. As a result, at least two small groups of civilians descended from that floor. Others remained on the floor to help the injured and move victims who were unable to walk to the stairwell to aid their rescue.[80]

Still others remained alive in the impact zone above the 78th floor. Damage was extensive, and conditions were highly precarious. The only survivor known to have escaped from the heart of the impact zone described the 81st floor-where the wing of the plane had sliced through his office-as a "demolition" site in which everything was "broken up" and the smell of jet fuel was so strong that it was almost impossible to breathe. This person escaped by means of an unlikely rescue, aided by a civilian fire warden descending from a higher floor, who, critically, had been provided with a flashlight.[81]

At least four people were able to descend stairwell A from the 81st floor or above. One left the 84th floor immediately after the building was hit. Even at that point, the stairway was dark, smoky, and difficult to navigate; glow strips on the stairs and handrails were a significant help. Several flights down, however, the evacuee became confused when he reached a smoke door that caused him to believe the stairway had ended. He was able to exit that stairwell and switch to another.[82]

Many civilians in and above the impact zone ascended the stairs. One small group reversed its descent down stairwell A after being advised by another civilian that they were approaching a floor "in flames."The only known survivor has told us that their intention was to exit the stairwell in search of clearer air. At the 91st floor, joined by others from intervening floors, they perceived themselves to be trapped in the stairwell and began descending again. By this time, the stairwell was "pretty black," intensifying smoke caused many to pass out, and fire had ignited in the 82nd-floor transfer hallway.[83]

Others ascended to attempt to reach the roof but were thwarted by locked doors. At approximately 9:30 a "lock release" order-which would unlock all areas in the complex controlled by the buildings' computerized security system, including doors leading to the roofs-was transmitted to the Security Command Center located on the 22nd floor of the North Tower. Damage to the software controlling the system, resulting from the impact of the plane, prevented this order from being executed.[84]

Others, attempting to descend, were frustrated by jammed or locked doors in stairwells or confused by the structure of the stairwell deviations. By the lower 70s, however, stairwells A and B were well-lit, and conditions were generally normal.[85]

Some civilians remained on affected floors, and at least one ascended from a lower point into the impact zone, to help evacuate colleagues or assist the injured.[86]

Within 15 minutes after the impact, debilitating smoke had reached at least one location on the 100th floor, and severe smoke conditions were reported throughout floors in the 90s and 100s over the course of the following half hour. By 9:30, a number of civilians who had failed to reach the roof remained on the 105th floor, likely unable to descend because of intensifying smoke in the stairwell. There were reports of tremendous smoke on that floor, but at least one area remained less affected until shortly before the building collapsed. There were several areas between the impact zone and the uppermost floors where conditions were better. At least a hundred people remained alive on the 88th and 89th floors, in some cases calling 911 for direction.[87]

The 911 system remained plagued by the operators' lack of awareness of what was occurring. Just as in the North Tower, callers from below and above the impact zone were advised to remain where they were and wait for help. The operators were not given any information about the inability to conduct rooftop rescues and therefore could not advise callers that they had essentially been ruled out. This lack of information, combined with the general advice to remain where they were, may have caused civilians above the impact not to attempt to descend, although stairwell A may have been passable.[88]

In addition, the 911 system struggled with the volume of calls and rigid standard operating procedures according to which calls conveying crucial information had to wait to be transferred to either EMS or FDNY dispatch.[89] According to one civilian who was evacuating down stairwell A from the heart of the impact zone and who stopped on the 31st floor in order to call 911, I told them when they answered the phone, where I was, that I had passed somebody on the 44th floor, injured-they need to get a medic and a stretcher to this floor, and described the situation in brief, and the person then asked for my phone number, or something, and they said-they put me on hold. "You gotta talk to one of my supervisors"-and suddenly I was on hold. And so I waited a considerable amount of time. Somebody else came back on the phone, I repeated the story. And then it happened again. I was on hold a second time, and needed to repeat the story for a third time. But I told the third person that I am only telling you once. I am getting out of the building, here are the details, write it down, and do what you should do.[90]

Very few 911 calls were received from floors below the impact, but at least one person was advised to remain on the 73rd floor despite the caller's protests that oxygen was running out. The last known 911 call from this location came at 9:52.[91]

Evidence suggests that the public-address system did not continue to function after the building was hit. A group of people trapped on the 97th floor, however, made repeated references in calls to 911 to having heard "announcements" to go down the stairs. Evacuation tones were heard in locations both above and below the impact zone.[92]

By 9:35, the West Street lobby level of the South Tower was becoming overwhelmed by injured people who had descended to the lobby but were having difficulty going on. Those who could continue were directed to exit north or east through the concourse and then out of the WTC complex.[93]

By 9:59, at least one person had descended from as high as the 91st floor of that tower, and stairwell A was reported to have been almost empty. Stairwell B was also reported to have contained only a handful of descending civilians at an earlier point in the morning. But just before the tower collapsed, a team of NYPD ESU officers encountered a stream of civilians descending an unidentified stairwell in the 20s. These civilians may have been descending from at or above the impact zone.[94]

North Tower. In the North Tower, civilians continued their evacuation. On the 91st floor, the highest floor with stairway access, all civilians but one were uninjured and able to descend. While some complained of smoke, heat, fumes, and crowding in the stairwells, conditions were otherwise fairly normal on floors below the impact. At least one stairwell was reported to have been "clear and bright" from the upper 80s down.[95]

Those who called 911 from floors below the impact were generally advised to remain in place. One group trapped on the 83rd floor pleaded repeatedly to know whether the fire was above or below them, specifically asking if 911 operators had any information from the outside or from the news. The callers were transferred back and forth several times and advised to stay put. Evidence suggests that these callers died.[96]

At 8:59, the Port Authority police desk at Newark Airport told a third party that a group of Port Authority civilian employees on the 64th floor should evacuate. (The third party was not at the WTC, but had been in phone contact with the group on the 64th floor.) At 9:10, in response to an inquiry from the employees themselves, the Port Authority police desk in Jersey City confirmed that employees on the 64th floor should "be careful, stay near the stairwells, and wait for the police to come up." When the third party inquired again at 9:31, the police desk at Newark Airport advised that they "absolutely" evacuate. The third party informed the police desk that the employees had previously received contrary advice from the FDNY, which could only have come via 911. These workers were not trapped, yet unlike most occupants on the upper floors, they had chosen not to descend immediately after impact. They eventually began to descend the stairs, but most of them died in the collapse of the North Tower.[97]

All civilians who reached the lobby were directed by NYPD and PAPD officers into the concourse, where other police officers guided them to exit the concourse and complex to the north and east so that they might avoid falling debris and victims.[98]

By 9:55, only a few civilians were descending above the 25th floor in stairwell B; these primarily were injured, handicapped, elderly, or severely overweight civilians, in some cases being assisted by other civilians.[99]

By 9:59, tenants from the 91st floor had already descended the stairs and exited the concourse. However, a number of civilians remained in at least stairwell C, approaching lower floors. Other evacuees were killed earlier by debris falling on the street.[100]

FDNY Response Increased Mobilization. Immediately after the second plane hit, the FDNY Chief of Department called a second fifth alarm.[101]

By 9:15, the number of FDNY personnel en route to or present at the scene was far greater than the commanding chiefs at the scene had requested. Five factors account for this disparity. First, while the second fifth alarm had called for 20 engine and 8 ladder companies, in fact 23 engine and 13 ladder companies were dispatched. Second, several other units self-dispatched. Third, because the attacks came so close to the 9:00 shift change, many firefighters just going off duty were given permission by company officers to "ride heavy" and became part of those on-duty teams, under the leadership of that unit's officer. Fourth, many off-duty firefighters responded from firehouses separately from the on-duty unit (in some cases when expressly told not to) or from home. The arrival of personnel in excess of that dispatched was particularly pronounced in the department's elite units. Fifth, numerous additional FDNY personnel-such as fire marshals and firefighters in administrative positions-who lacked a predetermined operating role also reported to the WTC.[102]

The Repeater System. Almost immediately after the South Tower was hit, senior FDNY chiefs in the North Tower lobby huddled to discuss strategy for the operations in the two towers. Of particular concern to the chiefs-in light of FDNY difficulties in responding to the 1993 bombing-was communications capability. One of the chiefs recommended testing the repeater channel to see if it would work.[103]

Earlier, an FDNY chief had asked building personnel to activate the repeater channel, which would enable greatly-enhanced FDNY portable radio communications in the high-rises. One button on the repeater system activation console in the North Tower was pressed at 8:54, though it is unclear by whom. As a result of this activation, communication became possible between FDNY portable radios on the repeater channel. In addition, the repeater's master handset at the fire safety desk could hear communications made by FDNY portable radios on the repeater channel. The activation of *transmission* on the master handset required, however, that a second button be pressed. That second button was never activated on the morning of September 11.[104]

At 9:05, FDNY chiefs tested the WTC complex's repeater system. Because the second button had not been activated, the chief on the master handset could not transmit. He was also apparently unable to hear another chief who was attempting to communicate with him from a portable radio, either because of a technical problem or because the volume was turned down on the console (the normal setting when the system was not in use). Because the repeater channel seemed inoperable-the master handset appeared unable to transmit or receive communications-the chiefs in the North Tower lobby decided not to use it. The repeater system was working at least partially, however, on portable FDNY radios, and firefighters subsequently used repeater channel 7 in the South Tower.[105]

FDNY North Tower Operations. Command and control decisions were affected by the lack of knowledge of what was happening 30, 60, 90, and 100 floors above. According to one of the chiefs in the lobby, "One of the most critical things in a major operation like this is to have information. We didn't have a lot of information coming in. We didn't receive any reports of what was seen from the [NYPD] helicopters. It was impossible to know how much damage was done on the upper floors, whether the stairwells were intact or not."[106] According to another chief present, "People watching on TV certainly had more knowledge of what was happening a hundred floors above us than we did in the lobby.... [W]ithout critical information coming in . . . it's very difficult to make informed, critical decisions[.]"[107]

As a result, chiefs in the lobby disagreed over whether anyone at or above the impact zone possibly could be rescued, or whether there should be even limited firefighting for the purpose of cutting exit routes through fire zones.[108]

Many units were simply instructed to ascend toward the impact zone and report back to the lobby via radio. Some units were directed to assist specific groups of individuals trapped in elevators or in offices well below the impact zone. One FDNY company successfully rescued some civilians who were trapped on the 22nd floor as a result of damage caused by the initial fireball.[109]

An attempt was made to track responding units' assignments on a magnetic board, but the number of units and individual firefighters arriving in the lobby made this an overwhelming task. As the fire companies were not advised to the contrary, they followed protocol and kept their radios on tactical channel 1, which would be monitored by the chiefs in the lobby. Those battalion chiefs who would climb would operate on a separate command channel, which also would be monitored by the chiefs in the lobby.[110]

Fire companies began to ascend stairwell B at approximately 9:07, laden with about 100 pounds of heavy protective clothing, self-contained breathing apparatuses, and other equipment (including hoses for engine companies and heavy tools for ladder companies).[111]

Firefighters found the stairways they entered intact, lit, and clear of smoke. Unbeknownst to the lobby command post, one battalion chief in the North Tower found a working elevator, which he took to the 16th floor before beginning to climb.[112]

In ascending stairwell B, firefighters were passing a steady and heavy stream of descending civilians. Firemen were impressed with the composure and total lack of panic shown by almost all civilians. Many civilians were in awe of the firefighters and found their mere presence to be calming.[113]

Firefighters periodically stopped on particular floors and searched to ensure that no civilians were still on it. In a few instances healthy civilians were found on floors, either because they still were collecting personal items or for no apparent reason; they were told to evacuate immediately. Firefighters deputized healthy civilians to be in charge of others who were struggling or injured.[114]

Climbing up the stairs with heavy protective clothing and equipment was hard work even for physically fit firefighters. As firefighters began to suffer varying levels of fatigue, some became separated from others in their unit.[115]

At 9:32, a senior chief radioed all units in the North Tower to return to the lobby, either because of a false report of a third plane approaching or because of his judgment about the deteriorating condition of the building. Once the rumor of the third plane was debunked, other chiefs continued operations, and there is no evidence that any units actually returned to the lobby. At the same time, a chief in the lobby was asked to consider the possibility of a rooftop rescue but was unable to reach FDNY dispatch by radio or phone. Out on West Street, however, the FDNY Chief of Department had already dismissed any rooftop rescue as impossible.[116]

As units climbed higher, their ability to communicate with chiefs on tactical 1 became more limited and sporadic, both because of the limited effectiveness of FDNY radios in high-rises and because so many units on tactical 1 were trying to communicate at once. When attempting to reach a particular unit, chiefs in the lobby often heard nothing in response.[117]

Just prior to 10:00, in the North Tower one engine company had climbed to the 54th floor, at least two other companies of firefighters had reached the sky lobby on the 44th floor, and numerous units were located between the 5th and 37th floors.[118]

FDNY South Tower and Marriott Hotel Operations. Immediately after the repeater test, a senior chief and a battalion chief commenced operations in the South Tower lobby. Almost at once they were joined by an OEM field responder. They were not, however, joined right away by a sizable number of fire companies, as units that had been in or en route to the North Tower lobby at 9:03 were not reallocated to the South Tower.[119]

A battalion chief and a ladder company found a working elevator to the 40th floor and from there proceeded to climb stairwell B. Another ladder company arrived soon thereafter, and began to rescue civilians trapped in an elevator between the first and second floors. The senior chief in the lobby expressed frustration about the lack of units he initially had at his disposal for South Tower operations.[120]

Unlike the commanders in the North Tower, the senior chief in the lobby and the ascending battalion chief kept their radios on repeater channel 7. For the first 15 minutes of the operations, communications among them and the ladder company climbing with the battalion chief worked well. Upon learning from a company security official that the impact zone began at the 78th floor, a ladder company transmitted this information, and the battalion chief directed an engine company staged on the 40th floor to attempt to find an elevator to reach that upper level.[121]

To our knowledge, no FDNY chiefs outside the South Tower realized that the repeater channel was functioning and being used by units in that tower. The senior chief in the South Tower lobby was initially unable to communicate his requests for more units to chiefs either in the North Tower lobby or at the outdoor command post.[122]

From approximately 9:21 on, the ascending battalion chief was unable to reach the South Tower lobby command post because the senior chief in the lobby had ceased to communicate on repeater channel 7. The vast majority of units that entered the South Tower did not communicate on the repeater channel.[123]

The first FDNY fatality of the day occurred at approximately 9:30, when a civilian landed on and killed a fireman near the intersection of West and Liberty streets.[124]

By 9:30, chiefs in charge of the South Tower still were in need of additional companies. Several factors account for the lag in response. First, only two units that had been dispatched to the North Tower prior to 9:03 reported immediately to the South Tower. Second, units were not actually sent until approximately five minutes after the FDNY Chief of Department ordered their dispatch. Third, those units that had been ordered at 8:53 to stage at the Brooklyn-Battery Tunnel-and thus very close to the WTC complex-were not dispatched after the plane hit the South Tower. Fourth, units parked further north on West Street, then proceeded south on foot and stopped at the overall FDNY command post on West Street, where in some cases they were told to wait. Fifth, some units responded directly to the North Tower. (Indeed, radio communications indicated that in certain cases some firemen believed that the South Tower was 1 WTC when in fact it was 2 WTC.) Sixth, some units couldn't find the staging area (at West Street south of Liberty) for the South Tower. Finally, the jumpers and debris that confronted units attempting to enter the South Tower from its main entrance on Liberty Street caused some units to search for indirect ways to enter that tower, most often through the Marriott Hotel, or simply to remain on West Street.[125]

A chief at the overall outdoor command post was under the impression that he was to assist in lobby operations of the South Tower, and in fact his aide already was in that lobby. But because of his lack of familiarity with the WTC complex and confusion over how to get to there, he instead ended up in the Marriott at about 9:35. Here he came across about 14 units, many of which had been trying to find safe access to the South Tower. He directed them to secure the elevators and conduct search-and-rescue operations on the upper floors of the Marriott. Four of these companies searched the spa on the hotel's top floor-the 22nd floor-for civilians, and found none.[126]

Feeling satisfied with the scope of the operation in the Marriott, the chief in the lobby there directed some units to proceed to what he thought was the South Tower. In fact, he pointed them to the North Tower. Three of the FDNY companies who had entered the North Tower from the Marriott found a working elevator in a bank at the south end of the lobby, which they took to the 23rd floor.[127]

In response to the shortage of units in the South Tower, at 9:37 an additional second alarm was requested by the chief at the West and Liberty streets staging area. At this time, the units that earlier had been staged on the Brooklyn side of the Brooklyn-Battery Tunnel were dispatched to the South Tower; some had gone through the tunnel already and had responded to the Marriott, not the South Tower.[128]

Between 9:45 and 9:58, the ascending battalion chief continued to lead FDNY operations on the upper floors of the South Tower. At 9:50, an FDNY ladder company encountered numerous seriously injured civilians on the 70th floor. With the assistance of a security guard, at 9:53 a group of civilians trapped in an elevator on the 78th-floor sky lobby were found by an FDNY company. They were freed from the elevator at 9:58. By that time the battalion chief had reached the 78th floor on stairwell A; he reported that it looked open to the 79th floor, well into the impact zone. He also reported numerous civilian fatalities in the area.[129]

FDNY Command and Control Outside the Towers. The overall command post consisted of senior chiefs, commissioners, the field communications van (Field Comm), numerous units that began to arrive after the South Tower was hit, and EMS chiefs and personnel.[130]

Field Comm's two main functions were to relay information between the overall operations command post and FDNY dispatch and to track all units operating at the scene on a large magnetic board. Both of these missions were severely compromised by the magnitude of the disaster on September 11. First, the means of transmitting information were unreliable. For example, while FDNY dispatch advised Field Comm that 100 people were reported via 911 to be trapped on the 105th floor of the North Tower, and Field Comm then attempted to convey that report to chiefs at the outdoor command post, this information did not reach the North Tower lobby. Second, Field Comm's ability to keep track of which units were operating where was limited, because many units reported directly to the North Tower, the South Tower, or the Marriott. Third, efforts to track units by listening to tactical 1 were severely hampered by the number of units using that channel; as many people tried to speak at once, their transmissions overlapped and often became indecipherable. In the opinion of one of the members of the Field Comm group, tactical 1 simply was not designed to handle the number of units operating on it that morning.[131]

The primary Field Comm van had access to the NYPD's Special Operations channel (used by NYPD Aviation), but it was in the garage for repairs on September 11. The backup van lacked that capability.[132]

The Chief of Department, along with civilian commissioners and senior EMS chiefs, organized ambulances on West Street to expedite the transport of injured civilians to hospitals.[133]

To our knowledge, none of the chiefs present believed that a total collapse of either tower was possible. One senior chief did articulate his concern that upper floors could begin to collapse in a few hours, and that firefighters thus should not ascend above floors in the 60s. That opinion was not conveyed to chiefs in the North Tower lobby, and there is no evidence that it was conveyed to chiefs in the South Tower lobby either.[134]

Although the Chief of Department had general authority over operations, tactical decisions remained the province of the lobby commanders. The highest-ranking officer in the North Tower was responsible for communicating with the Chief of Department. They had two brief conversations. In the first, the senior lobby chief gave the Chief of Department a status report and confirmed that this was a rescue, not firefighting, operation. In the second conversation, at about 9:45, the Chief of Department suggested that given how the North Tower appeared to him, the senior lobby chief might want to consider evacuating FDNY personnel.[135]

At 9:46, the Chief of Department called an additional fifth alarm, and at 9:54 an additional 20 engine and 6 ladder companies were sent to the WTC. As a result, more than one-third of all FDNY companies now had been dispatched to the WTC. At about 9:57, an EMS paramedic approached the FDNY Chief of Department and advised that an engineer in front of 7 WTC had just remarked that the Twin Towers in fact were in imminent danger of a total collapse.[136]

NYPD Response
Immediately after the second plane hit, the Chief of Department of the NYPD ordered a second Level 4 mobilization, bringing the total number of NYPD officers responding to close to 2,000.[137]

The NYPD Chief of Department called for Operation Omega, which required the protection of sensitive locations around the city. NYPD headquarters were secured and all other government buildings were evacuated.[138]

The ESU command post at Church and Vesey streets coordinated all NYPD ESU rescue teams. After the South Tower was hit, the ESU officer running this command post decided to send one ESU team (each with approximately six police officers) up each of the Twin Towers' stairwells. While he continued to monitor the citywide SOD channel, which NYPD helicopters were using, he also monitored the point-to-point tactical channel that the ESU teams climbing in the towers would use.[139]

The first NYPD ESU team entered the West Street-level lobby of the North Tower and prepared to begin climbing at about 9:15 A.M. They attempted to check in with the FDNY chiefs present, but were rebuffed. OEM personnel did not intervene. The ESU team began to climb the stairs. Shortly thereafter, a second NYPD ESU team entered the South Tower. The OEM field responder present ensured that they check in with the FDNY chief in charge of the lobby, and it was agreed that the ESU team would ascend and support FDNY personnel.[140]

A third ESU team subsequently entered the North Tower at its elevated mezzanine lobby level and made no effort to check in with the FDNY command post. A fourth ESU team entered the South Tower. By 9:59, a fifth ESU team was next to 6 WTC and preparing to enter the North Tower.[141]

By approximately 9:50, the lead ESU team had reached the 31st floor, observing that there appeared to be no more civilians still descending. This ESU team encountered a large group of firefighters and administered oxygen to some of them who were exhausted.[142]

At about 9:56, the officer running the ESU command post on Church and Vesey streets had a final radio communication with one of the ESU teams in the South Tower. The team then stated that it was ascending via stairs, was somewhere in the 20s, and was making slow progress because of the numerous descending civilians crowding the stairwell.[143]

Three plainclothes NYPD officers without radios or protective gear had begun ascending either stairwell A or C of the North Tower. They began checking every other floor above the 12th for civilians. Only occasionally did they find any, and in those few cases they ordered the civilians to evacuate immediately. While checking floors, they used office phones to call their superiors. In one phone call an NYPD chief instructed them to leave the North Tower, but they refused to do so. As they climbed higher, they encountered increasing smoke and heat. Shortly before 10:00 they arrived on the 54th floor.[144]

Throughout this period (9:03 to 9:59), a group of NYPD and Port Authority police officers, as well as two Secret Service agents, continued to assist civilians leaving the North Tower. They were positioned around the mezzanine lobby level of the North Tower, directing civilians leaving stairwells A and C to evacuate down an escalator to the concourse. The officers instructed those civilians who seemed composed to evacuate the complex calmly but rapidly. Other civilians exiting the stairs who were either injured or exhausted collapsed at the foot of these stairs; officers then assisted them out of the building.[145]

When civilians reached the concourse, another NYPD officer stationed at the bottom of the escalator directed them to exit through the concourse to the north and east and then out of the WTC complex. This exit route ensured that civilians would not be endangered by falling debris and people on West Street, on the plaza between the towers, and on Liberty Street.[146]

Some officers positioned themselves at the top of a flight of stairs by 5 WTC that led down into the concourse, going into the concourse when necessary to evacuate injured or disoriented civilians. Numerous other NYPD officers were stationed throughout the concourse, assisting burned, injured, and disoriented civilians, as well as directing all civilians to exit to the north and east. NYPD officers were also in the South Tower lobby to assist in civilian evacuation. NYPD officers stationed on Vesey Street between West Street and Church Street urged civilians not to remain in the area and instead to keep walking north.[147]

At 9:06, the NYPD Chief of Department instructed that no units were to land on the roof of either tower. At about 9:30, one of the helicopters present advised that a rooftop evacuation still would not be possible. One NYPD helicopter pilot believed one portion of the North Tower roof to be free enough of smoke that a hoist could be lowered in order to rescue people, but there was no one on the roof. This pilot's helicopter never attempted to hover directly over the tower. Another helicopter did attempt to do so, and its pilot stated that the severity of the heat from the jet fuel-laden fire in the North Tower would have made it impossible to hover low enough for a rescue, because the high temperature would have destabilized the helicopter.[148]

At 9:51, an aviation unit warned units of large pieces of debris hanging from the building. Prior to 9:59, no NYPD helicopter pilot predicted that either tower would collapse.[149]

Interaction of 911 Calls and NYPD Operations. At 9:37, a civilian on the 106th floor of the South Tower reported to a 911 operator that a lower floor-the "90-something floor"-was collapsing. This information was conveyed inaccurately by the 911 operator to an NYPD dispatcher. The dispatcher further confused the substance of the 911 call by telling NYPD officers at the WTC complex that "the 106th floor is crumbling" at 9:52, 15 minutes after the 911 call was placed. The NYPD dispatcher conveyed this message on the radio frequency used in precincts in the vicinity of the WTC and subsequently on the Special Operations Division channel, but not on City Wide channel 1.[150]

PAPD Response
Initial responders from outside PAPD commands proceeded to the police desk in 5 WTC or to the fire safety desk in the North Tower lobby. Some officers were then assigned to assist in stairwell evacuations; others were assigned to expedite evacuation in the plaza, concourse, and PATH station. As information was received of civilians trapped above ground-level floors of the North Tower, other PAPD officers were instructed to climb to those floors for rescue efforts. Still others began climbing toward the impact zone.[151]

At 9:11, the PAPD Superintendent and an inspector began walking up stairwell

B of the North Tower to assess damage near and in the impact zone. The PAPD Chief and several other PAPD officers began ascending a stairwell in order to reach the Windows on the World restaurant on the 106th floor, from which calls had been made to the PAPD police desk reporting at least 100 people trapped.[152]

Many PAPD officers from different commands responded on their own initiative. By 9:30, the PAPD central police desk requested that responding officers meet at West and Vesey and await further instructions. In the absence of a predetermined command structure to deal with an incident of this magnitude, a number of PAPD inspectors, captains, and lieutenants stepped forward at around 9:30 to formulate an on-site response plan. They were hampered by not knowing how many officers were responding to the site and where those officers were operating. Many of the officers who responded to this command post lacked suitable protective equipment to enter the complex.[153]

By 9:58, one PAPD officer had reached the 44th-floor sky lobby of the North Tower. Also in the North Tower, one team of PAPD officers was in the mid-20s and another was in the lower 20s. Numerous PAPD officers were also climbing in the South Tower, including the PAPD ESU team. Many PAPD officers were on the ground floors of the complex-some assisting in evacuation, others manning the PAPD desk in 5 WTC or assisting at lobby command posts.[154]

OEM Response
After the South Tower was hit, OEM senior leadership decided to remain in its "bunker" and continue conducting operations, even though all civilians had been evacuated from 7 WTC. At approximately 9:30, a senior OEM official ordered the evacuation of the facility, after a Secret Service agent in 7 WTC advised him that additional commercial planes were not accounted for. Prior to its evacuation, no outside agency liaisons had reached OEM. OEM field responders were stationed in each tower's lobby, at the FDNY overall command post, and, at least for some period of time, at the NYPD command post at Church and Vesey.[155]

Summary
The emergency response effort escalated with the crash of United 175 into the South Tower. With that escalation, communications as well as command and control became increasingly critical and increasingly difficult. First responders assisted thousands of civilians in evacuating the towers, even as incident commanders from responding agencies lacked knowledge of what other agencies and, in some cases, their own responders were doing.

From 9:59 until 10:28 A.M.
At 9:58:59, the South Tower collapsed in ten seconds, killing all civilians and emergency personnel inside, as well a number of individuals-both first responders and civilians-in the concourse, in the Marriott, and on neighboring streets. The building collapsed into itself, causing a ferocious windstorm and creating a

massive debris cloud. The Marriott hotel suffered significant damage as a result of the collapse of the South Tower.[156]

Civilian Response in the North Tower
The 911 calls placed from most locations in the North Tower grew increasingly desperate as time went on. As late as 10:28, people remained alive in some locations, including on the 92nd and 79th floors. Below the impact zone, it is likely that most civilians who were physically and emotionally capable of descending had exited the tower. The civilians who were nearing the bottom of stairwell C were assisted out of the building by NYPD, FDNY, and PAPD personnel. Others, who experienced difficulty evacuating, were being helped by first responders on lower floors.[157]

FDNY Response Immediate Impact of the Collapse of the South Tower. The FDNY overall command post and posts in the North Tower lobby, the Marriott lobby, and the staging area on West Street south of Liberty all ceased to operate upon the collapse of the South Tower, as did EMS staging areas, because of their proximity to the building.[158]

Those who had been in the North Tower lobby had no way of knowing that the South Tower had suffered a complete collapse. Chiefs who had fled from the overall command post on the west side of West Street took shelter in the underground parking garage at 2 World Financial Center and were not available to influence FDNY operations for the next ten minutes or so.[159]

When the South Tower collapsed, firefighters on upper floors of the North Tower heard a violent roar, and many were knocked off their feet; they saw debris coming up the stairs and observed that the power was lost and emergency lights activated. Nevertheless, those firefighters not standing near windows facing south had no way of knowing that the South Tower had collapsed; many surmised that a bomb had exploded, or that the North Tower had suffered a partial collapse on its upper floors.[160]

We do not know whether the repeater channel continued to function after 9:59.[161]

Initial Evacuation Instructions and Communications. The South Tower's total collapse was immediately communicated on the Manhattan dispatch channel by an FDNY boat on the Hudson River; but to our knowledge, no one at the site received this information, because every FDNY command post had been abandoned-including the overall command post, which included the Field Comm van. Despite his lack of knowledge of what had happened to the South Tower, a chief in the process of evacuating the North Tower lobby sent out an order within a minute of the collapse: "Command to all units in Tower 1, evacuate the building." Another chief from the North Tower lobby soon followed with an additional evacuation order issued on tactical 1.[162]

Evacuation orders did not follow the protocol for giving instructions when a building's collapse may be imminent-a protocol that includes constantly repeating "Mayday, Mayday, Mayday"-during the 29 minutes between the fall of the South Tower and that of the North Tower. In addition, most of the evacuation instructions did not mention that the South Tower had collapsed. However, at least three firefighters heard evacuation instructions which stated that the North Tower was in danger of "imminent collapse."[163]

FDNY Personnel above the Ground Floors of the North Tower. Within minutes, some firefighters began to hear evacuation orders over tactical 1. At least one chief also gave the evacuation instruction on the command channel used only by chiefs in the North Tower, which was much less crowded.[164]

At least two battalion chiefs on upper floors of the North Tower-one on the 23rd floor and one on the 35th floor-heard the evacuation instruction on the command channel and repeated it to everyone they came across. The chief on the 23rd floor apparently aggressively took charge to ensure that all firefighters on the floors in the immediate area were evacuating. The chief on the 35th floor also heard a separate radio communication stating that the South Tower had collapsed (which the chief on the 23rd floor may have heard as well). He subsequently acted with a sense of urgency, and some firefighters heard the evacuation order for the first time when he repeated it on tactical 1. This chief also had a bullhorn and traveled to each of the stairwells and shouted the evacuation order: "All FDNY, get the fuck out!" As a result of his efforts, many firefighters who had not been in the process of evacuating began to do so.[165]

Other firefighters did not receive the evacuation transmissions, for one of four reasons: First, some FDNY radios did not pick up the transmission because of the difficulties of radio communications in high-rises. Second, the numbers trying to use tactical 1 after the South Tower collapsed may have drowned out some evacuation instructions. According to one FDNY lieutenant who was on the 31st floor of the North Tower at the time, "[Tactical] channel 1 just might have been so bogged down that it may have been impossible to get that order through."[166] Third, some firefighters in the North Tower were off-duty and did not have radios. Fourth, some firefighters in the North Tower had been dispatched to the South Tower and likely were on the different tactical channel assigned to that tower.[167]

FDNY personnel in the North Tower who received the evacuation orders did not respond uniformly. Some units-including one whose officer knew that the South Tower had collapsed-either delayed or stopped their evacuation in order to assist nonambulatory civilians. Some units whose members had become separated during the climb attempted to regroup so they could descend together. Some units began to evacuate but, according to eyewitnesses, did not hurry. At least several firefighters who survived believed that they and others would have evacuated more urgently had they known of the South Tower's complete collapse. Other firefighters continued to sit and rest on floors while other companies descended past them and reminded them that they were supposed to evacuate. Some firefighters were determined not to leave the building while other FDNY personnel remained inside and, in one case, convinced others to remain with them. In another case, firefighters had successfully descended to the lobby, where another firefighter then persuaded them to reascend in order to look for specific FDNY personnel.[168]

Other FDNY personnel did not hear the evacuation order on their radio but were advised orally to leave the building by other firefighters and police who were themselves evacuating.[169]

By 10:24, approximately five FDNY companies reached the bottom of stairwell B and entered the North Tower lobby. They stood in the lobby for more than a minute, not certain what to do, as no chiefs were present. Finally, one firefighter-who had earlier seen from a window that the South Tower had collapsed-urged that they all leave, as this tower could fall as well. The units then proceeded to exit onto West Street. While they were doing so, the North Tower began its pancake collapse, killing some of these men.[170]

Other FDNY Personnel. The Marriott Hotel suffered significant damage in the collapse of the South Tower. Those in the lobby were knocked down and enveloped in the darkness of a debris cloud. Some were hurt but could walk. Others were more severely injured, and some were trapped. Several firefighters came across a group of about 50 civilians who had been taking shelter in the restaurant and assisted them in evacuating. Up above, at the time of the South Tower's collapse four companies were descending the stairs single file in a line of approximately 20 men. Four survived.[171]

At the time of the South Tower's collapse, two FDNY companies were either at the eastern side of the North Tower lobby, near the mall concourse, or actually in the mall concourse, trying to reach the South Tower. Many of these men were thrown off their feet by the collapse of the South Tower; they then attempted to regroup in the darkness of the debris cloud and evacuate civilians and themselves, not knowing that the South Tower had collapsed. Several of these firefighters subsequently searched the PATH station below the con-course-unaware that the PAPD had cleared the area of all civilians by 9:19.[172]

At about 10:15, the FDNY Chief of Department and the Chief of Safety, who had returned to West Street from the parking garage, confirmed that the South Tower had collapsed. The Chief of Department issued a radio order for all units to evacuate the North Tower, repeating it about five times. He then directed that the FDNY command post be moved further north on West Street and told FDNY units in the area to proceed north on West Street toward Chambers Street. At approximately 10:25, he radioed for two ladder companies to respond to the Marriott, where he was aware that both FDNY personnel and civilians were trapped.[173]

Many chiefs, including several of those who had been in the North Tower lobby, did not learn that the South Tower had collapsed until 30 minutes or more after the event. According to two eyewitnesses, however, one senior

FDNY chief who knew that the South Tower had collapsed strongly expressed the opinion that the North Tower would not collapse, because unlike the South Tower, it had not been hit on a corner.[174]

After the South Tower collapsed, some firefighters on the streets neighboring the North Tower remained where they were or came closer to the North Tower. Some of these firefighters did not know that the South Tower had collapsed, but many chose despite that knowledge to remain in an attempt to save additional lives. According to one such firefighter, a chief who was preparing to mount a search-and-rescue mission in the Marriott, "I would never think of myself as a leader of men if I had headed north on West Street after [the] South Tower collapsed." Just outside the North Tower on West Street one firefighter was directing others exiting the building, telling them when no jumpers were coming down and it was safe to run out. A senior chief had grabbed an NYPD bullhorn and was urging firefighters exiting onto West Street to continue running north, well away from the WTC. Three of the most senior and respected members of the FDNY were involved in attempting to rescue civilians and firefighters from the Marriott.[175]

NYPD Response

A member of the NYPD Aviation Unit radioed that the South Tower had collapsed immediately after it happened, and further advised that all people in the WTC complex and nearby areas should be evacuated. At 10:04, NYPD aviation reported that the top 15 stories of the North Tower "were glowing red" and that they might collapse. At 10:08, a helicopter pilot warned that he did not believe the North Tower would last much longer.[176]

Immediately after the South Tower collapsed, many NYPD radio frequencies became overwhelmed with transmissions relating to injured, trapped, or missing officers. As a result, NYPD radio communications became strained on most channels. Nevertheless, they remained effective enough for the two closest NYPD mobilization points to be moved further from the WTC at 10:06.[177]

Just like most firefighters, the ESU rescue teams in the North Tower had no idea that the South Tower had collapsed. However, by 10:00 the ESU officer running the command post at Church and Vesey ordered the evacuation of all ESU units from the WTC complex. This officer, who had observed the South Tower collapse, reported it to ESU units in the North Tower in his evacuation instruction.[178]

This instruction was clearly heard by the two ESU units already in the North Tower and the other ESU unit preparing to enter the tower. The ESU team on the 31st floor found the full collapse of the South Tower so unfathomable that they radioed back to the ESU officer at the command post and asked him to repeat his communication. He reiterated his urgent message.[179]

The ESU team on the 31st floor conferred with the FDNY personnel there to ensure that they, too, knew that they had to evacuate, then proceeded down stairwell B. During the descent, they reported seeing many firefighters who were resting and did not seem to be in the process of evacuating. They further reported advising these firefighters to evacuate, but said that at times they were not acknowledged. In the opinion of one of the ESU officers, some of these firefighters essentially refused to take orders from cops. At least one firefighter who was in the North Tower has supported that assessment, stating that he was not going to take an evacuation instruction from a cop that morning. However, another firefighter reports that ESU officers ran past him without advising him to evacuate.[180]

The ESU team on the 11th floor began descending stairwell C after receiving the evacuation order. Once near the mezzanine level-where stairwell C ended-this team spread out in chain formation, stretching from several floors down to the mezzanine itself. They used their flashlights to provide a path of beacons through the darkness and debris for civilians climbing down the stairs. Eventually, when no one else appeared to be descending, the ESU team exited the North Tower and ran one at a time to 6 WTC, dodging those who still were jumping from the upper floors of the North Tower by acting as spotters for each other. They remained in the area, conducting additional searches for civilians; all but two of them died.[181]

After surviving the South Tower's collapse, the ESU team that had been preparing to enter the North Tower spread into chain formation and created a path for civilians (who had exited from the North Tower mezzanine) to evacuate the WTC complex by descending the stairs on the north side of 5 and 6 WTC, which led down to Vesey Street. They remained at this post until the North Tower collapsed, yet all survived.[182]

The three plainclothes NYPD officers who had made it up to the 54th floor of the North Tower felt the building shake violently at 9:59 as the South Tower collapsed (though they did not know the cause). Immediately thereafter, they were joined by three firefighters from an FDNY engine company. One of the firefighters apparently heard an evacuation order on his radio, but responded in a return radio communication, "We're not fucking coming out!" However, the firefighters urged the police officers to descend because they lacked the protective gear and equipment needed to handle the increasing smoke and heat. The police officers reluctantly began descending, checking that the lower floors were clear of civilians. They proceeded down stairwell B, poking their heads into every floor and briefly looking for civilians.[183]

Other NYPD officers helping evacuees on the mezzanine level of the North Tower were enveloped in the debris cloud that resulted from the South Tower's collapse. They struggled to regroup in the darkness and to evacuate both themselves and civilians they encountered. At least one of them died in the collapse of the North Tower. At least one NYPD officer from this area managed to evacuate out toward 5 WTC, where he teamed up with a Port Authority police officer and acted as a spotter in advising the civilians who were still exiting when they could safely run from 1 WTC to 5 WTC and avoid being struck by people and debris falling from the upper floors.[184]

At the time of the collapse of the South Tower, there were numerous NYPD officers in the concourse, some of whom are believed to have died there. Those who survived struggled to evacuate themselves in darkness, assisting civilians as they exited the concourse in all directions.[185]

Port Authority Response
The collapse of the South Tower forced the evacuation of the PAPD command post on West and Vesey, compelling PAPD officers to move north. There is no evidence that PAPD officers without WTC Command radios received an evacuation order by radio. Some of these officers in the North Tower decided to evacuate, either on their own or in consultation with other first responders they came across. Some greatly slowed their own descent in order to assist nonambulatory civilians.[186]

After 10:28 A.M. The North Tower collapsed at 10:28:25 A.M., killing all civilians alive on upper floors, an undetermined number below, and scores of first responders. The FDNY Chief of Department, the Port Authority Police Department Superintendent, and many of their senior staff were killed. Incredibly, twelve firefighters, one PAPD officer, and three civilians who were descending stairwell B of the North Tower survived its collapse.[187]

On September 11, the nation suffered the largest loss of life-2,973-on its soil as a result of hostile attack in its history. The FDNY suffered 343 fatalities- the largest loss of life of any emergency response agency in history. The PAPD suffered 37 fatalities-the largest loss of life of any police force in history. The NYPD suffered 23 fatalities-the second largest loss of life of any police force in history, exceeded only by the number of PAPD officers lost the same day.[188]

Mayor Giuliani, along with the Police and Fire commissioners and the OEM director, moved quickly north and established an emergency operations command post at the Police Academy. Over the coming hours, weeks, and months, thousands of civilians and city, state, and federal employees devoted themselves around the clock to putting New York City back on its feet.[189]

9.3 EMERGENCY RESPONSE AT THE PENTAGON

If it had happened on any other day, the disaster at the Pentagon would be remembered as a singular challenge and an extraordinary national story. Yet the calamity at the World Trade Center that same morning included catastrophic damage 1,000 feet above the ground that instantly imperiled tens of thousands of people. The two experiences are not comparable. Nonetheless, broader lessons in integrating multiagency response efforts are apparent when we analyze the response at the Pentagon.

The emergency response at the Pentagon represented a mix of local, state, and federal jurisdictions and was generally effective. It overcame the inherent complications of a response across jurisdictions because the Incident Command System, a formalized management structure for emergency response, was in place in the National Capital Region on 9/11.[190]

Because of the nature of the event-a plane crash, fire, and partial building collapse-the Arlington County Fire Department served as incident commander. Different agencies had different roles. The incident required a major rescue, fire, and medical response from Arlington County at the U.S. military's headquarters-a facility under the control of the secretary of defense. Since it was a terrorist attack, the Department of Justice was the lead federal agency in charge (with authority delegated to the FBI for operational response). Additionally, the terrorist attack affected the daily operations and emergency management requirements of Arlington County and all bordering and surrounding jurisdictions.[191]

At 9:37, the west wall of the Pentagon was hit by hijacked American Airlines Flight 77, a Boeing 757. The crash caused immediate and catastrophic damage. All 64 people aboard the airliner were killed, as were 125 people inside the Pentagon (70 civilians and 55 military service members). One hundred six people were seriously injured and transported to area hospitals.[192]

While no emergency response is flawless, the response to the 9/11 terrorist attack on the Pentagon was mainly a success for three reasons: first, the strong professional relationships and trust established among emergency responders; second, the adoption of the Incident Command System; and third, the pursuit of a regional approach to response. Many fire and police agencies that responded had extensive prior experience working together on regional events and training exercises. Indeed, at the time preparations were under way at many of these agencies to ensure public safety at the annual meetings of the International Monetary Fund and the World Bank scheduled to be held later that month in Washington, D.C.[193]

Local, regional, state, and federal agencies immediately responded to the Pentagon attack. In addition to county fire, police, and sheriff's departments, the response was assisted by the Metropolitan Washington Airports Authority, Ronald Reagan Washington National Airport Fire Department, Fort Myer Fire Department, the Virginia State Police, the Virginia Department of Emergency Management, the FBI, FEMA, a National Medical Response Team, the Bureau of Alcohol, Tobacco, and Firearms, and numerous military personnel within the Military District of Washington.[194]

Command was established at 9:41. At the same time, the Arlington County Emergency Communications Center contacted the fire departments of Fairfax County, Alexandria, and the District of Columbia to request mutual aid.

The incident command post provided a clear view of and access to the crash site, allowing the incident commander to assess the situation at all times.[195]

At 9:55, the incident commander ordered an evacuation of the Pentagon impact area because a partial collapse was imminent; it occurred at 9:57, and no first responder was injured.[196]

At 10:15, the incident commander ordered a full evacuation of the command post because of the warning of an approaching hijacked aircraft passed along by the FBI. This was the first of three evacuations caused by reports of incoming aircraft, and the evacuation order was well communicated and well coordinated.[197]

Several factors facilitated the response to this incident, and distinguish it from the far more difficult task in New York. There was a single incident, and it was not 1,000 feet above ground. The incident site was relatively easy to secure and contain, and there were no other buildings in the immediate area. There was no collateral damage beyond the Pentagon.[198]

Yet the Pentagon response encountered difficulties that echo those experienced in New York. As the "Arlington County: After-Action Report" notes, there were significant problems with both self-dispatching and communications: "Organizations, response units, and individuals proceeding on their own initiative directly to an incident site, without the knowledge and permission of the host jurisdiction and the Incident Commander, complicate the exercise of command, increase the risks faced by bonafide responders, and exacerbate the challenge of accountability." With respect to communications, the report concludes: "Almost all aspects of communications continue to be problematic, from initial notification to tactical operations. Cellular telephones were of little value.... Radio channels were initially oversaturated.... Pagers seemed to be the most reliable means of notification when available and used, but most firefighters are not issued pagers."[199]

It is a fair inference, given the differing situations in New York City and Northern Virginia, that the problems in command, control, and communications that occurred at both sites will likely recur in any emergency of similar scale. The task looking forward is to enable first responders to respond in a coordinated manner with the greatest possible awareness of the situation.

9.4 ANALYSIS

Like the national defense effort described in chapter 1, the emergency response to the attacks on 9/11 was necessarily improvised. In New York, the FDNY, NYPD, the Port Authority, WTC employees, and the building occupants themselves did their best to cope with the effects of an unimaginable catastrophe-unfolding furiously over a mere 102 minutes-for which they were unprepared in terms of both training and mindset. As a result of the efforts of first responders, assistance from each other, and their own good instincts and goodwill, the vast majority of civilians below the impact zone were able to evacuate the towers.

The National Institute of Standards and Technology has provided a preliminary estimation that between 16,400 and 18,800 civilians were in the WTC complex as of 8:46 A.M. on September 11. At most 2,152 individuals died at the WTC complex who were not (1) fire or police first responders, (2) security or fire safety personnel of the WTC or individual companies, (3) volunteer civilians who ran to the WTC after the planes' impact to help others, or (4) on the two planes that crashed into the Twin Towers. Out of this total number of fatalities, we can account for the workplace location of 2,052 individuals, or 95.35 percent. Of this number, 1,942 or 94.64 percent either worked or were supposed to attend a meeting at or above the respective impact zones of the Twin Towers; only 110, or 5.36 percent of those who died, worked below the impact zone. While a given person's office location at the WTC does not definitively indicate where that individual died that morning or whether he or she could have evacuated, these data strongly suggest that the evacuation was a success for civilians below the impact zone.[200]

Several factors influenced the evacuation on September 11. It was aided greatly by changes made by the Port Authority in response to the 1993 bombing and by the training of both Port Authority personnel and civilians after that time. Stairwells remained lit near unaffected floors; some tenants relied on procedures learned in fire drills to help them to safety; others were guided down the stairs by fire safety officials based in the lobby. Because of damage caused by the impact of the planes, the capability of the sophisticated building systems may have been impaired. Rudimentary improvements, however, such as the addition of glow strips to the handrails and stairs, were credited by some as the reason for their survival. The general evacuation time for the towers dropped from more than four hours in 1993 to under one hour on September 11 for most civilians who were not trapped or physically incapable of enduring a long descent.

First responders also played a significant role in the success of the evacuation. Some specific rescues are quantifiable, such as an FDNY company's rescue of civilians trapped on the 22d floor of the North Tower, or the success of FDNY, PAPD, and NYPD personnel in carrying nonambulatory civilians out of both the North and South Towers. In other instances, intangibles combined to reduce what could have been a much higher death total. It is impossible to measure how many more civilians who descended to the ground floors would have died but for the NYPD and PAPD personnel directing them-via safe exit routes that avoided jumpers and debris-to leave the complex urgently but calmly. It is impossible to measure how many more civilians would have died but for the determination of many members of the FDNY, PAPD, and NYPD to continue assisting civilians after the South Tower collapsed. It is impossible to measure the calming influence that ascending firefighters had on descending civilians or whether but for the firefighters' presence the poor behavior of a very few civilians could have caused a dangerous and panicked mob flight. But the positive impact of the first responders on the evacuation came at a tremendous cost of first responder lives lost.[201]

Civilian and Private-Sector Challenges

The "first" first responders on 9/11, as in most catastrophes, were private-sector civilians. Because 85 percent of our nation's critical infrastructure is controlled not by government but by the private sector, private-sector civilians are likely to be the first responders in any future catastrophes. For that reason, we have assessed the state of private sector and civilian preparedness in order to formulate recommendations to address this critical need. Our recommendations grow out of the experience of the civilians at the World Trade Center on 9/11.

Lack of Protocol for Rooftop Rescues. Civilians at or above the impact zone in the North Tower had the smallest hope of survival. Once the plane struck, they were prevented from descending because of damage to or impassable conditions in the building's three stairwells. The only hope for those on the upper floors of the North Tower would have been a swift and extensive air rescue. Several factors made this impossible. Doors leading to the roof were kept locked for security reasons, and damage to software in the security command station prevented a lock release order from taking effect. Even if the doors had not been locked, structural and radiation hazards made the rooftops unsuitable staging areas for a large number of civilians; and even if conditions permitted general helicopter evacuations-which was not the case-only several people could be lifted at a time.

The WTC lacked any plan for evacuation of civilians on upper floors of the WTC in the event that all stairwells were impassable below.

Lack of Comprehensive Evacuation of South Tower Immediately after the North Tower Impact. No decision has been criticized more than the decision of building personnel not to evacuate the South Tower immediately after the North Tower was hit. A firm and prompt evacuation order would likely have led many to safety. Even a strictly "advisory" announcement would not have dissuaded those who decided for themselves to evacuate. The advice to stay in place was understandable, however, when considered in its context. At that moment, no one appears to have thought a second plane could hit the South Tower. The evacuation of thousands of people was seen as inherently dangerous. Additionally, conditions were hazardous in some areas outside the towers.[202]

Less understandable, in our view, is the instruction given to some civilians who had reached the lobby to return to their offices. They could have been held in the lobby or perhaps directed through the underground concourse.

Despite the initial advice given over its public-address system, the South Tower was ordered to be evacuated by the FDNY and PAPD within 12 minutes of the North Tower's being hit. If not for a second, unanticipated attack, the evacuation presumably would have proceeded.

Impact of Fire Safety Plan and Fire Drills on Evacuation. Once the South

Tower was hit, civilians on upper floors wasted time ascending the stairs instead of searching for a clear path down, when stairwell A was at least initially passable. Although rooftop rescues had not been conclusively ruled out, civilians were not informed in fire drills that roof doors were locked, that rooftop areas were hazardous, and that no helicopter evacuation plan existed.

In both towers, civilians who were able to reach the stairs and descend were also stymied by the deviations in the stairways and by smoke doors. This confusion delayed the evacuation of some and may have obstructed that of others. The Port Authority has acknowledged that in the future, tenants should be made aware of what conditions they will encounter during descent.

Impact of 911 Calls on Evacuation. The NYPD's 911 operators and FDNY dispatch were not adequately integrated into the emergency response. In several ways, the 911 system was not ready to cope with a major disaster. These operators and dispatchers were one of the only sources of information for individuals at and above the impact zone of the towers. The FDNY ordered both towers fully evacuated by 8:57, but this guidance was not conveyed to 911 operators and FDNY dispatchers, who for the next hour often continued to advise civilians not to self-evacuate, regardless of whether they were above or below the impact zones. Nor were 911 operators or FDNY dispatchers advised that rooftop rescues had been ruled out. This failure may have been harmful to civilians on the upper floors of the South Tower who called 911 and were not told that their only evacuation hope was to attempt to descend, not to ascend. In planning for future disasters, it is important to integrate those taking 911 calls into the emergency response team and to involve them in providing up-to-date information and assistance to the public.

Preparedness of Individual Civilians. One clear lesson of September 11 is that individual civilians need to take responsibility for maximizing the probability that they will survive, should disaster strike. Clearly, many building occupants in the World Trade Center did not take preparedness seriously. Individuals should know the exact location of every stairwell in their workplace. In addition, they should have access at all times to flashlights, which were deemed invaluable by some civilians who managed to evacuate the WTC on September 11.

Challenges Experienced by First Responders The Challenge of Incident Command. As noted above, in July 2001, Mayor Giuliani updated a directive titled "Direction and Control of Emergencies in the City of New York." The directive designated, for different types of emergencies, an appropriate agency as "Incident Commander"; it would be "responsible for the management of the City's response to the emergency." The directive also provided that where incidents are "so multifaceted that no one agency immediately stands out as the Incident Commander, OEM will assign the role of Incident Commander to an agency as the situation demands."[203]

To some degree, the Mayor's directive for incident command was followed on

9/11. It was clear that the lead response agency was the FDNY, and that the other responding local, federal, bistate, and state agencies acted in a supporting role. There was a tacit understanding that FDNY personnel would have primary responsibility for evacuating civilians who were above the ground floors of the Twin Towers, while NYPD and PAPD personnel would be in charge of evacuating civilians from the WTC complex once they reached ground level. The NYPD also greatly assisted responding FDNY units by clearing emergency lanes to the WTC.[204]

In addition, coordination occurred at high levels of command. For example, the Mayor and Police Commissioner consulted with the Chief of the Department of the FDNY at approximately 9:20.There were other instances of coordination at operational levels, and information was shared on an ad hoc basis. For example, an NYPD ESU team passed the news of their evacuation order to firefighters in the North Tower.[205]

It is also clear, however, that the response operations lacked the kind of integrated communications and unified command contemplated in the directive. These problems existed both within and among individual responding agencies.

Command and Control within First Responder Agencies. For a unified incident management system to succeed, each participant must have command and control of its own units and adequate internal communications. This was not always the case at the WTC on 9/11.

Understandably lacking experience in responding to events of the magnitude of the World Trade Center attacks, the FDNY as an institution proved incapable of coordinating the numbers of units dispatched to different points within the 16-acre complex. As a result, numerous units were congregating in the undamaged Marriott Hotel and at the overall command post on West Street by 9:30, while chiefs in charge of the South Tower still were in desperate need of units. With better understanding of the resources already available, additional units might not have been dispatched to the South Tower at 9:37.

The task of accounting for and coordinating the units was rendered difficult, if not impossible, by internal communications breakdowns resulting from the limited capabilities of radios in the high-rise environment of the WTC and from confusion over which personnel were assigned to which frequency. Furthermore, when the South Tower collapsed the overall FDNY command post ceased to operate, which compromised the FDNY's ability to understand the situation; an FDNY marine unit's immediate radio communication to FDNY dispatch that the South Tower had fully collapsed was not conveyed to chiefs at the scene. The FDNY's inability to coordinate and account for the different radio channels that would be used in an emergency of this scale contributed to the early lack of units in the South Tower, whose lobby chief initially could not communicate with anyone outside that tower.[206]

Though almost no one at 9:50 on September 11 was contemplating an imminent total collapse of the Twin Towers, many first responders and civilians were contemplating the possibility of imminent additional terrorist attacks throughout New York City. Had any such attacks occurred, the FDNY's response would have been severely compromised by the concentration of so many of its off-duty personnel, particularly its elite personnel, at the WTC.

The Port Authority's response was hampered by the lack of both standard operating procedures and radios capable of enabling multiple commands to respond in unified fashion to an incident at the WTC. Many officers reporting from the tunnel and airport commands could not hear instructions being issued over the WTC Command frequency. In addition, command and control was complicated by senior Port Authority Police officials becoming directly involved in frontline rescue operations.

The NYPD experienced comparatively fewer internal command and control and communications issues. Because the department has a history of mobilizing thousands of officers for major events requiring crowd control, its technical radio capability and major incident protocols were more easily adapted to an incident of the magnitude of 9/11. In addition, its mission that day lay largely outside the towers themselves. Although there were ESU teams and a few individual police officers climbing in the towers, the vast majority of NYPD personnel were staged outside, assisting with crowd control and evacuation and securing other sites in the city. The NYPD ESU division had firm command and control over its units, in part because there were so few of them (in comparison to the number of FDNY companies) and all reported to the same ESU command post. It is unclear, however, whether non-ESU NYPD officers operating on the ground floors, and in a few cases on upper floors, of the WTC were as well coordinated.

Significant shortcomings within the FDNY's command and control capabilities were painfully exposed on September 11. To its great credit, the department has made a substantial effort in the past three years to address these. While significant problems in the command and control of the PAPD also were exposed on September 11, it is less clear that the Port Authority has adopted new training exercises or major incident protocols to address these shortcomings.[207]

Lack of Coordination among First Responder Agencies. Any attempt to establish a unified command on 9/11 would have been further frustrated by the lack of communication and coordination among responding agencies. Certainly, the FDNY was not "responsible for the management of the City's response to the emergency," as the Mayor's directive would have required. The command posts were in different locations, and OEM headquarters, which could have served as a focal point for information sharing, did not play an integrating role in ensuring that information was shared among agencies on 9/11, even prior to its evacuation. There was a lack of comprehensive coordination between FDNY, NYPD, and PAPD personnel climbing above the ground floors in the Twin Towers.

Information that was critical to informed decisionmaking was not shared among agencies. FDNY chiefs in leadership roles that morning have told us that their decision making capability was hampered by a lack of information from NYPD aviation. At 9:51 A.M., a helicopter pilot cautioned that "large pieces" of the South Tower appeared to be about to fall and could pose a danger to those below. Immediately after the tower's collapse, a helicopter pilot radioed that news. This transmission was followed by communications at 10:08, 10:15, and 10:22 that called into question the condition of the North Tower. The FDNY chiefs would have benefited greatly had they been able to communicate with personnel in a helicopter.

The consequence of the lack of real-time intelligence from NYPD aviation should not be overstated. Contrary to a widely held misperception, no NYPD helicopter predicted the fall of either tower before the South Tower collapsed, and no NYPD personnel began to evacuate the WTC complex prior to that time. Furthermore, the FDNY, as an institution, was in possession of the knowledge that the South Tower had collapsed as early as the NYPD, as its fall had been immediately reported by an FDNY boat on a dispatch channel. Because of internal breakdowns within the department, however, this information was not disseminated to FDNY personnel on the scene.

The FDNY, PAPD, and NYPD did not coordinate their units that were searching the WTC complex for civilians. In many cases, redundant searches of specific floors and areas were conducted. It is unclear whether fewer first responders in the aggregate would have been in the Twin Towers if there had been an integrated response, or what impact, if any, redundant searches had on the total number of first responder fatalities.

Whether the lack of coordination between the FDNY and NYPD on September 11 had a catastrophic effect has been the subject of controversy. We believe that there are too many variables for us to responsibly quantify those consequences. It is clear that the lack of coordination did not affect adversely the evacuation of civilians. It is equally clear, however, that the Incident Command System did not function to integrate awareness among agencies or to facilitate interagency response.[208]

If New York and other major cities are to be prepared for future terrorist attacks, different first responder agencies within each city must be fully coordinated, just as different branches of the U.S. military are. Coordination entails a unified command that comprehensively deploys all dispatched police, fire, and other first responder resources.

In May 2004, New York City adopted an emergency response plan that expressly contemplates two or more agencies jointly being lead agency when responding to a terrorist attack but does not mandate a comprehensive and unified incident command that can deploy and monitor all first responder resources from one overall command post. In our judgment, this falls short of an optimal response plan, which requires clear command and control, common training, and the trust that such training creates. The experience of the military suggests that integrated into such a coordinated response should be a unified field intelligence unit, which should receive and combine information from all first responders-including 911 operators. Such a field intelligence unit could be valuable in large and complex incidents.

Radio Communication Challenges: The Effectiveness and Urgency of Evacuation Instructions. As discussed above, the location of the NYPD ESU command post was crucial in making possible an urgent evacuation order explaining the South Tower's full collapse. Firefighters most certainly would have benefited from that information.

A separate matter is the varied success at conveying evacuation instructions to personnel in the North Tower after the South Tower's collapse. The success of NYPD ESU instruction is attributable to a combination of (1) the strength of the radios, (2) the relatively small numbers of individuals using them, and (3) use of the correct channel by all.

The same three factors worked against successful communication among FDNY personnel. First, the radios' effectiveness was drastically reduced in the high-rise environment. Second, tactical channel 1 was simply overwhelmed by the number of units attempting to communicate on it at 10:00. Third, some firefighters were on the wrong channel or simply lacked radios altogether.

It is impossible to know what difference it made that units in the North Tower were not using the repeater channel after 10:00. While the repeater channel was at least partially operational before the South Tower collapsed, we do not know whether it continued to be operational after 9:59.

Even without the repeater channel, *at least* 24 of the *at most* 32 companies who were dispatched to and actually in the North Tower received the evacuation instruction-either via radio or directly from other first responders. Nevertheless, many of these firefighters died, either because they delayed their evacuation to assist civilians, attempted to regroup their units, lacked urgency, or some combination of these factors. In addition, many other firefighters not dispatched to the North Tower also died in its collapse. Some had their radios on the wrong channel. Others were off-duty and lacked radios. In view of these considerations, we conclude that the technical failure of FDNY radios, while a contributing factor, was not the primary cause of the many firefighter fatalities in the North Tower.[209]

The FDNY has worked hard in the past several years to address its radio deficiencies. To improve radio capability in high-rises, the FDNY has internally developed a "post radio" that is small enough for a battalion chief to carry to the upper floors and that greatly repeats and enhances radio signal strength.[210]

The story with respect to Port Authority police officers in the North Tower is less complicated; most of them lacked access to the radio channel on which the

Port Authority police evacuation order was given. Since September 11, the Port Authority has worked hard to integrate the radio systems of their different commands.

The lesson of 9/11 for civilians and first responders can be stated simply: in the new age of terror, they-we-are the primary targets. The losses America suffered that day demonstrated both the gravity of the terrorist threat and the commensurate need to prepare ourselves to meet it.

The first responders of today live in a world transformed by the attacks on 9/11. Because no one believes that every conceivable form of attack can be prevented, civilians and first responders will again find themselves on the front lines. We must plan for that eventuality. A rededication to preparedness is perhaps the best way to honor the memories of those we lost that day.

> National Commission on Terrorist Attacks Upon the United States
> The Commission closed on August 21, 2004. This site is archived.

In 1986, the world watched and waited nervously as news slowly emerged from the then Soviet Union that the unthinkable may have happened at one of its nuclear power facilities. It had become apparent that a critical incident had occurred at the Chernobyl nuclear power plant in the Ukraine. A core reactor fire had led to a containment failure, releasing a deadly radiological cloud to the open environment. Thousands were evacuated from their homes, crops and water supplies were contaminated, fears about birth and developmental defects affecting the children and pregnant women soared, and scores of courageous firefighters and power plant workers lost their lives in a heroic and fatal struggle to quickly contain the release. Nations as far away as Sweden and Norway anxiously tracked the enormous plume of radioactive material as winds swept across Western Europe.

To this day the fears of Chernobyl have proved to be founded. Homes and whole communities evacuated in the immediate aftermath still remain uninhabitable. Water and soil in the region are considered radioactive and unusable and indeed there has been a profound, unmistakable spike in birth defects, developmental disorders, and cancer rates among the people who were exposed to the released radiation. A nuclear release is considered among the most terrifying catastrophic events because of its ability to harm so many, so quickly and for such a long time afterwards.

Finally, this chapter will conclude with an examination of a real-world, existing catastrophic event response plan. The *Federal Radiological Emergency Response Plan* has been developed in order to anticipate response needs and responder priorities, in the nightmarish event of a radiological or nuclear emergency impacting a community. The FRERP is representative of the type of planning and consideration that catastrophic event managers must consider in the face of the threats inherent in the modern world.

Federal Radiological Emergency Response Plan (FRERP)—Operational Plan[3]

Section Contents

I. Introduction and Background
- A. Introduction
- B. Participating Federal Agencies
- C. Scope
- D. Plan Considerations
  1. Public and Private Sector Response
  2. Coordination by Federal Agencies
  3. Federal Agency Authorities
  4. Federal Agency Resource Commitments
  5. Requests for Federal Assistance
  6. Reimbursement
- E. Training and Exercises
- F. Relationship to the Federal Response Plan (FRP)
  1. Without a Stafford Act Declaration
  2. With a Stafford Act Declaration
- G. Authorities

II. Concept of Operations
- A. Introduction
- B. Determination of Lead Federal Agency (LFA)
  1. Nuclear Facility
     a. Licensed by Nuclear Regulatory Commission (NRC) or an Agreement State
     b. Owned or Operated by DOD or DOE
     c. Not Licensed, Owned, or Operated by a Federal Agency or an Agreement State
  2. Transportation of Radioactive Materials
     a. Shipment of Materials Licensed by NRC or an Agreement State
     b. Materials Shipped by or for DOD or DOE

c. Shipment of Materials Not Licensed or Owned by a Federal Agency or an Agreement State

   3. Satellites Containing Radioactive Materials

   4. Impact from Foreign or Unknown Source

   5. Other Types of Emergencies

C. Radiological Sabotage and Terrorism

D. Response Functions and Responsibilities

   1. Onscene Coordination

   2. Onsite Management

   3. Radiological Monitoring and Assessment

      a. Role of Department of Energy (DOE)

      b. Role of the Environmental Protection Agency (EPA)

      c. Role of the Lead Federal Agency (LFA)

      d. Role of Other Federal Agencies

   4. Protective Action Recommendations

      a. Role of the Lead Federal Agency (LFA)

      b. Role of the Advisory Team for Environment, Food, and Health

   5. Other Federal Resource Support

      a. Role of the Federal Emergency Management Agency (FEMA)

      b. Role of Other Federal Agencies

   6. Public Information Coordination

      a. Role of the Lead Federal Agency (LFA)

      b. Role of the Federal Emergency Management Agency (FEMA)

      c. Role of Other Participating Agencies

   7. Congressional and White House Coordination

      a. Congressional Coordination

      b. White House Coordination

   8. International Coordination

   9. Response Function Overview

E. Stages of the Federal Response

   1. Notification

      a. Role of the Lead Federal Agency (LFA)

      b. Role of Federal Emergency Management Agency (FEMA)

   2. Activation and Deployment

      a. Role of the Lead Federal Agency (LFA)

      b. Role of Federal Emergency Management Agency (FEMA)

      c. Role of Other Federal Agencies

   3. Response Operations

      a. Joint Operations Center (JOC)

      b. Disaster Field Office (DFO)

      c. Federal Radiological Monitoring and Assessment Center (FRMAC)

      d. Advisory Team for Environment, Food, and Health

      e. Joint Information Center (JIC)

   4. Response Deactivation

   5. Recovery

Appendix A: Acronyms

Appendix B: Definitions

Appendix C: Federal Agency Response Missions, Capabilities and Resources, References, and Authorities

   A. Department of Agriculture

      1. Summary of Response Mission

      2. Capabilities and Resources

      3. USDA References

      4. USDA Specific Authorities.

   B. Department of Commerce

      1. Summary of Response Mission

      2. Capabilities and Resources

      3. DOC References

      4. DOC Specific Authorities

   C. Department of Defense

      1. Summary of Response Mission

      2. Capabilities and Resources

      3. DOD References.

      4. DOD Specific Authorities.

   D. Department of Energy

      1. Summary of Response Mission

      2. Capabilities and Resources

      3. DOE References

4. DOE Specific Authorities
E. Department of Health and Human Services
   1. Summary of Response Mission
   2. Capabilities and Resources
   3. HHS References
   4. HHS Specific Authorities
F. Department of Housing and Urban Development
   1. Summary of Response Mission
   2. Capabilities and Resources
   3. HUD References
   4. HUD Specific Authorities
G. Department of the Interior
   1. Summary of Response Mission
   2. Capabilities and Resources
   3. DOI References
   4. DOI Specific Authorities
H. Department of Justice
   1. Summary of Response Mission
   2. Capabilities and Resources
   3. DOJ References
   4. DOJ Specific Authorities
I. Department of State
   1. Summary of Response Mission
   2. Capabilities and Resources
   3. DOS References
   4. DOS Specific Authorities
J. Department of Transportation
   1. Summary of Response Mission
   2. Capabilities and Resources
   3. DOT References
   4. DOT Specific Authorities
K. Department of Veterans Affairs
   1. Summary of Response Mission

2. Capabilities and Resources

   3. VA References

   4. VA Specific Authorities

L. Environmental Protection Agency

   1. Summary of Response Mission

   2. Capabilities and Resources

   3. EPA References

   4. EPA Specific Authorities

M. Federal Emergency Management Agency

   1. Summary of Response Mission

   2. Capabilities and Resources

   3. FEMA References

   4. FEMA Specific Authorities

N. General Services Administration

   1. Summary of Response Mission

   2. Capabilities and Resources

   3. Funding

   4. GSA References

   5. GSA Specific Authorities

O. National Aeronautics and Space Administration

   1. Summary of Response Mission

   2. Capabilities and Resources

   3. NASA References

   4. NASA Specific Authorities

P. National Communications System

   1. Summary of Response Mission

   2. Capabilities and Resources

   3. NCS References

   4. NCS Specific Authorities

Q. Nuclear Regulatory Commission

   1. Summary of Response Mission

   2. Capabilities and Resources

3. NRC References

4. NRC Specific Authorities

Publication Information

Published Date: May 1, 1996

This plan, signed by FEMA Director James L. Witt on May 1, 1996, was originally published in the Federal Register as a Notice, dated May 8, 1996, Part III, pp. 20944-20970. A subsequent Correction to Notice, dated June 5, 1996, pp. 28583-28584, included the two figures appearing on Page II-19 that were inadvertently omitted in the original Notice.

Figures

Figure II-1. Notification Process.
Figure II-2. Onscene Response Operations Structure.

I. Introduction and Background

A. Introduction

The objective of the Federal Radiological Emergency Response Plan (FRERP) is to establish an organized and integrated capability for timely, coordinated response by Federal agencies to peacetime radiological emergencies.

The FRERP:

1. Provides the Federal Government's concept of operations based on specific authorities for responding to radiological emergencies

2. Outlines Federal policies and planning considerations on which the concept of operations of this Plan and Federal agency specific response plans are based and

3. Specifies authorities and responsibilities of each Federal agency that may have a significant role in such emergencies.

There are two Sections in this Plan. Section I contains background, considerations, and scope. Section II describes the concept of operations for response.

B. Participating Federal Agencies

Each participating agency has responsibilities and/or capabilities that pertain to various types of radiological emergencies. The following Federal agencies participate in the FRERP:

1. Department of Agriculture (USDA)
2. Department of Commerce (DOC)
3. Department of Defense (DOD)

4. Department of Energy (DOE)
5. Department of Health and Human Services (HHS)
6. Department of Housing and Urban Development (HUD)
7. Department of the Interior (DOI)
8. Department of Justice (DOJ)
9. Department of State (DOS)
10. Department of Transportation (DOT)
11. Department of Veterans Affairs (VA)
12. Environmental Protection Agency (EPA)
13. Federal Emergency Management Agency (FEMA)
14. General Services Administration (GSA)
15. National Aeronautics and Space Administration (NASA)
16. National Communications System (NCS) and
17. Nuclear Regulatory Commission (NRC).

C. Scope

The FRERP covers any peacetime radiological emergency that has actual, potential, or perceived radiological consequences within the United States, its Territories, possessions, or territorial waters and that could require a response by the Federal Government. The level of the Federal response to a specific emergency will be based on the type and/or amount of radioactive material involved, the location of the emergency, the impact on or the potential for impact on the public and environment, and the size of the affected area. Emergencies occurring at fixed nuclear facilities or during the transportation of radioactive materials, including nuclear weapons, fall within the scope of the Plan regardless of whether the facility or radioactive materials are publicly or privately owned, Federally regulated, regulated by an Agreement State, or not regulated at all. (Under the Atomic Energy Act of 1954 [Subsection 274.b.], the NRC has relinquished to certain States its regulatory authority for licensing the use of source, byproduct, and small quantities of special nuclear material.)

D. Plan Considerations

1. Public and Private Sector Response

For an emergency at a fixed nuclear facility or a facility not under the control of a Federal agency, State and local governments have primary responsibility for determining and implementing measures to protect life, property, and the environment in areas outside the facility boundaries. The owner or operator of a nuclear facility has primary responsibility for actions within the boundaries of that facility, for providing notification and advice to offsite officials, and for minimizing the radiological hazard to the public.

For emergencies involving an area under Federal control, the responsibility for onsite actions belongs to a Federal agency, while offsite actions are the responsibility of the State or local government.

For all other emergencies, the State or local government has the responsibility for taking emergency actions both onsite and offsite, with support provided, upon request, by Federal agencies as designated in Section II of this plan.

2. Coordination by Federal Agencies

This Plan describes how the Federal response to a radiological emergency will be organized. It includes guidelines for notification of Federal agencies and States, coordination and leadership of Federal response activities onscene, and coordination of Federal public information activities and Congressional relations by Federal agencies. The Plan suggests ways in which the State, local, and Federal agencies can most effectively integrate their actions. The degree to which the Federal response is merged or to which activities are adjusted will be based upon the requirements and priorities set by the State.

Appropriate independent emergency actions may be taken by the participating Federal agencies within the limits of their own statutory authority to protect the public, minimize immediate hazards, and gather information about the emergency that might be lost by delay.

3. Federal Agency Authorities

Some Federal agencies have authority to respond to certain situations affecting public health and safety with or without a State request. Appendix C of this Plan cites relevant legislative and executive authorities. This Plan does not create any new authorities nor change any existing ones.

A response to radiological emergencies on or affecting Federal lands not occupied by a government agency should be coordinated with the agency responsible for managing that land to ensure that response activities are consistent with Federal statutes governing the use and occupancy of these lands. This coordination is necessary in the case of Indian tribal lands because Federally recognized Indian tribes have a special relationship with the U.S. Government, and the State and local governments may have limited or no authority on their reservations.

In the event of an offsite radiological accident involving a nuclear weapon, special nuclear material, classified components, or all three, the owner (either DOD, DOE, or NASA) will declare a National Defense Area (NDA) or National Security Area (NSA), respectively, and this area will become "onsite" for the purposes of this plan. NDAs and NSAs are established to safeguard classified information, and/or restricted data, or equipment and material. Establishment of these areas places non-Federal lands under Federal control and results only from an emergency event. It is possible that radioactive contamination would extend beyond the boundaries of these areas.

In accordance with appropriate national security classification directives, information may be classified concerning nuclear weapons, special nuclear materials at reactors, and certain fuel cycle facilities producing military fuel.

4. Federal Agency Resource Commitments

Agencies committing resources under this Plan do so with the understanding that the duration of the commitment will depend on the nature and extent of the emergency and the State and local resources available. Should another emergency occur that is more serious or of higher priority (such as one that may jeopardize national security), Federal agencies will reassess resources committed under this Plan.

5. Requests for Federal Assistance

State and local government requests for assistance, as well as those from owners and operators of radiological facilities or activities, may be made directly to the Federal agencies listed in Table II-1, FEMA, or to other Federal agencies with whom they have preexisting arrangements or relationships.

6. Reimbursement

The cost of each Federal agency's participation in support of the FRERP is the responsibility of that agency, unless other agreements or reimbursement mechanisms exist. GSA will be reimbursed for supplies and services provided under this Plan in accordance with prior interagency agreements.

E. Training and Exercises

Federal agencies, in conjunction with State and local governments, will periodically exercise the FRERP. Each agency will coordinate its exercises with the Federal Radiological Preparedness Coordinating Committee's (FRPCC's) Subcommittee on Federal Response to avoid duplication and to invite participation by other Federal agencies.

Federal agencies will assist other Federal agencies and State and local governments with planning and training activities designed to improve response capabilities. Each agency should coordinate its training programs with the FRPCC's Subcommittee on Training to avoid duplication and to make its training available to other agencies.

F. Relationship to the Federal Response Plan (FRP)

1. Without a Stafford Act Declaration

Federal agencies will respond to radiological emergencies using the FRERP, each agency in accordance with existing statutory authorities and funding resources. The LFA has responsibility for coordination of the overall Federal response to the emergency. FEMA is responsible for coordinating non-radiological support using the structure of the Federal Response Plan (FRP).

2. With a Stafford Act Declaration

When a major disaster or emergency is declared under the Stafford Act and an associated radiological emergency exists, the functions and responsibilities of the FRERP remain the same. The LFA coordinates the management of the radiological response with the Federal Coordinating Officer (FCO). Although the direction of the radiological response remains the same with the LFA, the FCO has the overall responsibility for the coordination of Federal assistance in support of State and local governments using the FRP.

G. Authorities

The following authorities are the basis for the development of this Plan:

1. Nuclear Regulatory Commission Authorization, Public Law 96-295, June 30, 1980, Section 304. This authorization requires the President to prepare and publish a "National Contingency Plan" (subsequently renamed the FRERP) to provide for expeditious, efficient, and coordinated action by appropriate Federal agencies to protect the public health and safety in case of accidents at commercial nuclear power plants.

2. Executive Order (E.O.) 12241, National Contingency Plan, September 29, 1980. This E.O. delegates to the Director of FEMA the responsibility for publishing the National Contingency Plan (i.e., the FRERP) for accidents at nuclear power facilities and requires that it be published from time to time in the Federal Register. Executive Order 12241 has been amended by Executive Order 12657, FEMA Assistance in Emergency Preparedness Planning at Commercial Nuclear Power Plants.

Authorities for the activities of individual Federal agencies appear in Appendix C.

II. Concept of Operations

A. Introduction

The concept of operations for a response provides for the designation of one agency as the Lead Federal Agency (LFA) and for the establishment of onscene, interagency response centers. The FRERP describes both the responsibilities of the LFA and other Federal agencies that may be involved and the functions of each of the on scene centers.

The concept of operations recognizes the preeminent role of State and local governments for determining and implementing any measures to protect life, property, and the environment in areas not under the control of a Federal agency.

B. Determination of Lead Federal Agency (LFA)

The agency that is responsible for leading and coordinating all aspects of the Federal response is referred to as the LFA and is determined by the type of emergency. In situations where a Federal agency owns, authorizes, regulates, or is otherwise deemed responsible for the facility or radiological activity causing the emergency and has authority to conduct and manage Federal actions onsite, that agency normally will be the LFA.

The following identifies the LFA for each specified type of radiological emergency.

1. Nuclear Facility

a. Licensed by Nuclear Regulatory Commission (NRC) or an Agreement State

The NRC is the LFA for an emergency that occurs at a fixed facility or regarding an activity licensed by the NRC or an Agreement State. These include, but are not limited to, commercial nuclear power reactors, fuel cycle facilities, DOE-owned gaseous diffusion facilities that are operating under NRC regulatory oversight, and radiopharmaceutical manufacturers.

b. Owned or Operated by DOD or DOE

The LFA is either DOD or DOE, depending on which agency owns or authorizes operation of the facility. These emergencies may involve reactor operations, nuclear material and weapons production, radioactive material from nuclear weapons, or other radiological activities.

c. Not Licensed, Owned, or Operated by a Federal Agency or an Agreement State

The EPA is the LFA for an emergency that occurs at a facility not licensed, owned, or operated by a Federal agency or an Agreement State. These include facilities that possess, handle, store, or process radium or accelerator-produced radioactive materials.

2. Transportation of Radioactive Materials

a. Shipment of Materials Licensed by NRC or an Agreement State

The NRC is the LFA for an emergency that involves radiological material licensed by the NRC or an Agreement State.

b. Materials Shipped by or for DOD or DOE

The LFA is either DOD or DOE depending on which of these agencies has custody of the material at the time of the accident.

c. Shipment of Materials Not Licensed or Owned by a Federal Agency or an

Agreement State

The EPA is the LFA for an emergency that involves radiological material not licensed or owned by a Federal agency or an Agreement State.

3. Satellites Containing Radioactive Materials

NASA is the LFA for NASA spacecraft missions. DOD is the LFA for DOD spacecraft missions. DOE and EPA provide technical assistance to DOD and NASA.

In the event of an emergency involving a joint U.S. Government and foreign government spacecraft venture containing radioactive sources and/or classified components, the LFA will be DOD or NASA, as appropriate. A joint U.S./foreign venture is defined as an activity in which the U.S. Government has an ongoing interest in the successful completion of the mission and is intimately involved in mission operations. A joint venture is not created by simply selling or supplying material to a foreign country for use in their spacecraft. DOE and EPA will provide technical support and assistance to the LFA.

4. Impact from Foreign or Unknown Source

The EPA is the LFA for an emergency that involves radioactive material from a foreign or unknown source that has actual, potential, or perceived radiological consequences in the United States, its Territories, possessions, or territorial waters. The foreign or unknown source may be a reactor (e.g., Chernobyl), a spacecraft containing radioactive material, radioactive fallout from atmospheric testing of nuclear devices, imported radioactively contaminated material, or a shipment of foreign-owned radioactive material. Unknown sources of radioactive material refers to that material whose origin and/or radiological nature is not yet established. These types of sources include contaminated scrap metal or abandoned radioactive material. DOD, DOE, NASA, and NRC provide technical assistance to EPA.

5. Other Types of Emergencies

In the event of an unforeseen type of emergency not specifically described in this Plan or a situation where conditions exist involving overlapping responsibility that could cause confusion regarding LFA role and responsibilities, DOD, DOE, EPA, NASA, and NRC will confer upon receipt of notification of the emergency to determine which agency is the LFA.

Table II-1.-Identification of Lead Federal Agency for Radiological Emergencies

Type of emergency	Lead Federal agency
1. Nuclear Facility:	
a. Licensed by NRC or an Agreement State	NRC

b. Owned or Operated by DOD or DOE	DOD or DOE
c. Not Licensed, Owned, or Operated by a Federal Agency or an Agreement State	EPA
2. Transportation of Radioactive Materials:	
a. Shipment of Materials Licensed by NRC or an NRC. Agreement State	NRC
b. Materials Shipped by or for DOD or DOE	DOD or DOE
c. Shipment of Materials Not Licensed or Owned by a Federal Agency or an Agreement State	EPA
3. Satellites Containing Radioactive Materials	NASA or DOD
4. Impact from Foreign or Unknown Source	EPA
5. Other Types of Emergencies	LFAs confer

C. Radiological Sabotage and Terrorism

For fixed facilities and materials in transit, responses to radiological emergencies generally do not depend on the initiating event. The coordinated response to contain or mitigate a threatened or actual release of radioactive material would be essentially the same whether it resulted from an accidental or deliberate act. For malevolent acts involving improvised nuclear or radiation dispersal devices, the response is further complicated by the magnitude of the threat and the need for specialized technical expertise/actions. Therefore, sabotage and terrorism are not treated as separate types of emergencies rather, they are considered a complicating dimension of the types listed in Table II-1.

The Atomic Energy Act directs the Federal Bureau of Investigation (FBI) to investigate all alleged or suspected criminal violations of the Act. Additionally, the FBI is legally responsible for locating any nuclear weapon, device, or material and for restoring nuclear facilities to their rightful custodians. In view of its unique responsibilities under the Atomic Energy Act (amended by the Energy Reorganization Act), the FBI has concluded formal agreements with the LFAs that provide for interface, coordination, and technical assistance in support of the FBI's mission.

Generally, for fixed facilities and materials in transit, the designated LFA and supporting agencies will perform the functions delineated in this plan and provide technical support and assistance to the FBI in the performance of its mission. It would be difficult to outline all the possible scenarios arising from criminal or terrorist activity. As a result, the Federal response will be tailored to the specific circumstances of the event at hand. For those emergencies where an LFA is not specifically designated (e.g., improvised nuclear device), the Federal response will be guided by the established interagency agreements and contingency plans. In accordance with these agreements and plans, the signatory agency(ies) supporting the FBI will coordinate and manage the technical portion of the response and activate/request assistance under the FRERP for measures to protect the public health and safety. In all cases, the FBI will manage and direct the law enforcement and intelligence aspects of the response coordinating activities with appropriate Federal, State, and local agencies within the framework of the FRERP and/or as provided for in established interagency agreements or plans.

D. Response Functions and Responsibilities

1. Onscene Coordination

The LFA will lead and coordinate all Federal onscene actions and assist State and local governments in determining measures to protect life, property, and the environment. The LFA will ensure that FEMA and other Federal agencies assist the State and local government agencies in implementing protective actions, if requested by the State and local government agencies.

The LFA will coordinate Federal response activities from an onscene location, referred to as the Joint Operations Center (JOC). Until the LFA has established its base of operations in a JOC, the LFA will accomplish that coordination from another LFA facility, usually a Headquarters operations center.

In the absence of existing agreements for radiological emergencies occurring on or with possible consequences to Indian tribal lands, DOI will provide liaison between federally recognized Indian tribal governments and LFA, State, and local agencies for coordination of response and protective action efforts. Additionally, DOI will advise and assist the LFA on economic, social, and political matters in the United States insular areas should a radiological emergency occur.

2. Onsite Management

The LFA will oversee the onsite response monitor and support owner or operator activities (when there is an owner or operator) provide technical support to the owner or operator, if requested and serve as the principal Federal source of information about onsite conditions. The LFA will provide a hazard assessment of onsite conditions that might have significant offsite impact and ensure onsite measures are taken to mitigate offsite consequences.

3. Radiological Monitoring and Assessment

DOE has the initial responsibility for coordinating the offsite Federal radiological monitoring and assessment assistance during the response to a radiological emergency. In a prolonged response, EPA will assume the responsibility for coordinating the assistance at some mutually agreeable time, usually after the emergency phase.

Some of the participating Federal agencies may have radiological planning and emergency responsibilities as part of their statutory authority, as well as established working relationships with State counterpart agencies. The monitoring and assessment activity, coordinated by DOE, does not alter those responsibilities but complements them by providing for coordination of the initial Federal radiological monitoring and assessment response activity.

Activities will:

- Support the monitoring and assessment programs of the States
- Respond to the assessment needs of the LFA and
- Meet statutory responsibilities of participating Federal agencies.

Federal offsite monitoring and assessment activities will be coordinated with those of the State. Federal agency plans and procedures for implementing this monitoring and assessment activity are designed to be compatible with the radiological emergency planning requirements for State, local governments, specific facilities, and existing memoranda of understanding and interagency agreements.

DOE may respond to a State or LFA request for assistance by dispatching a Radiological Assistance Program (RAP) team. If the situation requires more assistance than a RAP team can provide, DOE will alert or activate additional resources. These resources may include the establishment of a Federal Radiological Monitoring and Assessment Center (FRMAC) to be used as an onscene coordination center for Federal radiological assessment activities. Federal and State agencies are encouraged to collocate their radiological assessment activities.

Federal radiological monitoring and assessment activities will be activated as a component of an FRERP response or pursuant to a direct request from State or local governments, other Federal agencies, licensees for radiological materials, industries, or the general public after evaluating the magnitude of the problem and coordinating with the State(s) involved.

DOE and other participating Federal agencies may learn of an emergency when they are alerted to a possible problem or receive a request for radiological assistance. DOE will maintain national and regional coordination offices as points of access to Federal radiological emergency assistance. Requests for Federal radiological monitoring and assessment assistance will generally be directed to the appropriate DOE radiological assistance Regional Coordinating Office. Requests also can go directly to DOE's Emergency Operations Center (EOC) in Washington, DC. When other agencies receive requests for Federal radiological monitoring and assessment assistance, they will promptly notify the DOE EOC.

a. Role of Department of Energy (DOE)

(1) Initial Response Coordination Responsibility. DOE, as coordinator, has the

following responsibilities:

   (a) Coordinate Federal offsite radiological environmental monitoring and assessment activities

   (b) Maintain technical liaison with State and local agencies with monitoring and assessment responsibilities

   (c) Maintain a common set of all offsite radiological monitoring data, in an accountable, secure, and retrievable form, and ensure the technical integrity of the FRMAC data

   (d) Provide monitoring data and interpretations, including exposure rate contours, dose projections, and any other requested radiological assessments, to the LFA, and to the States

   (e) Provide, in cooperation with other Federal agencies, the personnel and equipment needed to perform radiological monitoring and assessment activities

   (f) Request supplemental assistance and technical support from other Federal agencies as needed and

   (g) Arrange consultation and support services through appropriate Federal agencies to all other entities (e.g., private contractors) with radiological monitoring functions and capabilities, and technical and medical advice on handling radiological contamination and population monitoring.

(2) Transition of Response Coordination Responsibility. The DOE FRMAC Director will work closely with the Senior EPA representative to facilitate a smooth transition of the Federal radiological monitoring and assessment coordination responsibility to EPA at a mutually agreeable time and after consultation with the States and LFA. The following conditions are intended to be met prior to this transfer:

   (a) The immediate emergency condition has been stabilized

   (b) Offsite releases of radioactive material have ceased, and there is little or no potential for further unintentional offsite releases

   (c) The offsite radiological conditions have been characterized and the immediate consequences have been assessed

   (d) An initial long-range monitoring plan has been developed in conjunction with the affected States and appropriate Federal agencies and

   (e) EPA has received adequate assurances from the other Federal agencies that they will commit the required resources, personnel, and funds for the duration of the Federal response.

b. Role of the Environmental Protection Agency (EPA)

Prior to assuming responsibility for the FRMAC, EPA will:

(1) Provide resources, including personnel, equipment, and laboratory support (including mobile laboratories), to assist DOE in monitoring radioactivity levels in the environment

(2) Assume coordination of Federal radiological monitoring and assessment responsibilities from DOE after the transition

(3) Assist in the development and implementation of a long- term monitoring plan and

(4) Provide nationwide environmental monitoring data from the Environmental Radiation Ambient Monitoring Systems for assessing the national impact of the accident.

c. Role of the Lead Federal Agency (LFA)

(1) Ensure that State's needs are addressed.

(2) Approve the release of official Federal offsite monitoring data and assessments.

(3) Provide other available radiological monitoring data to the State and to the FRMAC.

d. Role of Other Federal Agencies

Agencies carrying out responsibilities related to radiological monitoring and assessment during a Federal response also will coordinate their activities with FRMAC. This coordination will not limit the normal working relationship between a Federal agency and its State counterparts nor restrict the flow of information from that agency to the States. The radiological monitoring and assessment responsibilities of the other Federal agencies include:

(1) Department of Agriculture (USDA)

>(a) Inspect meat and meat products, poultry and poultry products, and egg products identified for interstate and foreign commerce to assure that they are safe for human consumption.

>(b) Assist, in conjunction with HHS, in monitoring the production, processing, storage, and distribution of food through the wholesale level to eliminate contaminated product or to reduce the contamination in the product to a safe level.

>(c) Collect agricultural samples within the Ingestion Exposure Pathway Emergency Planning Zone. Assist in the evaluation and assessment of data to determine the impact of the emergency on agriculture.

(2) Department of Commerce (DOC)

>(a) Prepare operational weather forecasts tailored to support emergency response activities.

>(b) Prepare and disseminate predictions of plume trajectories, dispersion, and deposition.

(c) Archive, as a special collection, the meteorological data from national observing systems applicable to the monitoring and assessment of the response.

(d) Ensure that marine fishery products available to the public are not contaminated.

(e) Provide assistance and reference material for calibrating radiological instruments.

(3) Department of Defense (DOD)

(a) Provide radiological resources to include trained response personnel, specialized radiation instruments, mobile instrument calibration, repair capabilities, and expertise in site restoration.

(b) Perform special sampling of airborne contamination on request.

(4) Department of Health and Human Services (HHS)

(a) In conjunction with USDA, inspect production, processing, storage, and distribution facilities for human food and animal feeds, which may be used in interstate commerce, to assure protection of the public health.

(b) Collect samples of agricultural products to monitor and assess the extent of contamination as a basis for recommending or implementing protective actions.

(5) Department of the Interior (DOI)

(a) Provide hydrologic advice and assistance, including monitoring personnel, equipment, and laboratory support.

(b) Advise and assist in evaluating processes affecting radioisotopes in soils, including personnel, equipment, and laboratory support.

(c) Advise and assist in the development of geographical information systems (GIS) databases to be used in the analysis and assessment of contaminated areas including personnel, equipment, and databases.

(6) Nuclear Regulatory Commission (NRC)

(a) Provide assistance in Federal radiological monitoring and assessment activities during incidents.

(b) Provide, where available, continuous measurement of ambient radiation levels around NRC licensed facilities, primarily power reactors using thermoluminescent dosimeters (TLD).

4. Protective Action Recommendations

Federal protective action recommendations provide advice to State and local governments on measures that they should take to avoid or reduce exposure of the public to radiation from a release of radioactive material. This includes advice on emergency actions such as sheltering, evacuation, and prophylactic use of stable iodine. It also includes longer term measures to avoid or minimize exposure to residual radiation or exposure through the ingestion pathway such as restriction of food, temporary relocation, and permanent resettlement.

a. Role of the Lead Federal Agency (LFA)

The LFA will assist State and local authorities, if requested, by advising them on protective actions for the public. The development or evaluation of protective action recommendations will be based upon the Protective Action Guides (PAGs) issued by EPA and HHS. In providing such advice, the LFA will use advice from other Federal agencies with technical expertise on those matters whenever possible. The LFA's responsibilities for the development, evaluation, and presentation of protective action recommendations are to:

(1) Respond to requests from State and local governments for technical information and assistance

(2) Consult with representatives from EPA, HHS, USDA, and other Federal agencies as needed to provide advice to the LFA on protective actions

(3) Review all recommendations made by other Federal agencies exercising statutory authorities related to protective actions to ensure consistency

(4) Prepare a coordinated Federal position on protective action recommendations whenever time permits and

(5) Present the Federal assessment of protective action recommendations, in conjunction with FEMA and other Federal agencies when practical, to State or other offsite authorities.

b. Role of the Advisory Team for Environment, Food, and Health

Advice on environment, food, and health matters will be provided to the LFA through the Advisory Team for Environment, Food, and Health (Advisory Team) consisting of representatives of EPA, HHS, and USDA supported by other Federal agencies, as warranted by the circumstances of the emergency. The Advisory Team provides direct support to the LFA and has no independent authority. The Advisory Team will not release information or make recommendations to the public unless authorized to do so by the LFA. The Advisory Team will select a chair for the Team. The Advisory Team will normally collocate with the FRMAC.

For emergencies with potential for causing widespread radiological contamination where no onscene FRMAC is established, the functions of the Advisory Team may be accomplished in the LFA response facility in Washington, DC.

The primary role of the Advisory Team is to provide a mechanism for timely, interagency coordination of advice to the LFA, States, and other Federal agencies concerning matters related to the following areas:

(1) Environmental assessments (field monitoring) required for developing recommendations

(2) PAGs and their application to the emergency

(3) Protective action recommendations using data and assessment from the FRMAC

(4) Protective actions to prevent or minimize contamination of milk, food, and water and to prevent or minimize exposure through ingestion

(5) Recommendations regarding the disposition of contaminated livestock and poultry

(6) Recommendations for minimizing losses of agricultural resources from radiation effects

(7) Availability of food, animal feed, and water supply inspection programs to assure wholesomeness

(8) Relocation, reentry, and other radiation protection measures prior to recovery

(9) Recommendations for recovery, return, and cleanup issues

(10) Health and safety advice or information for the public and for workers

(11) Estimate effects of radioactive releases on human health and environment

(12) Guidance on the use of radioprotective substances (e.g., thyroid blocking agents), including dosage and projected radiation doses that warrant the use of such drugs and

(13) Other matters, as requested by the LFA.

5. Other Federal Resource Support

FEMA will coordinate the provision of non-radiological (i.e., not related to radiological monitoring and assessment) Federal resources and assistance to affected State and local governments. The Federal non-radiological resource and assistance coordination function will be performed at the Disaster Field Office (DFO) (or other appropriate location established by FEMA).

a. Role of the Federal Emergency Management Agency (FEMA)

(1) Monitor the status of the Federal response to requests for non-radiological assistance from the affected States and provide this information to the States.

(2) Keep the LFA informed of requests for assistance from the State and the status of the Federal response.

(3) Identify and inform Federal agencies of actual or apparent omissions, redundancies, or conflicts in response activity.

(4) Establish and maintain a source of integrated, coordinated information about

the status of all non-radiological resource support activities.

(5) Provide other non-radiological support to Federal agencies responding to the emergency.

b. Role of Other Federal Agencies

In order to properly coordinate activities, Federal agencies responding to requests for non- radiological support or directly providing such support under statutory authorities will provide liaison personnel to the DFO. The following indicates types of assistance that may be provided by Federal agencies as needed or requested:

(1) Department of Agriculture (USDA)

(a) Provide emergency food coupon assistance in officially designated disaster areas, if a need is determined by officials and if the commercial food system is sufficient to accommodate the use of food coupons.

(b) Provide for placement of USDA donated food supplies from warehouses, local schools, and other outlets to emergency care centers. These are foods donated to various outlets through USDA food programs.

(c) Provide lists that identify locations of alternate sources of food and livestock feed.

(d) Assist in providing temporary housing for evacuees.

(e) Assess damage to crops, soil, livestock, poultry, and processing facilities and incorporate findings in a damage assessment report.

(f) Provide emergency communications assistance to the agricultural community through the State Research, Education, and Extension Services" electronic mail system.

(2) Department of Commerce (DOC)

Provide radiation shielding materials.

(3) Department of Defense (DOD)-DOD may provide assistance in the form of personnel, logistics and telecommunications, advice on proper medical treatment of personnel exposed to or contaminated by radioactive materials, and assistance, including airlift services, when available, upon the request of the LFA or FEMA. Requests for assistance must be directed to the National Military Command Center or through channels established by prior agreements.

(4) Department of Energy (DOE)-Provide advice on proper medical treatment of personnel exposed to or contaminated by radioactive materials.

(5) Department of Health and Human Services (HHS)

(a) Ensure the availability of health and medical care and other human services (especially for the aged, poor, infirm, blind, and others most in need).

(b) Assist in providing crisis counseling to victims in affected geographic areas.

(c) Provide guidance to State and local health officials on disease control measures and epidemiological surveillance and study of exposed populations.

(d) Provide advice on proper medical treatment of personnel exposed to or contaminated by radioactive materials.

(e) Provide advice and guidance in assessing the impact of the effects of radiological incidents on the health of persons in the affected area.

(6) Department of Housing and Urban Development (HUD)

(a) Review and report on available housing for disaster victims and displaced persons.

(b) Assist in planning for and placing homeless victims in available housing.

(c) Provide staff to support emergency housing within available resources.

(d) Provide housing assistance and advisory personnel.

(7) Department of the Interior (DOI)-Advise and assist in assessing impacts to economic, social, and political issues relating to natural resources, including fish and wildlife, subsistence uses, public lands, Indian Tribal lands, land reclamation, mining, minerals, and water resources.

(8) Department of Transportation (DOT)

(a) Support State and local governments by identifying sources of civil transportation on request and when consistent with statutory responsibilities.

(b) Coordinate the Federal civil transportation response in support of emergency transportation plans and actions with State and local governments. (This may include provision of Federally controlled transportation assets and the controlling of airspace or transportation routes to protect commercial transportation and to facilitate the movement of response resources to the scene.)

(c) Provide Regional Emergency Transportation Coordinators and staff to assist State and local authorities in planning and response.

(d) Provide technical advice and assistance on the transportation of radiological materials and the impact of the incident on the transportation system.

(9) Department of Veterans Affairs (VA)

(a) Provide medical assistance using Medical Emergency Radiological Response Teams (MERRTs).

(b) Provide temporary housing.

(10) General Services Administration (GSA)

(a) Provide acquisition and procurement of floor space, telecommunications and automated data processing services, supplies, services, transportation, computers, contracting, equipment, and material as well as specified logistical services that exceed the capabilities of other Federal agencies.

(b) Activate the Regional Emergency Communications Planner (RECP) and a Federal Emergency Communications Coordinator (FECC). RECP will provide technical support and accept guidance from the FEMA Regional Director during the pre-deployment phase of a telecommunications emergency.

(c) Upon request, will dispatch the FECC to the scene to expedite the provision of the telecommunications services.

(11) National Communications System (NCS)-Acting through its operational element, the National Coordinating Center for Telecommunications (NCC), the NCS will ensure the provision of adequate telecommunications support to Federal FRERP operations.

6. Public Information Coordination

Public information coordination is most effective when the owner/operator, Federal, State, local, and other relevant information sources participate jointly. The primary location for linking these sources is the Joint Information Center (JIC).

Prior to the establishment of Federal operations at the JIC, it may be necessary to release Federal information regarding public health and safety. In these instances, Federal agencies will coordinate with the LFA and the State in advance or as soon as possible after the information has been released.

This coordination will accomplish the following: compile information about the status of the emergency, response actions, and instructions for the affected population coordinate all information from various sources with the other Federal, State, local, and non-governmental response organizations allow various sources to work cooperatively, yet maintain their independence in disseminating information disseminate timely, consistent, and accurate information to the public and the news media and establish coordinated arrangements for dealing with citizen inquiries.

a. Role of the Lead Federal Agency (LFA)

The LFA is responsible for information on the status of the overall Federal response, specific LFA response activities, and the status of onsite conditions.

The LFA will:

(1) Develop joint information procedures for providing Federal information to and for obtaining information from all Federal agencies participating in the response

(2) Work with the owner/operator and State and local government information officers to develop timely coordinated public information releases

(3) Inform the media that the JIC is the primary source of onscene public information and news from facility, local, State, and Federal spokespersons

(4) Establish and manage Federal public information operations at the JIC and

(5) Coordinate Federal public information among the various media centers.

b. Role of the Federal Emergency Management Agency (FEMA)

FEMA will assist the LFA in coordinating non-radiological information among Federal agencies and with the State. When mutually agreeable, FEMA may assume responsibility from the LFA for coordinating Federal public information. Should this occur, it will usually be after the onsite situation has been stabilized and recovery efforts have begun.

c. Role of Other Participating Agencies

All Federal agencies with an operational response role under the FRERP will coordinate public information activities at the JIC. Each Federal agency will provide information on the status of its response and on technical information.

7. Congressional and White House Coordination

a. Congressional Coordination

Federal agencies will coordinate their responses to Congressional requests for information with the LFA. Points of contact for this function are the Congressional Liaison Officers. All Federal agency Congressional Liaison Officers and Congressional staffs seeking site-specific information about the emergency should contact the LFA headquarters Congressional Affairs Office. Congress may request information directly from any Federal agency. Any agency responding to such requests should inform the LFA as soon as feasible.

b. White House Coordination

The LFA will report to the President and keep the White House informed on all aspects of the emergency. The White House may request information directly from any Federal agency. Any agency responding to such requests should inform the LFA as soon as feasible. The LFA will submit reports to the White House. The initial report should cover, if possible, the nature of and prognosis for the radiological situation causing the emergency and the actual or potential offsite radiological impact. Subsequent reports by the LFA should cover the status of mitigation, corrective actions, protective measures, and overall Federal response to the emergency. Federal agencies should provide information related to the technical and radiological aspects of the response directly to the LFA. FEMA will compile information related to the non-radiological resource support aspects of the response and submit to the LFA for inclusion in the report(s).

8. International Coordination

In the event of an environmental impact or potential impact upon the United States, its possessions, Territories, or territorial waters from a radiological emergency originating on foreign soil or, conversely, a domestic incident with an actual or potential foreign impact, the LFA will immediately inform DOS (which has responsibility for official interactions with foreign governments). The LFA will keep DOS informed of all Federal response activities. The DOS will coordinate notification and information gathering activities with foreign governments, except in cases where existing bilateral agreements permit direct communication. Where the LFA has existing bilateral agreements that permit direct exchange of information, those agencies should keep DOS informed of consultations with their foreign counterparts. Agency officials should take care that consultations do not exceed the scope of the relevant agreement(s). The LFA will ensure that any offers of assistance to or requests from foreign governments are coordinated with DOS.

9. Response Function Overview

Table II-2 provides an overview of the responsible Federal agencies for major response functions.

Table II-2.-Response Function Overview

Response action	Responsible agency
(1) Maintain cognizance of the Federal response conduct and manage Federal onsite actions.	LFA
(2) Coordinate Federal offsite radiological monitoring and assessment:	
Initial Response	DOE.
Intermediate and Long-Term Response	EPA
(3) Develop and evaluate recommendations for offsite protective actions for the public.	LFA, in coordination with other agencies.
(4) Present recommendations for offsite protective actions to the appropriate State and/or local officials	LFA, FEMA, in conjunction with other Federal agencies when practical.

(5) Coordinate Federal offsite non-radiological resource support.	FEMA
(6) Coordinate release of Federal information to the LFA after public. mutual agreement.	FEMA
(7) Coordinate release of Federal information to LFA.	Congress
(8) Provide reports to the President and keep the White House informed on all aspects of the emergency.	LFA
(9) Coordinate international aspects and make DOS as required international notifications. appropriate.	LFA
(10) Coordinate the law enforcement aspects of a criminal act involving radioactive material.	DOJ/FB

E. Stages of the Federal Response

The Federal response is divided into five stages: Notification, Activation and Deployment, Response Operations, Response Deactivation, and Recovery.

1. Notification

The owner or operator of the facility or radiological activity is generally the first to become aware of a radiological emergency and is responsible for notifying the State and local authorities and the LFA. The notification should include:

- Location and nature of the accident,
- An assessment of the severity of the problem,
- Potential and actual offsite consequences, and
- Initial response actions.

If any Federal agency receives notification from any source other than FEMA or the LFA, the agency will notify the LFA. See Figure II-1 for the notification process.

a. Role of the Lead Federal Agency (LFA)

(1) Verify accuracy of notification

(2) Notify FEMA and advisory team agencies and provide information

(3) Verify that other Federal agencies have been notified and

(4) Verify that the State has been notified.

b. Role of Federal Emergency Management Agency (FEMA)

(1) Verify that the State has been notified of the emergency and

(2) Notify other Federal agencies as appropriate.

2. Activation and Deployment

Once notified, each agency will respond according to its plan. The LFA will assess the technical response requirements and cause the activation and deployment of response components. FEMA, in conjunction with the LFA, will coordinate the non-radiological assistance in support of State and local governments. Initially, the LFA, FEMA, and other Federal agencies will coordinate response actions from their headquarters locations, usually from their respective headquarters EOCs.

a. Role of the Lead Federal Agency (LFA)

(1) Deploy LFA response personnel to the scene and provide liaison to the State and local authorities as appropriate

(2) Designate a Federal Onscene Commander (OSC) at the scene of the emergency to manage onsite activities and coordinate the overall Federal response to the emergency

(3) Establish bases of Federal operation, such as the JOC and the JIC

(4) Coordinate the Federal response with the owner/operator and

(5) Provide advice on the radiological hazard to the Federal responders.

b. Role of Federal Emergency Management Agency (FEMA)

(1) Activate a Regional Operations Center (ROC) to monitor the situation

(2) Establish contact with the LFA and the affected State to determine the status of non-radiological response requirements

(3) Designate a Senior FEMA Official (SFO) to coordinate activities with the LFA and

(4) Coordinate the provision of non-radiological Federal resources and assistance.

c. Role of Other Federal Agencies

(1) Designate an onscene Senior Agency Official

(2) Activate agency emergency response personnel and deploy them to the scene

(3) Deploy FRMAC assets

(4) Deploy Advisory Team representatives

(5) Keep the LFA and FEMA informed of status of response activities and

(6) Coordinate all State requests and offsite activities with the LFA and FEMA,

3. Response Operations

The following describes the general operational structure for meeting Federal agency roles and responsibilities in response to a radiological emergency. At the headquarters level, the LFA, FEMA, and other Federal agencies (OFAs) will generally exchange liaison personnel and maintain staffs at their EOCs to support their respective onscene operations. Federal agencies may also activate a regional or field office EOC in support of the emergency. Figure II-2 provides a graphic depiction of the onscene structure.

a. Joint Operations Center (JOC)

The JOC$^{(1)}$ is established by the LFA under the operational control of the Federal OSC as the focal point for management and direction of onsite activities, establishment of State requirements and priorities, and coordination of the overall Federal response. The JOC may be established in a separate onscene location or collocated with an existing emergency operations facility. The following elements may be represented in the JOC:

(1) LFA staff and onsite liaison

(2) FEMA/DFO liaison

(3) FRMAC liaison

(4) Advisory Team liaison

(5) Other Federal agency liaison, as needed

(6) LFA Public information liaison

(7) LFA Congressional liaison and

(8) State and local liaison.

b. Disaster Field Office (DFO)

The DFO is established by FEMA as the focal point for the coordination and provision of non-radiological resource support based on coordinated State requirements/priorities. The DFO is established at an onscene location in coordination with State and local authorities and other Federal agencies. The following elements may be represented in the DFO:

(1) LFA liaison

(2) Other appropriate Federal agency personnel

(3) State and local liaison

(4) Public information liaison and

(5) Congressional liaison.

c. Federal Radiological Monitoring and Assessment Center (FRMAC)

The FRMAC is established by DOE (with subsequent transfer to EPA for intermediate and long-term actions) for the coordination of Federal radiological monitoring and assessment activities with that of State and local agencies. The FRMAC is established at an onscene location in coordination with State and local authorities and other Federal agencies. The following elements may be represented in the FRMAC:

(1) DOE/DOE contractor technical staff and capabilities

(2) EPA/EPA contractor technical staff and capabilities

(3) DOC technical staff and capabilities

(4) LFA technical liaison

(5) DOE public information liaison

(6) Other Federal agency liaisons, as needed

(7) State and local liaison and

(8) DFO liaison.

d. Advisory Team for Environment, Food, and Health

The Advisory Team is established by representatives from EPA, USDA, HHS, and other Federal agencies as needed for the provision of interagency coordinated advice and recommendations to the State and LFA concerning environmental, food, and health matters. For the ease of transfer of radiological monitoring and assessment data and coordination with Federal, State, and local representatives, the Advisory Team is normally collocated with the FRMAC.

e. Joint Information Center (JIC)

The JIC[2] is established by the LFA, under the operational control of the LFA-designated Public Information Officer, as a focal point for the coordination and provision of information to the public and media concerning the Federal response to the emergency. The JIC is established at an onscene location in coordination with State and local agencies and other Federal agencies. The following elements should be represented at the JIC:

(1) LFA Public Information Officer and staff

(2) FEMA Public Information Officer and staff

(3) Other Federal agency Public Information, as needed

(4) State and local Public Information Officers and

(5) Owner/Operator Public Information Officers and staff.

4. Response Deactivation

a. Each agency will discontinue emergency response operations when advised that Federal assistance is no longer required from their agency or when its statutory responsibilities have been fulfilled. Prior to discontinuing its response operation, each agency should discuss its intent to do so with the LFA, FEMA, and the State.

b. The LFA will consult with participating Federal agencies and the State and local government to determine when the Federal information coordination operations at the JIC should be terminated. This will occur normally at a time when the rate of information generated and coordinated by the LFA has decreased to the point where it can be handled through the normal day-to-day coordination process. The LFA will inform the other participants of their intention to deactivate Federal information coordination operations at the JIC and advise them of the procedures for continued coordination of information pertinent to recovery from the radiological emergency.

c. FEMA will consult with the LFA, other Federal agencies, and the State(s) as to when the onscene coordination of non- radiological assistance is no longer required. Prior to ending operations at the DFO, FEMA will inform all participating organizations of the schedule for doing so.

d. The LFA will terminate JOC operations and the Federal response after consulting with FEMA, other participating Federal agencies, and State and local officials, and after determining that onscene Federal assistance is no longer required.

e. The agency managing the FRMAC will consult with the LFA, FEMA, other participating Federal agencies, and State and local officials to determine when a formal FRMAC structure and organization is no longer required. Normally, this will occur when operations move into the recovery phase and extensive Federal multi-agency resources are no longer required to augment State and local radiological monitoring and assessment activities.

5. Recovery

a. The State or local governments have the primary responsibility for planning the recovery of the affected area. (The term recovery as used here encompasses any action dedicated to the continued protection of the public and resumption of normal activities in the affected area.) Recovery planning will be initiated at the request of the States, but it will generally not take place until after the initiating conditions of the emergency have stabilized and immediate actions to protect public health and safety and property have been accomplished. The Federal Government will, on request, assist the State and local governments in developing offsite recovery plans, prior to the deactivation of the Federal response. The LFA will coordinate the overall activity of Federal agencies involved in the recovery process.

b. The radiological monitoring and assessment activities will be terminated when the EPA, after consultation with the LFA and other participating Federal agencies, and State and local officials, determines that:

(1) There is no longer a threat to the public health and safety or to the environment,

(2) State and local resources are adequate for the situation, and

(3) There is mutual agreement of the agencies involved to terminate the response.

Appendix A: Acronyms

CFR	Code of Federal Regulations
DFO	Disaster Field Office
DOC	Department of Commerce
DOD	Department of Defense
DOE	Department of Energy
DOI	Department of the Interior
DOJ	Department of Justice
DOS	Department of State
DOT	Department of Transportation
EICC	Emergency Information and Coordination Center
EO	Executive Order
EOC	Emergency Operations Center
EPA	Environmental Protection Agency
ERT	Emergency Response Team
ERT-A	Emergency Response Team-Advance Element
FBI	Federal Bureau of Investigation
FCO	Federal Coordinating Officer
FECC	Federal Emergency Communications Coordinator
FEMA	Federal Emergency Management Agency
FRERP	Federal Radiological Emergency Response Plan
FRMAC	Federal Radiological Monitoring and Assessment Center
FRP	Federal Response Plan
FRPCC	Federal Radiological Preparedness Coordinating Committee
GIS	Geographical Information Systems

GSA	General Services Administration
HHS	Department of Health and Human Services
HUD	Department of Housing and Urban Development
JIC	Joint Information Center
JOC	Joint Operations Center
LFA	Lead Federal Agency
MERRT	Medical Emergency Radiological Response Team
NASA	National Aeronautics and Space Administration
NCC	National Coordinating Center for Telecommunications
NCS	National Communications System
NDA	National Defense Area
NOAA	National Oceanic and Atmospheric Administration (DOC)
NRC	Nuclear Regulatory Commission
NSA	National Security Area
OSC	Onscene Commander
PAG	Protective Action Guide
PIO	Public Information Officer
RAP	Radiological Assistance Program (DOE)
RECP	Regional Emergency Communications Planner
SCO	State Coordinating Officer
SFO	Senior FEMA Official
TLD	Thermoluminescent dosimeter
USDA	United States Department of Agriculture
VA	Department of Veterans Affairs

Appendix B: Definitions

Advisory Team for Environment, Food, and Health An interagency team, consisting of representatives from EPA, HHS, USDA, and representatives from other Federal agencies as necessary, that provides advice to the LFA and States, as requested on matters associated with environment, food, and health issues during a radiological emergency.

Agreement State A State that has entered into an Agreement under the Atomic Energy Act of 1954, as amended, in which NRC has relinquished to such States the majority of its regulatory authority over source, byproduct, and special nuclear material in quantities not sufficient to form a critical mass.

Assessment The evaluation and interpretation of radiological measurements and other information to provide a basis for decision- making. Assessment can include projections of offsite radiological impact.

Coordinate To advance systematically an exchange of information among principals who have or may have a need to know certain information in order to carry out their role in a response.

Disaster Field Office (DFO) A center established in or near the designated area from which the Federal Coordinating Officer (FCO) and representatives of Federal response agencies will interact with State and local government representatives to coordinate non-technical resource support.

Emergency Any natural or man-caused situation that results in or may result in substantial injury or harm to the population or substantial damage to or loss of property.

Emergency Response Team (ERT) A team of Federal interagency personnel headed by FEMA and deployed to the site of an emergency to serve as the FCO's key staff and assist with accomplishing FEMA responsibilities at the DFO.

Federal Coordinating Officer (FCO) The Federal official appointed in accordance with the provisions of P.L. 93-288, as amended, to coordinate the overall response and recovery activities under a major disaster or emergency declaration. The FCO represents the President as provided by Section 302 of P.L. 93-288, as amended, for the purpose of coordinating the administration of Federal relief activities in the designated area. Additionally, the FCO is delegated responsibilities and performs those for the FEMA Director as outlined in Executive Order 12148, and those responsibilities delegated to the FEMA Regional Director in Title 44 Code of Federal Regulations, Part 206.

Federal Radiological Monitoring and Assessment Center (FRMAC) An operations center usually established near the scene of a radiological emergency from which the Federal field monitoring and assessment assistance is directed and coordinated.

Federal Radiological Preparedness Coordinating Committee (FRPCC) An interagency committee, created under 44 CFR Part 351, to coordinate Federal radiological planning and training.

Federal Response Plan (FRP) The plan designed to address the consequences of any disaster or emergency situation in which there is a need for Federal assistance under the authorities of the Robert T. Stafford Disaster Relief and Emergency Assistance Act, 42 U.S.C. 5121 et seq.

FRMAC Director The person designated by DOE or EPA to manage operations in the FRMAC.

Joint Information Center (JIC) A center established to coordinate the Federal public information activities onscene. It is the central point of contact for all news media at the scene of the incident. Public information officials from all participating Federal agencies should collocate at the JIC. Public information officials from participating State and local agencies also may collocate at the JIC.

Joint Operations Center (JOC) Established by the LFA under the operational control of the OSC, as the focal point for management and direction of onsite activities, coordination/establishment of State requirements/priorities, and coordination of the overall Federal response.

Joint U.S. Government/Foreign Government Space Venture Any space venture conducted jointly by the U.S. Government (DOD or NASA) with a foreign government or foreign governmental entity that is characterized by an ongoing U.S. Government interest in the successful completion of the mission, active involvement in mission operations, and uses radioactive sources and/or classified components, regardless of which country owns or provides said sources or components, within the space vehicle. For the purposes of this plan, in a situation whereby the U.S. Government simply sells or supplies radioactive material to a foreign country for use in a space vehicle and otherwise has no active mission involvement, it shall not be considered a joint venture.

Lead Federal Agency (LFA) The agency that is responsible for leading and coordinating all aspects of the Federal response is referred to as the LFA and is determined by the type of emergency. In situations where a Federal agency owns, authorizes, regulates, or is otherwise deemed responsible for the facility or radiological activity causing the emergency and has authority to conduct and manage Federal actions onsite, that agency normally will be the LFA.

License An authorization issued to a facility owner or operator by the NRC pursuant to the conditions of the Atomic Energy Act of 1954, as amended, or issued by an Agreement State pursuant to appropriate State laws. NRC licenses certain activities under section 170(a) of that Act.

Local Government Any county, city, village, town, district, or political subdivision of any State, and Indian tribe or authorized tribal organization, or Alaska Native village or organization, including any rural community or unincorporated town or village or any other public entity.

Monitoring The use of sampling and radiation detection equipment to determine the levels of radiation.

National Defense Area (NDA) An area established on non-Federal lands located within the United States, its possessions or its territories, for safeguarding classified defense information or protecting DOD equipment and/or material. Establishment of a National Defense Area temporarily places such non-Federal lands under the effective control of the Department of Defense and results only from an emergency event. The senior DOD representative at the scene shall define the boundary, mark it with a physical barrier, and post warning signs. The landowner's consent and cooperation shall be obtained whenever possible however, military necessity shall dictate the final location, shape, and size of the NDA.

National Security Area (NSA) An area established on non-Federal lands located within the United States, its possessions or territories, for safeguarding classified information, and/or restricted data or equipment and material belonging to DOE or NASA. Establishment of a National Security Area temporarily places such non-Federal lands under the effective control of DOE or NASA and results only from an emergency event. The senior DOE or NASA representative having custody of the material at the scene shall define the boundary, mark it with a physical barrier, and post warning signs. The landowner's consent and cooperation shall be obtained whenever possible however, operational necessity shall dictate the final location, shape, and size of the NSA.

Nuclear Facilities Nuclear installations that use or produce radioactive materials in their normal operations.

Offsite The area outside the boundary of the onsite area. For emergencies occurring at fixed nuclear facilities, "offsite" generally refers to the area beyond the facility boundary. For emergencies that do not occur at fixed nuclear facilities and for which no physical boundary exists, the circumstances of the emergency will dictate the boundary of the offsite area. Unless a Federal agency has the authority to define and control a restricted area, the State or local government will define an area as "onsite" at the time of the emergency, based on required response activities.

Offsite Federal Support Federal assistance in mitigating the offsite consequences of an emergency and protecting the public health and safety, including assistance with determining and implementing public protective action measures.

Onscene The area directly affected by radiological contamination and environs. Onscene includes onsite and offsite areas.

Onscene Commander (OSC) The lead official designated at the scene of the emergency to manage onsite activities and coordinate the overall Federal response to the emergency.

Onsite The area within (a) the boundary established by the owner or operator of a fixed nuclear facility, or (b) the area established by the LFA as a National Defense Area or National Security Area, or (c) the area established around a downed/ditched U.S. spacecraft, or (d) the boundary established at the time of the emergency by the State or local government with jurisdiction for a transportation accident not occurring at a fixed nuclear facility and not involving nuclear weapons.

Onsite Federal Support Federal assistance that is the primary responsibility of the Federal agency that owns, authorizes, regulates, or is otherwise deemed responsible for the radiological facility or material being transported, i.e., the LFA. This response supports State and local efforts by supporting the owner or operator's efforts to bring the incident under control and thereby prevent or minimize offsite consequences.

Owner or Operator The organization that owns or operates the nuclear facility or carrier or cargo that causes the radiological emergency. The owner or operator may be a Federal agency, a State or local government, or a private business.

Protective Action Guide (PAG) A radiation exposure or contamination level or range established by appropriate Federal or State agencies at which protective actions should be considered.

Protective Action Recommendation (Federal) Federal advice to State and local governments on measures that they should take to avoid or reduce exposure of the public to radiation from an accidental release of radioactive material. This includes emergency actions such as sheltering, evacuation, and prophylactic use of stable iodine. It also includes longer term measures to avoid or minimize exposure to residual radiation or exposure through the ingestion pathway such as restriction of food, temporary relocation, and permanent resettlement.

Public Information Officer (PIO) Official at headquarters or in the field responsible for preparing and coordinating the dissemination of public information in cooperation with other responding Federal, State, and local agencies.

Radiological Assistance Program (RAP) Team A response team dispatched to the site of a radiological incident by the U.S. Department of Energy (DOE) regional coordinating office responding to a radiological incident. RAP Teams are located at DOE operations offices and national laboratories and some area offices.

Radiological Emergency A radiological incident that poses an actual, potential, or perceived hazard to public health or safety or loss of property.

Recovery Recovery, in this document, includes all types of emergency actions dedicated to the continued protection of the public or to promoting the resumption of normal activities in the affected area.

Recovery Plan A plan developed by each State, with assistance from the responding Federal agencies, to restore the affected area.

Regional Operations Center (ROC) The temporary operations facility for the coordination of Federal response and recovery activities, located at the FEMA Regional Office (or at the Federal Regional Center) and led by the FEMA Regional Director or Deputy Regional Director until the DFO becomes operational.

Senior FEMA Official (SFO) Official appointed by the Director of FEMA, or his representative, to initially direct the FEMA response at the scene of a radiological emergency. Also, acts as the Team Leader for the Advance Element of the Emergency Response Team (ERT-A).

State Coordinating Officer (SCO) An official designated by the Governor of the affected State to work with the LFA's Onscene Commander and Senior FEMA Official or Federal Coordinating Officer in coordinating the response efforts of Federal, State, local, volunteer, and private agencies.

Subcommittee on Federal Response A subcommittee of the Federal Radiological Preparedness Coordinating Committee formed to develop and test the Federal Radiological Emergency Response Plan. Most agencies that will participate in the Federal radiological emergency response are represented on this subcommittee.

Transportation Emergency For the purposes of this plan, any emergency that involves a transportation vehicle or shipment containing radioactive materials outside the boundaries of a facility.

Transportation of Radioactive Materials The loading, unloading, movement, or temporary storage en route of radioactive materials.

Appendix C: Federal Agency Response Missions, Capabilities and Resources, References, and Authorities

Each Federal agency develops and maintains a plan that describes a detailed concept of operations for implementing this Plan. This section contains summary information about the following Federal agencies:

Department of Agriculture (USDA)
Department of Commerce (DOC)
Department of Defense (DOD)
Department of Energy (DOE)
Department of Health and Human Services (HHS)
Department of Housing and Urban Development (HUD)
Department of the Interior (DOI)
Department of Justice (DOJ)
Department of State (DOS)
Department of Transportation (DOT)
Department of Veterans Affairs (VA)
Environmental Protection Agency (EPA)
Federal Emergency Management Agency (FEMA)
General Services Administration (GSA)
National Aeronautics and Space Administration (NASA)
National Communications System (NCS)
Nuclear Regulatory Commission (NRC)

Summary information for each agency contains: (1) a response mission statement, (2) a description of the agency's response capabilities and resources, (3) agency response plan and procedures references, and (4) sources of agency authority.

A. Department of Agriculture

1. Summary of Response Mission

The United States Department of Agriculture (USDA) provides assistance to State and local governments in developing agricultural protective action recommendations and in providing agricultural damage assessments. USDA will actively participate with EPA and HHS on the Advisory Team for Environment, Food, and Health when convened. USDA regulatory responsibilities for the inspection of meat, meat products, poultry, poultry products, and egg products are essential uninterruptible functions that would continue during an emergency.

2. Capabilities and Resources

USDA can provide assistance to State and local governments through emergency response personnel located at its Washington, DC, headquarters and from USDA State and County Emergency Board representatives located throughout the country. USDA Emergency Board representatives have knowledge of local agriculture and can provide specific advice to the local agricultural community. In addition, USDA State and County Emergency Boards can assist in the collection of agricultural samples during a radiological emergency. USDA actively participates with EPA and HHS on the Advisory Team when convened.

The functions and capabilities of the USDA to provide assistance in the event of a radiological emergency include the following:

a. Provide assistance through regular USDA programs, if legally adaptable to radiological emergencies

b. Provide emergency food coupon assistance in officially designated disaster areas, if a need is determined by officials and if the commercial food system is sufficient to accommodate the use of food coupons

c. Assist in reallocation of USDA-donated food supplies from warehouses, local schools, and other outlets to emergency care centers. These are foods donated to various outlets through USDA food programs

d. Provide lists that identify locations of alternate sources of food and livestock feed and arrange for transportation of the food and feed if requested

e. Provide advice to State and local officials regarding the disposition of livestock and poultry contaminated by radiation

f. Inspect meat and meat products, poultry and poultry products, and egg products identified for interstate and foreign commerce to assure that they are safe for human consumption

g. Assist State and local officials, in coordination with HHS and EPA, in the recommendation and implementation of protective actions to limit or prevent the ingestion of contaminated food

h. Assist, in conjunction with HHS, in monitoring the production, processing, storage, and distribution of food through the wholesale level to eliminate contaminated product or to reduce the contamination in the product to a safe level

i. Assess damage to crops, soil, livestock, poultry, and processing facilities and incorporate findings into a damage assessment report

j. Provide advice to State and local officials on minimizing losses to agricultural resources from radiation effects

k. Provide information and assistance to farmers, food processors, and distributors to aid them in returning to normal after a radiological emergency

l. Provide a liaison to State agricultural agencies if requested

m. Assist DOE at the FRMAC in collecting agricultural samples within the Ingestion Exposure Pathway Emergency Planning Zone. Assist in the evaluation and assessment of data to determine the impact of the emergency on agriculture

n. Assist in providing temporary housing for evacuees who have been displaced from their homes due to a radiological emergency and

o. Provide emergency communications assistance to the agricultural community through the Cooperative Extension System, an electronic mail system.

3. USDA References

USDA Radiological Emergency Response Plan, January 1988.

4. USDA Specific Authorities.

a. Title 7, U.S.C. 241-273.
b. Title 7, U.S.C. 341-349.
c. Title 7, U.S.C. 612 C.
d. Title 7, U.S.C. 612 C Note.
e. Title 7, U.S.C. 1431.
f. Title 7, U.S.C. 1622.
g. Title 7, U.S.C. 2014(h).
h. Title 7, U.S.C. 2204.
i. Title 16, U.S.C. 590 a-f.
j. Title 21, U.S.C. 451 et seq.
k. Title 21, U.S.C. 601 et seq.
l. Title 21, U.S.C. 1031-1056.
m. Title 42, U.S.C. 1480.
n. Title 42, U.S.C. 3271-3274.
o. Title 50, U.S.C. Appendix 2251 et seq.
p. Title 7, CFR 2.51 (a)(30).
q. E.O. 12656, November 18, 1988.
r. DR 1800-1, March 5, 1993.

B. Department of Commerce

1. Summary of Response Mission

The National Oceanic and Atmospheric Administration (NOAA) is the primary agency within the Department of Commerce (DOC) responsible for providing assistance to the Federal, State, and local organizations responding to a radiological emergency. Other assistance may be provided by the National Institute of Standards and Technology. DOC's responsibilities include:

a. Acquiring and disseminating weather data and providing weather forecasts in direct support of the emergency response operation

b. Preparing and disseminating predictions of plume trajectories, dispersion, and deposition of radiological material released into the atmosphere

c. Providing local meteorological support as needed to assure the quality of these predictions

d. Organizing and maintaining a special data archive for meteorological information related to the emergency and its assessment

e. Ensuring that marine fishery products available to the public are not contaminated

f. Providing assistance and reference material for calibrating radiological instruments and

g. Providing radiation shielding materials.

2. Capabilities and Resources

NOAA is the principal DOC participant in the response to a radiation accident. NOAA prepares both routine and special weather forecasts, and makes use of these forecasts to predict atmospheric transport and dispersion. NOAA's forecasts may be the basis for all public announcements on the movement of contamination from accidents occurring outside U.S. territory or during domestic accidents when any released radioactive material is expected to be carried off-site. NOAA has capabilities to do the following:

a. Provide current and forecast meteorological information as needed to guide aerial monitoring and sampling, and to predict the transport and dispersion of radioactive materials (gases, liquids, and particles).

b. Routinely forecast the atmospheric transport, dispersion, and deposition of the radioactive materials, and disseminate the results of these computations via automatic facsimile to all relevant parties, twice per day.

c. Produce (and archive) special high-resolution meteorological data sets for providing an improved capability to predict atmospheric transport and dispersion of radioactive materials in the atmosphere.

d. Augment routine and special upper atmosphere and surface meteorological observation systems, as required to improve the quality of these predictions.

e. Evaluate NOAA's transport and dispersion forecast products in conjunction with those of other nations" weather services responding to the emergency, to provide a more internationally consistent product.

Additionally, DOC may provide support to HHS at its request, through the National Marine Fisheries Service, in order to avoid human consumption of contaminated commercial fishery products (marine area only). The National Institute of Standards and Technology can assist in calibrating radiological instruments by comparison with national standards or by providing standard reference materials for calibration, as well as making extensive data on the physical properties of materials available. The National Institute of Standards and Technology can also supply temporary radiation shielding materials.

3. DOC References

National Plan for Radiological Emergencies at Commercial Nuclear Power Plants. Federal Coordinator for Meteorological Services and Supporting Research, National Oceanic and Atmospheric Administration, November 1982.

4. DOC Specific Authorities

Department of Commerce Organization Order 25-5B, as amended, June 18, 1987.

C. Department of Defense

1. Summary of Response Mission

The Department of Defense (DOD) is charged with the safe handling, storage, maintenance, assembly, and transportation of nuclear weapons and other radioactive materials in DOD custody, and with the safe operation of DOD nuclear facilities. Inherent in this responsibility is the requirement to protect life and property from any health or safety hazards that could ensue from an accident or significant incident associated with these materials or activities.

The DOD role in a Federal response will depend on the circumstances of the emergency. DOD will be the LFA if the emergency involves one of its facilities or a nuclear weapon in its custody. Within DOD, the military service or agency responsible for the facility, ship, or area is responsible for the onsite response. The military service or agency having custody of the material outside an installation boundary is responsible for the onsite response. For emergencies occurring under circumstances for which DOD is not responsible, DOD will not be the LFA, but will support and assist in the Federal response.

2. Capabilities and Resources

Offsite authority and responsibility at a nuclear accident rest with State and local officials. It is important to recognize that for nuclear weapons or weapon component accidents, land may be temporarily placed under effective Federal control by the establishment of a National Defense Area or National Security Area to protect U.S. Government classified materials. These lands will revert to State control upon disestablishment of the National Defense Area or National Security Area.

DOD has a trained and equipped nuclear response organization to deal with accidents at its facilities or involving materials in its custody. Radiological resources include trained response personnel, specialized radiation instruments, and mobile instrument calibration and repair capabilities. DOD also may perform special sampling of airborne contamination on request. Descriptions of the capabilities and assets of DOD response teams can be found in DOD 5100.52M.

DOD may provide assistance in the form of personnel, logistics and telecommunications, assistance and expertise in site restoration, including airlift services, when available, upon the request of the LFA or FEMA. Requests for assistance must be directed to the National Military Command Center or through channels established by prior agreements.

3. DOD References.

   a. DOD Directive 5100.52, DOD Response to an Accident or Significant Incident Involving Radiological Materials.

   b. DOD Directive 5230.16, Nuclear Accident and Incident Public Affairs Guidance.

   c. DOD Directive 3025.1, Military Support to Civil Authorities.

   d. DOD Directive 3025.12, Military Assistance for Civil Disturbances.

   e. DOD Directive 3150.5, DOD Response to Improvised Nuclear Device (IND) Incident.

   f. DOD 5100.52M, Nuclear Weapon Accident Response Procedures (NARP) Manual.

   g. Joint Federal Bureau of Investigation, Department of Energy, and Department of Defense Agreement for Response to Improvised Nuclear Device Incidents.

4. DOD Specific Authorities.

   a. The Atomic Energy Act of 1954, as amended, 42 U.S.C. 2011- 2284.

   b. Public Law 97 351, "Convention on the Physical Protection of Nuclear Material Implementation Act of 1982."

   c. Department of Defense, Department of Energy, Federal Emergency Management Agency Memorandum of Agreement on Response to Nuclear Weapon Accidents and Nuclear Weapon Significant Incidents, 1983.

D. Department of Energy

1. Summary of Response Mission

The Department of Energy (DOE) owns and operates a variety of radiological activities throughout the United States. These activities include: fixed nuclear sites the use, storage, and shipment of a variety of radioactive materials the shipment of spent reactor fuel the production, assembly, and shipment of nuclear weapons and special nuclear materials the production and shipment of radioactive sources for space ventures and the storage and shipment of radioactive and mixed waste. DOE is responsible for the safe operation of these activities and should an emergency occur at one of its sites or an activity under its control, DOE will be the LFA for the Federal response.

Due to its technical capabilities and resources, the DOE may perform other roles within the Federal response to a radiological emergency. With extensive, field-based radiological resources throughout the United States available for emergency deployment, the DOE responds to requests for offsite radiological monitoring and assessment assistance and serves as the initial coordinator of all such Federal assistance (to include initial management of the FRMAC) to State and local governments. With other specialized, deployable assets, DOE assists other Federal agencies responding to malevolent nuclear emergencies, accidents involving nuclear weapons not under DOE custody, emergencies caused by satellites containing radioactive sources, and other radiological incidents as appropriate.

2. Capabilities and Resources

DOE has trained personnel, radiological instruments, mobile laboratories, and radioanalytical facilities located at its national laboratories, production, and other facilities throughout the country. Through eight Regional Coordinating Offices, these resources form the basis for the Radiological Assistance Program, which can provide technical assistance in any radiological emergency. DOE can provide specialized radiation detection instruments and support for both its response as LFA and as initial coordinator of Federal radiological monitoring and assessment assistance. Some of the specialized resources and capabilities include:

a. Aerial monitoring capability for tracking dispersion of radioactive material and mapping ground contamination

b. A computer-based, emergency preparedness and response predictive capability that provides rapid predictions of the transport, diffusion, and deposition of radionuclides released to the atmosphere and dose projections to people and the environment

c. Specialized equipment and instruments and response teams for locating radioactive materials and handling damaged nuclear weapons

d. Medical experts on radiation effects and the treatment of exposed or contaminated patients and

e. Support facilities for DOE response, including command post supplies, communications systems, generators, and portable video and photographic capabilities.

3. DOE References

a. DOE Order 5500.1B, Emergency Management System, April 1991.

b. DOE Order 5500.2B, Emergency Categories, Classes, and Notification and Reporting Requirements, April 1991.

c. DOE Order 5500.3A, Planning and Preparedness for Operational Emergencies, April 1991.

d. DOE Order 5500.4A, Public Affairs Policy and Planning Requirements for Emergencies, June 1992.

e. DOE Order 5530.1A, Accident Response Group, September 1991.

f. DOE Order 5530.2, Nuclear Emergency Search Team, September 1991.

g. DOE Order 5530.3, Radiological Assistance Program, January 1992.

h. DOE Order 5530.4, Aerial Measuring System, September 1991.

i. DOE Order 5530.5, Federal Radiological Monitoring and Assessment Center, July 1992.

4. DOE Specific Authorities

a. Atomic Energy Act of 1954, as amended, 42 U.S.C. 2011- 2284.

b. Energy Reorganization Act of 1974, 42 U.S.C. 5801 et seq.

c. Department of Energy Organization Act of 1977, 42 U.S.C. 7101 et seq.

d. Nuclear Waste Policy Act of 1982, 42 U.S.C. 10101 et seq.

e. Title 44, Code of Federal Regulations, Part 351, Radiological Emergency Planning and Preparedness, 351.24, The Department of Energy.

E. Department of Health and Human Services

1. Summary of Response Mission

In a radiological emergency, the Department of Health and Human Services (HHS) assists with the assessment, preservation, and protection of human health and helps ensure the availability of essential health/medical and human services. Overall, the Office of Public Health and Science, Office of Emergency Preparedness, coordinates the HHS emergency response. HHS provides technical and nontechnical assistance in the form of advice, guidance, and resources to Federal, State, and local governments. The principal HHS response comes from the U.S. Public Health Service. HHS actively participates with EPA and USDA on the Advisory Team for Environment, Food, and Health when convened.

2. Capabilities and Resources

HHS has personnel located at headquarters, regional offices, and at laboratories and other facilities who can provide assistance in radiological emergencies. The agency can provide the following kinds of advice, guidance, and assistance:

a. Assist State and local government officials in making evacuation and relocation decisions

b. Ensure the availability of health and medical care and other human services (especially for the aged, the poor, the infirm, the blind, and others most in need)

c. Provide advice and guidance in assessing the impact of the effects of radiological incidents on the health of persons in the affected area

d. Assist in providing crisis counseling to victims in affected geographic areas

e. Provide guidance on the use of radioprotective substances (e.g., thyroid blocking agents), including dosage, and also projected radiation doses that warrant the use of such drugs

f. In conjunction with DOE and DOD, advise medical personnel on proper medical treatment of people exposed to or contaminated by radioactive materials

g. Recommend Protective Action Guides for food and animal feed and assist in developing technical recommendations on protective measures for food and animal feed and

h. Provide guidance to State and local health officials on disease control measures and epidemiological surveillance and study of exposed populations.

3. HHS References

a. 55 FR 2879, January 29, 1990-Delegations of authority to the Assistant Secretary for Health for department-wide emergency preparedness functions.

b. 55 FR 2885, January 29, 1990-Statement of organization, functions and delegations of authority to the Office of Emergency Preparedness.

c. Federal Response Plan, Emergency Support Functions #8 (Health and Medical Services), April 1992.

d. Disaster Response Guides, Operating Divisions, Various Dates.

4. HHS Specific Authorities

a. Public Health Service Act, as amended, 42 U.S.C. 201 et seq.

b. Federal Food, Drug, and Cosmetic Act of 1938, as amended, 21 U.S.C. 301-392.

c. Snyder Act, 25 U.S.C. 13 (1921).

d. Transfer Act, 42 U.S.C. 2004b.

e. Indian Health Care Improvement Act, 25 U.S.C. 1601 et seq.

f. The Robert T. Stafford Disaster Relief and Emergency Assistance Act, as amended, Title VI, 42 U.S.C. 5195 et seq.

g. Comprehensive Environmental Response, Compensation, and Liability Act of 1980 (SUPERFUND), 42 U.S.C. 9601 et seq., as amended by the SUPERFUND Amendments and Reauthorization Act of 1986 (Public Law 99-499) (1986).

h. 42 U.S.C. 3030-Section 310 of the Older Americans Act.

i. 42 U.S.C. 601 et seq.-Section 401 et seq. of the Social Security Act.

j. 45 CFR 233.120-Emergency Community Services Homeless Grant Program.

k. 45 CFR 233.120-AFDC Emergency Assistance Program.

l. 45 CFR 233.20(a)(2)(v)-AFDC Special Needs Allowance.

m. Runaway and Homeless Youth Act, as amended, Section 366(0).

n. Omnibus Budget Reconciliation Act of 1981, Title XXVI (as amended by Public Laws 98-558, 99-425, 101-501, 101-517)- Low Income Home Energy Assistance Program.

o. E.O. 12656, National Security Emergency Preparedness-Part 8.

F. Department of Housing and Urban Development

1. Summary of Response Mission

The Department of Housing and Urban Development (HUD) provides information on available housing for disaster victims or displaced persons. HUD assists in planning for and placing homeless victims by providing emergency housing and technical support staff within available resources.

2. Capabilities and Resources

HUD has capabilities to do the following:

a. Review and report on available housing for disaster victims and displaced persons

b. Assist in planning for and placing homeless victims in available housing

c. Provide staff to support emergency housing within available resources and

d. Provide technical housing assistance and advisory personnel.

3. HUD References

HUD Handbook 3200.02, REV-3, "Disaster Response and Assistance."

4. HUD Specific Authorities

HUD housing programs provide the Department some discretion, to the extent permissible by law, in granting waivers of eligibility requirements to disaster-displaced families. These programs provide rental housing assistance, HUD/FHA-insured loans to repair and rebuild homes, and HUD/FHA-insured loans to purchase new or existing housing, under the following authorities:

a. National Housing Act, as amended, 12 U.S.C. 1701 et seq.

b. United States Housing Act of 1977, as amended, 42 U.S.C. 1437c et seq.

c. Housing and Community Development Act of 1974, as amended, 42 U.S.C. 5301 et seq.

d. National Affordable Housing Act of 1990 (P.L. 101-625), as amended.

G. Department of the Interior

1. Summary of Response Mission

The Department of the Interior (DOI) manages over 500 million acres of Federal lands and thousands of Federal natural resources facilities and is responsible for these lands and facilities, as well as other natural resources such as endangered and threatened species, migratory birds, anadromous fish, and marine mammals, when they are threatened by a radiological emergency. In addition, DOI coordinates emergency response plans for DOI-managed refuges, parks, recreation areas, monuments, public lands, and Indian trust lands with State and local authorities operates its water resources projects to protect municipal and agricultural water supplies in cases of radiological emergencies and provides advice and assistance concerning hydrologic and natural resources, including fish and wildlife, to Federal, State, and local governments upon request. DOI also administers the Federal Government's trust responsibility for 512 Federally recognized Indian tribes and villages, and about 50 million acres of Indian lands. The Bureau of Indian Affairs of the Department of the Interior is available to assist other agencies in consulting with these tribes about radiological emergency preparedness and responses to emergencies. DOI also has certain responsibilities for the United States insular areas.

2. Capabilities and Resources

DOI has personnel at headquarters and in regional offices with technical expertise to do the following:

a. Advise and assist in assessing the nature and extent of radioactive releases to water resources including support of monitoring personnel, equipment, and laboratory analytical capabilities.

b. Advise and assist in evaluating processes affecting radioisotopes in soils, including personnel, equipment, and laboratory support.

c. Advise and assist in the development of geographical information systems (GIS) databases to be used in the analysis and assessment of contaminated areas including personnel, equipment, and databases.

d. Provide hydrologic advice and assistance, including monitoring personnel, equipment, and laboratory support.

e. Advise and assist in assessing and minimizing offsite consequences on natural resources, including fish and wildlife, subsistence uses, land reclamation, mining, and mineral expertise.

f. Advise and assist the United States insular areas on economic, social, and political matters.

g. Coordinate and provide liaison between Federal, State, and local agencies and Federally recognized Indian tribal governments on questions of radiological emergency preparedness and responses to incidents.

3. DOI References

a. 910 DM 5 (Draft)-Interior Emergency Operations, Federal Radiological Emergency Response Plan.

b. 296 DM 3 (Draft)-Interior Emergency Delegations, Radiological Emergencies.

4. DOI Specific Authorities

a. Organic Act of 1879 providing for "surveys, investigations, and research covering the topography, geology, hydrology, and the mineral and water resources of the United States," 43 U.S.C. 31 (USGS).

b. Appropriations Act of 1894 providing for gaging streams and assessment of water supplies of the U.S., 28 Stat. 398 (USGS).

c. OMB Circular A-67 (1964) giving DOI (USGS) responsibility"* * * for the design and operation of the national network for acquiring data on the quantity and quality of surface ground waters * * *" (USGS).

d. The Reclamation Act of 1902, as amended, 43 U.S.C. 391, and project authorization acts (BuRec).

e. National Park Service Act of 1916, 16 U.S.C. 1 et seq., and park enabling acts (NPS).

f. The Snyder Act of 1921, as amended, 25 U.S.C. 13. DOI shall direct, supervise, and expend such monies appropriated by Congress for the benefit, care, and assistance of Indians throughout the United States for such purposes as the relief of distress, and conservation of health, for improvement of operation and maintenance of existing Indian irrigation and water supply systems * * * etc. (BIA).

g. National Wildlife Refuge System Administration Act of 1966, as amended, 16 U.S.C. 668dd, and refuge enabling acts (FWS).

h. Federal Land Policy and Management Act of 1976, 43 U.S.C. 1701 et seq. (BLM).

i. Endangered Species Act (1973), as amended, 16 U.S.C. 1531 et seq. Federal agencies may not jeopardize the continued existence of endangered or threatened species (FWS).

j. Migratory Bird Treaty Act (1918), as amended, 16 U.S.C. 703 et seq. Prohibits the taking of migratory birds without permits (FWS).

k. Anadromous Fish Conservation Act, as amended, 16 U.S.C. 757a et seq. Reestablishes anadromous fish habitat (FWS).

l. Marine Mammal Protection Act (1972), as amended, 16 U.S.C. 1361 et seq. Conserves marine mammals with management of certain species vested in DOI (FWS).

H. Department of Justice

1. Summary of Response Mission

The Department of Justice (DOJ) is the lead agency for coordinating the Federal response to acts of terrorism in the United States and U.S. territories. Within the DOJ, the Federal Bureau of Investigation (FBI) will manage the law enforcement aspect of the Federal response to such incidents. The FBI also is responsible for investigating all alleged or suspected criminal violations of the Atomic Energy Act of 1954, as amended.

2. Capabilities and Resources

The FBI will coordinate all law enforcement operations including intelligence gathering, hostage negotiations, and tactical operations.

3. DOJ References

a. Memorandum of Understanding between DOJ, DOD, and DOE for Responding to Domestic Malevolent Nuclear Weapons Emergencies.

b. Federal Bureau of Investigation Nuclear Incident Response Plan.

c. Memorandum of Understanding between DOE and the FBI for Responding to Nuclear Threat Incidents.

   d. Memorandum of Understanding between the FBI and the NRC Regarding Nuclear Threat Incidents Involving NRC-Licensed Facilities, Materials, or Activities.

   e. Memorandum of Understanding between DOE, FBI, White House Military Office, and the U.S. Secret Service Regarding Nuclear Incidents Concerning the Office of the President and Vice President of the United States.

   f. Joint Federal Bureau of Investigation, Department of Energy, and Department of Defense Agreement for Response to Improvised Nuclear Device Incidents.

4. DOJ Specific Authorities

   a. Atomic Energy Act of 1954, 42 U.S.C. 2011-2284.

   b. 18 U.S.C. 831 (Prohibited Transactions Involving Nuclear Materials).

I. Department of State

1. Summary of Response Mission

The Department of State (DOS) is responsible for the conduct of relations between the U.S. Government and other governments and international organizations and for the protection of U.S. interests and citizens abroad.

In a radiological emergency outside the United States, DOS is responsible for coordinating U.S. Government actions concerning the event in the country where it occurs (including evacuation of U.S. citizens, if necessary) and internationally. Should the FRERP be invoked due to the need for domestic action, DOS will continue to hold this role within the FRPCC structure. Specifically, DOS will coordinate foreign information-gathering activities and, in particular, conduct all contacts with foreign governments except in cases where existing bilateral agreements permit direct agency-to-agency cooperation. In the latter situation, the U.S. agency will keep DOS fully informed of all communications.

In a domestic radiological emergency with potential international trans-boundary consequences, DOS will coordinate all contacts with foreign governments and agencies except where existing bilateral agreements provide for direct exchange of information. DOS is responsible for conveying the U.S. Government response to foreign offers of assistance.

2. Capabilities and Resources

The State Department maintains embassies, missions, interest sections (in countries where the United States does not have diplomatic relations), and consulates throughout the world. The State Department Operations Center is capable of secure, immediate, around-the-clock communications with diplomatic posts. The diplomatic personnel stationed at a post are knowledgeable of local factors important to clear and concise communication, and frequently speak the local language. The Ambassador is the President's personal representative to the host government, and his country team is responsible for coordinating official contacts between the U.S. Government and the host government or international organization.

3. DOS References

Task Force Manual for Crisis Management (rev. 11 January 1990).

4. DOS Specific Authorities

   a. Presidential Directive/NSC-27 (PD-27) of January 19, 1978.
   b. 22 U.S.C. 2656.
   c. 22 U.S.C. 2671(a)(92)(A).

J. Department of Transportation

1. Summary of Response Mission

The Department of Transportation (DOT) Radiological Emergency Response Plan for Non-Defense Emergencies provides assistance to State and local governments when a radiological emergency adversely affects one or more transportation modes and the States or local jurisdictions requesting assistance have inadequate technical and logistical resources to meet the demands created by a radiological emergency.

2. Capabilities and Resources

DOT can assist Federal, State, and local governments with emergency transportation needs and contribute to the response by assisting with the control and protection of transportation near the area of the emergency. DOT has capabilities to do the following:

   a. Support State and local governments by identifying sources of civil transportation on request and when consistent with statutory responsibilities.

   b. Coordinate the Federal civil transportation response in support of emergency transportation plans and actions with State and local governments. (This may include provision of Federally controlled transportation assets and the controlling of transportation routes to protect commercial transportation and to facilitate the movement of response resources to the scene.)

   c. Provide Regional Emergency Transportation Coordinators and staff to assist State and local authorities in planning and response.

   d. Provide technical advice and assistance on the transportation of radiological materials and the impact of the incident on the transportation system.

e. Provide exemptions from normal transportation hazardous materials regulations if public interest is best served by allowing shipments to be made in variance with the regulations. Most exemptions are issued following public notice procedures, but if emergency conditions exist, DOT can issue emergency exemptions by telephone.

f. Control airspace, including the imposition of Temporary Flight Restrictions and issuance of Notices to Airmen (NOTAMS), both to give priority to emergency flights and protect aircraft from contaminated airspace.

DOT is responsible for dealing with the International Atomic Energy Agency and foreign Competent Authorities on issues related to packaging and other standards for the international transport of radioactive materials. If a transport accident involves international shipments of radioactive materials, DOT will be the point of contact for working with the transportation authorities of the foreign country that offered the material for transport in the United States.

3. DOT References

a. Department of Transportation Radiological Emergency Response Plan for Non-Defense Emergencies, August 1985.

b. DOT Order 1900.8, Department of Transportation Civil Emergency Preparedness Policies and Program(s).

c. DOT Order 1900.7D, Crisis Action Plan.

d. Transportation Annex (Emergency Support Function #1), Federal Response Plan.

4. DOT Specific Authorities

a. 49 U.S.C. 301.

b. 44 CFR 351, Radiological Emergency Planning and Preparedness, 351.25, The Department of Transportation.

K. Department of Veterans Affairs

1. Summary of Response Mission

The Department of Veterans Affairs (VA) can assist other Federal agencies, State and local governments, and individuals in an emergency by providing immediate and long-term medical care, including management of radiation trauma, as well as first aid, at its facilities or elsewhere. VA can make available repossessed VA mortgaged homes to be used for housing for affected individuals. VA can manage a system of disposing of the deceased. VA can provide medical, biological, radiological, and other technical guidance for response and recovery reactions. Generally, none of these actions will be taken unilaterally but at the request of a responsible senior Federal official and with appropriate external funding.

2. Capabilities and Resources

In addition to the capabilities listed above, VA:

a. Operates almost 200 full-facility hospitals and outpatient clinics throughout the United States

b. Has almost 200,000 employees with broad medical, scientific, engineering and design, fiscal, and logistical capabilities

c. Manages the National Cemetery System in 38 States

d. May have a large inventory of repossessed homes (this inventory varies according to economic trends)

e. Is one of the Federal managers of the National Disaster Medical System

f. Is a participant in the VA/DOD contingency plan for Medical Backup in times of national emergency

g. Has the capability to manage the medical effects of radiation trauma using the VA's Medical Emergency Radiological Response Teams (MERRTs) and

h. Has a fully equipped emergency center with multi-media communications at the Emergency Medical Preparedness Office (EMPO).

3. VA References

MP-1, Part II, Chapter 13 (Emergency Preparedness Plan), March 20, 1985, as revised.

4. VA Specific Authorities

a. The Robert T. Stafford Disaster Relief and Emergency Assistance Act, as amended, Title VI, 42 U.S.C. 5195 et seq.

b. National Security Decision Directive Number 47 (NSDD-47), July 22, 1982, Emergency Mobilization Preparedness.

c. National Security Decision Directive Number 97 (NSDD-97), June 13, 1982, National Security Telecommunications Policy.

d. National Plan of Action for Emergency Mobilization Preparedness.

e. Veterans Administration and Department of Defense Health Resources Sharing and Emergency Operations Act, 38 U.S.C. 5001 et seq.

f. E.O. 11490, Assignment of Preparedness Functions to Federal Departments and Agencies, October 28, 1969, as amended, 3 CFR, 1966-1970 Comp., p. 820.

g. E.O. 12656, Assignment of Emergency Preparedness Responsibilities,

November 18, 1988, 3 CFR, 1988 Comp., p. 585.

h. E.O. 12657, Federal Emergency Management Agency Assistance, Emergency Preparedness Planning at Commercial Nuclear Power Plants, November 23, 1988, 3 CFR, 1988 Comp., p. 611.

L. Environmental Protection Agency

1. Summary of Response Mission

The Environmental Protection Agency (EPA) assists Federal, State, and local governments during radiological emergencies by providing environmental and water supply monitoring, recommending protective actions, and assessing the consequences of radioactivity releases to the environment. These services may be provided at the request of the Federal or State Government, or EPA may respond to an emergency unilaterally in order to fulfill its statutory responsibility. EPA actively participates with USDA and HHS on the Advisory Team when convened.

2. Capabilities and Resources

EPA can provide personnel, resources, and equipment (including mobile monitoring laboratories) from its facilities in Montgomery, AL, and Las Vegas, NV, and technical support from Headquarters and regional offices. EPA has capability to do the following:

a. Direct environmental monitoring activities and assess the environmental consequences of radioactivity releases.

b. Develop Protective Action Guides.

c. Recommend protective actions and other radiation protection measures.

d. Recommend acceptable emergency levels of radioactivity and radiation in the environment.

e. Prepare health and safety advice and information for the public.

f. Assist in the preparation of long-term monitoring and area restoration plans and recommend clean-up criteria.

g. Estimate effects of radioactive releases on human health and environment.

h. Provide nationwide environmental monitoring data from the Environmental Radiation Ambient Monitoring Systems for assessing the national impact of the emergency.

3. EPA References

a. U.S. Environmental Protection Agency Radiological Emergency Response Plan, Office of Radiation Programs, December 1986.

b. Letter of Agreement between DOE and EPA for Notification of Accidental Radioactivity Releases into the Environment from DOE Facilities, January

8, 1978.

c. Letter of Agreement between NRC and EPA for Notification of Accidental Radioactivity Releases to the Environment from NRC Licensed Facilities, July 28, 1982.

d. Manual of Protective Action Guides and Protective Actions for Nuclear Incidents, Office of Radiation Programs, January 1990.

e. Memorandum of Understanding Between the Federal Emergency Management Agency and the Environmental Protection Agency Concerning the Use of High Frequency Radio for Radiological Emergency Response 1981, Office of Radiation Programs, EPA.

4. EPA Specific Authorities

a. Atomic Energy Act of 1954, as amended, 42 U.S.C. 2011 et seq. (1970), and Reorganization Plan #3 of 1970.

b. Public Health Service Act, as amended, 42 U.S.C. 241 et seq. (1970).

c. Safe Drinking Water Act, as amended, 42 U.S.C. 300f et seq. (1974).

d. Clean Air Act, as amended, 42 U.S.C. 7401 et seq. (1977).

e. Comprehensive Environmental Response, Compensation, and Liability Act of 1980 (SUPERFUND), 42 U.S.C. 9601 et seq., as amended by the Superfund Amendments and Reauthorization Act of 1986 (Public Law 99-499) (1986).

f. E.O. 12656, Assignment of Emergency Preparedness Responsibilities, November 18, 1988, 3 CFR, 1988 Comp., p. 585.

M. Federal Emergency Management Agency

1. Summary of Response Mission

The Federal Emergency Management Agency (FEMA) is responsible for coordinating offsite Federal response activities and Federal assistance to State and local governments for functions other than radiological monitoring and assessment. FEMA's coordination role is to promote an effective and efficient response by Federal agencies at both the national level and at the scene of the emergency. FEMA coordinates the activities of Federal, State, and local agencies at the national level through the use of its Emergency Support Team and at the scene of the emergency with its Emergency Response Team.

2. Capabilities and Resources

FEMA will provide personnel who are experienced in disaster assistance to establish and operate the DFO public information officials to coordinate public information activities personnel to coordinate reporting to the White House and liaison with the Congress and personnel experienced in information support for the Federal response. FEMA personnel are familiar with the capabilities of other Federal agencies and can aid the States and other Federal agencies in obtaining the assistance they need. FEMA will:

a. Coordinate assistance to State and local governments among the Federal agencies

b. Coordinate Federal agency response activities, except those pertaining to the FRMAC, and coordinate these with the activities of the LFA

c. Work with the LFA to coordinate the dissemination of public information concerning Federal emergency response activities. Promote the coordination of public information releases with State and local governments, appropriate Federal agencies, and appropriate private sector authorities and

d. Help obtain logistical support for Federal agencies.

3. FEMA References

a. Federal Response Plan, April, 1992, and subsequent changes.

b. Emergency Response Team Plans for FEMA Regions I, II, III, IV, V, VI, VII, VIII, IX, and X, various dates.

c. NRC/FEMA Operational Response Procedures for Response to a Commercial Nuclear Reactor Accident (NUREG-0981/FEMA-51), Rev. 1, February 1985.

d. Memorandum of Understanding for Incident Response between the Federal Emergency Management Agency and the Nuclear Regulatory Commission, October 22, 1980.

e. Department of Defense, Department of Energy, Federal Emergency Management Agency Memorandum of Agreement for Response to Nuclear Weapon Accidents and Nuclear Weapon Significant Incidents, 1983.

f. Memorandum of Understanding, GSA and FEMA, February 1989.

4. FEMA Specific Authorities

a. The Robert T. Stafford Disaster Relief and Emergency Assistance Act, P.L. 93-288, as amended, 42 U.S.C. 5121 et seq.

b. E.O. 12148 of July 20, 1979, Federal Emergency Management, 3 CFR, 1979 Comp., p. 412.

c. E.O. 12241 of September 29, 1980, National Contingency Plan, 3 CFR, 1980 Comp., p. 282.

d. E.O. 12472 of April 3, 1984, Assignment of National Security and

Emergency Preparedness Telecommunications Functions, 3 CFR, 1984 Comp., p. 193.

e. E.O. 12656 of November 18, 1988, Assignment of Emergency Preparedness Responsibilities, 3 CFR, 1988 Comp., p. 585.

f. E.O. 12657 of November 18, 1988, Federal Emergency Management Agency Assistance in Emergency Preparedness Planning at Commercial Nuclear Power Plants, 3 CFR, 1988 Comp., p. 611.

g. 44 CFR 351, Radiological Emergency Planning and Preparedness.

h. 44 CFR 352, Commercial Nuclear Power Plants: Emergency Preparedness Planning.

N. General Services Administration

1. Summary of Response Mission

The General Services Administration (GSA) is responsible to direct, coordinate, and provide logistical support of other Federal agencies. GSA, in accordance with the National Plan for Telecommunications Support During Non-Wartime Emergencies, manages the provision and operations of telecommunications and automated data processing services. A GSA employee, the Federal Emergency Communications Coordinator (FECC), in accordance with appropriate regulations and plans, is appointed to perform communications management functions.

2. Capabilities and Resources

GSA provides acquisition and procurement of floor space, telecommunications and automated data processing services, transportation, supplies, equipment, material it also provides specified logistical services that exceed the capabilities of other Federal agencies. GSA also provides contracted advisory and support services to Federal agencies and provides security services on Federal property leased by or under the control of GSA. GSA will identify a Regional Emergency Communications Planner (RECP) and FECC, when required, for each of the 10 standard Federal regions. GSA will authorize the RECP to provide technical support and to accept guidance from the FEMA Regional Director during the pre-deployment phase of a telecommunications emergency. The GSA Regional Emergency Coordinator will coordinate all the services provided. Upon request of the Senior FEMA Official (SFO) through the Regional Emergency Coordinator, GSA will dispatch the FECC to the disaster site to expedite the provision of the telecommunications services.

3. Funding

GSA is not funded by Congressional appropriations. All requests for support are funded by the requestor in accordance with normal procedures or existing agreements.

4. GSA References

a. Memorandum of Understanding between GSA and FEMA Pertaining to Disaster Assistance Programs, Superfund Relocation Program, and Federal Radiological Emergency Response Plan Programs, February 2, 1989.

b. GSA Orders in the 2400 Series (Emergency Management).

c. National Communications System Plan for Telecommunications Support to Non-Wartime Emergencies, January 1992.

d. National Telecommunications System Telecommunication Procedures Manuals.

5. GSA Specific Authorities

a. The Federal Property and Administrative Services Act of 1947, as amended, 40 U.S.C. 471 et seq.

b. The Communications Act of 1934, 47 U.S.C. 390 et seq.

c. The Defense Production Act of 1950, as amended, 50 App. 2061 et seq.

d. E.O. 12472 of April 3, 1984, Assignment of National Security and Emergency Preparedness Telecommunications Functions, 3 CFR, 1984 Comp., p. 193.

e. Federal Acquisition Regulations, 48 CFR 1.

f. The General Services Administration Acquisition Regulations, 41 CFR 5.

g. Federal Property Management Regulations, 41 CFR 101.

h. Federal Travel Regulations, 41 CFR 301-304.

O. National Aeronautics and Space Administration

1. Summary of Response Mission

The role of the National Aeronautics and Space Administration (NASA) in a Federal response will depend on the circumstances of the emergency. NASA will be the LFA and will coordinate the initial response and support of other agencies as agreed to in specific interagency agreements when the launch vehicle or payload carrying the nuclear source is a NASA responsibility.

2. Capabilities and Resources

NASA has launch facilities and the ability to provide launch vehicle and space craft telemetry data through its tracking and data network. NASA also has the capability to provide limited radiological monitoring and emergency response from its field centers in Florida, Alabama, Maryland, Virginia, Ohio, Texas, and California.

3. NASA References

a. KHB 1860.1B KSC Ionizing Radiation Protection Program.

b. Memorandum of Understanding between the Department of Energy and the National Aeronautics and Space Administration concerning Radioisotope Power Systems for Space Missions, dated July 26, 1991, as supplemented.

4. NASA Specific Authorities

a. National Aeronautics and Space Act of 1958, as amended, 42 U.S.C. 2451 et seq.

b. NASA Policy Directives (NPDs), as applicable.

P. National Communications System

1. Summary of Response Mission

Under the National Plan for Telecommunications Support in Non-Wartime Emergencies, the Manager, National Communications System (NCS), is responsible for adequate telecommunications support to the Federal response and recovery operations. The Manager, NCS, will identify, upon the request of the Senior FEMA Official, a Communications Resource Manager from the NCS/National Coordinating Center (NCC) staff when any of the following conditions exist: (1) when local telecommunications vendors are unable to satisfy all telecommunications service requirements (2) when conflicts between multiple Federal Emergency Communications Coordinators occur or (3) if the allocation of available resources cannot be fully accomplished at the field level. The Manager, NCC, will monitor all extraordinary situations to determine that adequate national security emergency preparedness telecommunications services are being provided to support the Federal response and recovery operations.

2. Capabilities and Resources

NCS can provide the expertise and authority to coordinate the communications for the Federal response and to assist appropriate State agencies in meeting their communications requirements.

3. NCS References

a. National Plan for Telecommunications Support in Non-Wartime Emergencies, September 1987.

b. Memorandum of Understanding, GSA and FEMA, February 1989.

c. E.O. 12046 (relates to the transfer of telecommunications functions), the White House, March 27, 1978, 3 CFR, 1978 comp., p. 158.

4. NCS Specific Authorities

a. E.O. 12472, Assignment of National Security and Emergency Preparedness Telecommunications Functions, April 3, 1984, 3 CFR, 1984 Comp., p. 193.

b. E.O. 11490, October 30, 1969, 3 CFR, 1966-1970 Comp., p. 820.

c. E.O. 12046, March 27, 1978, 3 CFR, 1978 Comp., p. 158.

d. White House Memorandum, National Security and Emergency Preparedness: Telecommunications and Management and Coordination Responsibilities, July 5, 1978.

Q. Nuclear Regulatory Commission

1. Summary of Response Mission

The U.S. Nuclear Regulatory Commission (NRC) regulates the use of byproduct, source, and special nuclear material, including activities at commercial and research nuclear facilities. If an incident involving NRC- regulated activities poses a threat to the public health or safety or environmental quality, the NRC will be the LFA. In such an incident, the NRC is responsible for monitoring the activities of the licensee to ensure that appropriate actions are being taken to mitigate the consequences of the incident and to ensure that appropriate protective action recommendations are being made to offsite authorities in a timely manner. In addition, the NRC will support its licensees and offsite authorities, including confirming the licensee's recommendations to offsite authorities.

Consistent with NRC's agreement to participate in FRMAC, the NRC may also be called upon to assist in Federal radiological monitoring and assessment activities during incidents for which it is not the LFA.

2. Capabilities and Resources

a. The NRC has trained personnel who can assess the nature and extent of the radiological emergency and its potential offsite effects on public health and safety and provide advice, when requested, to the State and local agencies with jurisdiction based on this assessment.

b. The NRC can assess the facility operator's recommendations and, if needed, develop Federal recommendations on protective actions for State and local governments with jurisdiction that consider, as required, all substantive views of other Federal agencies.

c. The NRC has a system of thermoluminescent dosimeters (TLD) established around every commercial nuclear power reactor in the country. The NRC can retrieve and exchange these TLDs promptly and obtain immediate readings onscene.

3. NRC References

a. NRC Incident Response Plan Revision 2 (NUREG-0728), NRC Office for Analysis and Evaluation of Operational Data, June 1987.

b. Regions I through V Supplements to NUREG-0845, 1990.

c. NRC/FEMA Operational Response Procedures for Response to a

Commercial Nuclear Reactor Accident, (NUREG-0981 FEMA- 51), Rev. 1, February 1985.

d. Operational Response Procedures Developed between NRC, EPA, HHS, DOE, and USDA, January 1991.

e. Memorandum of Understanding for Incident Response between the Federal Emergency Management Agency and the Nuclear Regulatory Commission, October 22, 1980.

f. Memorandum of Understanding Between the FBI and the NRC Regarding Nuclear Threat Incidents Involving NRC-Licensed Facilities, Materials, and Activities, March 13, 1991.

g. NUREG/BR-0150, "Response Technical Manual," November 1993.

h. NUREG-1442 (Rev. 1)/FEMA-REP-17 (Rev. 1), "Emergency Response Resources Guide," July 1992.

i. NUREG-1467, "Federal Guide for a Radiological Response: Supporting the Nuclear Regulatory Commission During the Initial Hours of a Serious Accident," November 1993.

j. NUREG-1471, "U.S. NRC Concept of Operations," February 1994.

k. NUREG-1210, "Pilot Program NRC Severe Reactor Accident Incident Response Training Manual," February 1987.

4. NRC Specific Authorities

a. Atomic Energy Act of 1954, as amended, 42 U.S.C. 2011- 2284.

b. Energy Reorganization Act of 1974, 42 U.S.C. 5841 et seq.

c. 10 CFR Parts 0 to 199.

[FR Doc. 96-11313 Filed 5-7-96 8:45 am] BILLING CODE 6718-02-P

The Contents entry for this article reads as follows:

Federal radiological emergency response plan, 20944

Notes:

1. For NRC reactor licensees, the JOC is within the Emergency Operations Facility (EOF). The EOF would be staffed in accordance with the owner/operator's site-specific Emergency Plan.

2. For NRC licensees, the Federal JIC is within the JIC established by the owner/operator.

Endnotes

[1] ttp://www.wildfirelessons.net/Hot_Tips/IMT_Shuttle_Response_FINAL.pdf

[2] http://www.9-11commission.gov/report/911Report_Ch9.htm

[3] http://www.fas.org/nuke/guide/usa/doctrine/national/frerp.htm

THREE

PLANNING WITH OTHER AGENCIES

Courtesy of Corbis Images.

3

Overview

If a community-wide catastrophe occurs, a flood, for instance or a wildfire, or a terrorist attack, then a multi-agency response involving hundreds of trained and qualified first responders and a variety of equipment and resources would certainly seem like a logical conclusion. What if, however, the combined assets of multiple agencies resulted in confusion regarding who should function as the lead agency, or who should be incident commander? What if jurisdictional 'turf wars' resulted in discord and uncertainty about who was in charge? What if communications systems and basic equipment could not properly interoperate, degrading the efficacy of the resources? These questions stem from the real-world experience of multi-agency catastrophic event responses. In California, for example, during the Loma Prieta earthquake of 1989 or the Oakland Hills wildfires of 1991, an inability of agencies to communicate and jurisdictional conflicts resulted in reduced response efficiency and hampered the ability to stem losses and coordinate efforts.

These conditions led to the development of complex incident command systems. The most current, which the State of California has now signed into law as the standard for multi-agency responses, is known as SEMS or Standardized Emergency Management System. Recognizing that modern catastrophic event response presents complex scene management challenges and coordination, protocols such as the Incident Command System (ICS) and later SEMS were developed to establish responsibilities, streamline decision-making, and prescribe standards for interaction and interoperability.

City of Los Angeles—Multi-Agency Coordination[1]

The following document, which is a part of the City of Los Angeles' general emergency plan, provides an excellent insight into the concepts and considerations of multi-agency coordination. Pay close attention to the key multi-agency planning concepts that are included in this excerpt, such as mutual aid agreements, joint operations provisions, impasse agreements, and plans for state and Federal assistance.

Revised 9/96 Multi-Agency Coordination

Part 5 - Multi-Agency Coordination

This part describes the use of mutual aid, City/County joint operational procedures, multi- and interagency coordination, and the use of volunteers and private sector resources during emergencies.

5.1 Mutual Aid

Mutual aid is support rendered by one jurisdiction to another during declared emergencies. The purpose of mutual aid is to provide personnel and logistical support to meet the immediate requirement of an emergency situation, when the resources normally available to that jurisdiction or agency are insufficient.

Mutual aid assistance provided to or by the City of Los Angeles will be made in accordance with the California Disaster and Civil Defense Master Mutual Aid Agreement and comply with the provisions set forth in this section.

5.1.1 Mutual Aid Activation

All mutual aid rendered under California's Disaster and Civil Defense Master Mutual Aid Agreement is based on an incremental and progressive system of mobilization. Under normal conditions, mutual aid plans are activated in ascending order, i.e., local, operational area, region and state.

- Local resources include those available through mutual aid agreements with neighboring jurisdictions, including the resources of the private sector. Local mutual aid resources are activated by requests to participating agencies.

- Operational area resources are mobilized by the appropriate Operational Area Coordinator in response to requests for assistance from an authorized local official.

- Depending on the type of mutual aid, regional resources are mobilized by the Governor's Office of Emergency Services (OES) Regional Manager or a discipline specific Regional Mutual Aid Coordinator in response to requests for assistance from an Operational Area Coordinator.

- Inter-regional mutual aid is mobilized through regional coordinators or OES regional managers, in response to requests made by a mutual aid region to the State Operations Center (SOC).

- During major emergencies, state government resources are mobilized through OES in response to requests received through regional mutual aid coordinators.

5.1.2 Mutual Aid Authority

Mutual aid assistance may be provided to or by the City under one or more of the following authorities:

- City mutual aid agreements
- California Master Mutual Aid Agreement

- California Fire and Rescue Emergency Plan
- California Law Enforcement Mutual Aid Plan
- California Coroners' Mutual Aid Plan
- Medical Mutual Aid Plan
- Public Works Mutual Aid Agreement
- Volunteer Engineers Safety Assessment Plan
- Federal Disaster Relief Act of 1974 (Public Law 93-288), as amended

5.1.3 Requesting and Using Mutual Aid Resources

Each EOO division shall prepare mutual aid plans and procedures to obtain support in fulfilling that division's emergency operations responsibilities.

Mutual aid will always be requested through established channels. The established channel will vary depending upon the mutual aid system being used. The City will reasonably exhaust its own resources before calling for outside assistance and will respond to requests for mutual aid only when to do so would not unreasonably deplete its own resources.

All mutual aid requests related to the City must be coordinated through the Mayor's Office prior to sending resources out of the City, in accordance with Executive Directive No. 58. Departmental general managers and bureau directors or their designees shall immediately notify the Mayor and the Deputy Mayor for Public Safety upon receipt of any request for mutual aid from the OES or an outside jurisdiction.

Financial reimbursement for mutual aid costs may become available as a result of state and/or federal disaster declarations. Departments can contact the CAO EPD to verify whether a formal declaration of emergency or disaster has been made.

Detailed procedures for requesting and using mutual aid resources vary by functional discipline. Fire, law enforcement and disaster medical mutual aid systems have established discipline specific Operational Area Mutual Aid Coordinators which will be the primary City contact. Specific mutual aid request procedures are contained in departmental plans and procedures.

5.2 City/County Joint Emergency Operations Procedures

5.2.1 The Operational Area

Under SEMS, an operational area is defined as an intermediate level of the state emergency organization, consisting of a county and all political subdivisions within the county. The operational area is one of the five organizational levels within SEMS (field, local government, operational area, region, state).

The operational area organization serves as an intermediate link in the lines of communication and coordination between local jurisdictions and the state emergency organization. Operational area mutual aid coordinators will in some cases function from different facilities.

5.2.2 County Government Assistance

The City of Los Angeles relies on the following agencies of the County of Los Angeles to provide appropriate disaster/emergency related services, as authorized by law.

- Department of Children's Services
- Coroner - Chief Medical Examiner
- District Attorney
- Health Services Department
- Medical Disaster Care Committee
- Mental Health Services
- Municipal Courts
- Probation Department
- Public Defender
- Department of Public Social Services
- Public Works
- Sheriff
- Southern California Hospital Council
- Superior Courts

5.2.3 City Agreements with the County Operational Area

The City and County of Los Angeles have entered into a joint agreement regarding procedures to be followed during an emergency. Procedures are listed below and briefly discussed in the following sections.

- Requesting EOC activation
- Exchange of EOC liaison personnel
- City requests for County support
- County requests for City support
- Media and/or public information announcements
- Establishment of Multi-Agency coordination
- Processing of intelligence and situation reports
- Emergency Management Mutual Aid (EMMA) requests
- Operational Area Satellite Information System (OASIS) back-up
- County-wide Integrated Radio System (CWIRS) radio talk groups
- Emergency Management Information System (EMIS) work station SOPs

- Joint Agreement procedures
- Impasse resolution

5.2.4 Requesting City EOC Activation

The City will activate its EOC if requested to do so by the County EOC. The level of activation and lead agency will be determined by the size and type of event.

5.2.5 Exchange of EOC Liaison Personnel

Upon activation of the County and City EOCs, the City will provide EOC liaison personnel to the County EOC who possess a comprehensive knowledge of the City's overall capabilities and resources, and who have immediate and direct access to the City's EOC command/management. The designated liaison will depend on the type of emergency and the lead department during activation.

Exchange of liaison personnel will be required upon an activation of the County's EOC and a Level II or higher activation of the City's EOC. This procedure may be modified upon the concurrence of both EOC Directors.

5.2.6 City Requests for County Support

Existing Mutual Aid Agreement requests - Departments may receive and/or make requests directly to their County counterpart, if existing mutual aid agreements are in place. All City requests for mutual aid, or requests for the City to provide mutual aid must be coordinated with the Mayor's Office.

Other assistance requests - Requests for assistance that are not covered by formal, preestablished Mutual Aid Agreements will normally be made over the County's EMIS, providing the system is operational. These requests, made by the City's EOC Director, or his/her designee, will be directed to the Operations Section Chief in the County EOC for processing.

The City EOC will be notified of the disposition of their request over EMIS. If EMIS is not operational, the Operations Section Coordinator will inform the City's EOC Director by the most appropriate means of communication available. Final action should always be documented using EMIS.

5.2.7 County Requests for City Support

Existing Mutual Aid Agreement requests - If existing Mutual Aid Agreements are in place. Departments may receive requests directly from their County counterpart.

Other assistance requests - Requests for assistance that are not covered by formal, preestablished Mutual Aid Agreements will normally be made over EMIS, providing the system is operational. These requests will be directed to the Logistics Section in the City EOC for processing. The County EOC will be notified of the disposition of its request over EMIS. If EMIS is not operational, the Logistics Section Coordinator will inform the Operations Section Coordinator in the County EOC by the most appropriate means of communication available. Final action should always be documented using EMIS.

5.2.8 Media/Public Information Announcements

The Public Information Section Coordinator in the City's EOC will coordinate with the

County's PIO on any news releases related to a disaster which includes information about the County.

5.2.9 Establishment of Multi-Agency Coordination

The County EOC Plans and Intelligence Section Coordinator will contact the City EOC Planning/Intelligence Section Coordinator at least each morning after the County's daily planning meeting to communicate priorities for the next operational period. The County will coordinate any local government mutual aid requests within the operational area.

5.2.10 Processing of Intelligence and Situation Reports

Los Angeles City reports for the operational period will be attached to County intelligence and situation reports without alteration.

5.2.11 Emergency Management Mutual Aid (EMMA) Requests

Requests for EMMA personnel support from the City to the County will be directed to and coordinated by the County EOC Operations Section.

Requests for EMMA support from the County to the City will be directed to the City EOC Director. The City's EOC Director or Deputy Director will contact the Mayor's Office for permission to respond to a request for assistance from an outside jurisdiction, per Executive Directive No. 58.

5.2.12 OASIS Back-Up Procedures

Communications over OASIS, as representing the operational area, will originate from the County EOC. In the event the City loses OASIS capability, the City EOC Logistics Section Coordinator shall contact the County EOC Logistics Section to request OASIS support.

5.2.13 County-wide Integrated Radio System (CWIRS) Talk Groups

The County has provided the City with a CWIRS radio. The City EOC Liaison Section Coordinator will ensure that the County liaison to the City EOC has access to a CWIRS radio with equivalent capability. In the event of a failure of both the phone and EMIS systems, the County liaison will be able to communicate with the County EOC by CWIRS radio.

5.2.14 Emergency Management Information System (EMIS) Support

The County has provided the City with an EMIS work station at the City EOC for use by the County liaison to the City EOC as well as the City's EOC staff. If EMIS is operational it will serve as the primary means of recorded communication between the county and City EOCs. The City will send situation reports and intelligence reports (as well as requests for assistance) over EMIS. The County will use EMIS to respond to City queries and requests for assistance, as well as to provide the City with information copies of all reports sent to the State.

5.2.15 Additional Joint Agreement Procedures

Transportation System Restoration and Transportation Allocation Plan

This plan provides a means for augmenting participating agencies' transportation resources during emergencies.

Sheltering and Mass Care

This plan provides a means for avoiding redundant shelters in adjacent Los Angeles City/County areas. It also provides consideration for such special populations as unaccompanied minors, frail or medically dependant elderly, and injured or disabled individuals.

Structural Evaluation and Demolition and Mass Debris Removal

This plan provides for:

- An updated inventory of LA City/County resources and contact lists, and a contact list of the Associated General Contractors of California.
- Ways to establish uniform policies on building re-entry and possession recovery.
- Coordination for identification to allow volunteers to cross law enforcement perimeters.
- Methods of minimizing hazards associated with demolition of damaged buildings.

A proposal to develop a Mass Debris Removal Plan.

Mass Fatalities Management

This plan provides for the continued development of resource lists and operational plans to support temporary morgues and fatality collection points. It also provides for City assistance in coordination with funeral directors to implement the County Mass Fatality Management Plan.

Disaster-Related Medical/Health Services

This plan provides for common disaster triage operational plans and procedures for predesignation of casualty collection points.

5.2.16 City/County Impasse Resolution

If the County is unable to fill a request for support, or to provide that support in a timely manner, the City EOC Director will contact the County EOC Director to attempt to resolve the situation. If a mutually agreeable resolution is not forthcoming, the City may request issue resolution assistance through the City's OES Liaison. Any decision on the part of the City to involve the City OES Liaison will be followed by immediate notification of the County EOC.

5.3 Multi-Agency or Inter-Agency Coordination

Multi-agency or inter-agency coordination is the participation of agencies and disciplines working together in a coordinated effort to facilitate decisions for overall emergency response activities, including the sharing of critical resources and the prioritization of incidents. Multi-agency or inter-agency coordination is recommended under California Government Code 8607 at all levels of SEMS.

Multi-agency coordination involves a mix of agencies, e.g., federal, state, local, etc., that may be working together to solve a particular problem. Inter-agency coordination is used to describe a mix of agencies working together from within a single jurisdiction, e.g., city fire, police and public works.

The City of Los Angeles routinely utilizes both multi-agency and inter-agency coordination at the incident (field), DOC and EOC levels. Ad-hoc groups can be formed at every level to facilitate decisions and resolve problems.

5.4 State and Federal Coordination

5.4.1 State Government Assistance

Various agencies of state government and OES provide the following disaster/emergency related services:

- Pre-event review and approval of emergency plans;
- Receiving and disseminating emergency situation information;
- Receiving, evaluating and disseminating emergency situation information;
- Preparing emergency proclamations and orders for the Governor and disseminating them;
- Preparing and maintaining the California Emergency Plan and associated readiness programs, and coordinating these with local governments;
- Coordinating the emergency activities of all state agencies;
- Processing and acting on mutual aid requests;
- Coordination of the regional response to a disaster, including collection and evaluation of situation information and allocation of available resources;
- Forwarding situation reports and resource requests to the OES SOC;
- Maintaining liaison with local, state and federal emergency response agencies;
- Providing assistance and guidance to the City in preparing emergency plans and procedures, and in conducting emergency exercises; and
- Coordinating local and state mutual aid activity.

5.4.2 Federal Government Assistance

Various federal agencies provide, as authorized by law, the following disaster/emergency related services:

- Search and rescue aerial surveys;
- Disaster loans for homeowners and businesses;

- Emergency conservation measures;
- Emergency loans for agriculture;
- Flood protection;
- Information about available assistance;
- Health and welfare;
- Repairs to federal aid System roads; and
- Tax refunds.

5.5 American Red Cross Assistance

The American Red Cross provides the following disaster/emergency related services:

- Emergency shelter and meals for disaster victims;
- Meals for disaster workers, if no other source is available;
- When appropriate, Red Cross provides meals for people in the impacted area who are living without power or who are cleaning up damage to their residences;
- Emergency assistance to individuals and families affected by a disaster with verifiable disaster caused needs including: groceries, new clothing, rental assistance, furniture and other household items, replacement medications, work related supplies and emergency home repairs;
- Physical and mental health services for disaster workers and disaster victims;
- Disaster welfare inquiry for family members who are unable to locate relatives in the disaster area; and
- Blood and blood products to hospitals.

All Red Cross assistance is based on need and is free. Proof of citizenship is not requested or required. When appropriate and requested, the American Red Cross assigns an agency representative to the City EOC for level two and three activations. The Red Cross liaison sits in the Public Welfare and Sheltering Division of the Operations Section and works with representatives from Recreation and Parks and other members of the division to coordinate shelter and feeding services for disaster victims.

5.6 Volunteer and Private Sector Coordination

5.6.1 Responsibility for Volunteers

The Personnel and Recruitment Division of the EOO is responsible for the City's program to manage the use of non-organized volunteers during emergencies. This program provides for:

- The recruitment of volunteers through various resources;
- The registration of volunteers;

- The assignment of volunteers to the various Divisions; and
- Liaison with Council offices for use of volunteers.

Some EOO divisions utilize organized volunteer groups in their regular activities and have already registered these volunteers. They are normally utilized in a local emergency only by the divisions in which they have been pre-registered. Such programs currently exist in the Police, Fire, and Animal Regulation Departments.

5.6.2 Requests for Unregistered Volunteers

Divisions within the EOO may request volunteers during a local emergency through the Personnel and Recruitment Division Coordinator in the EOC. Assignment of volunteers will be based on operational needs, priorities and the availability of volunteers. Requests for volunteers must be specific in terms of: the number requested; required skills; and when, where and to whom the volunteers are to report. Registration and records of volunteers will be maintained by the EOO Personnel and Recruitment Division.

5.6.3 Disaster Service Workers

The Director of the EOO may require emergency service of any City officer or employee or any citizen. All public employees and registered volunteers of a jurisdiction, including the City of Los Angeles, having an accredited Disaster Council (EOB) are considered to be disaster service workers under the State Government Code. A disaster service worker also includes any unregistered person pressed into service during a state of emergency or state of war emergency by an official with that jurisdiction

5.6.4 Private Resources

Normally available resources may be insufficient during a major emergency affecting large portions of the City. Each EOO division shall develop contingency plans to obtain private resources that will assist in fulfilling that division's emergency operations responsibilities. Many privately owned resources are available for use during emergencies.

5.7 Payment for Emergency Services

Each officer, board, department and employee of the City shall render all possible assistance to the Mayor, the EOB and the Deputy Director (Chief of Police) in carrying out the provisions of the City's Emergency Operations Ordinance, including, but not limited to, planning, training and/or response to emergency incidents. While engaged in emergency services, officers and employees of the City shall be deemed to be engaged in their regular duties (LAAC, Section 8.73).

All other persons rendering services pursuant to the provisions of the City's Emergency Operations Ordinance shall serve without compensation from the City. While engaged in such services, they shall have the same immunities as officers and employees of the City performing similar duties (Administrative Code Section 8.74 a).

Standardized Emergency Management System

WHAT IS SEMS*?[2]

This is a very brief overview of the Standardized Emergency Management System, which is a series of organizational systems for use in the response to any disaster that enables emergency personnel to respond to emergency or disaster situations with a clear and consistent organizational structure. SEMS provides government organizations with clear and consistent strategies for responding to emergencies and disasters, and a table of organization that's consistent among all California governments. SEMS can be used in any emergency, including earthquakes, fires, mudslides, toxic accidents, civil unrest and others. Go to www.oes.ca.gov for more information.

WHO USES SEMS ?

All governments in California are expected to use SEMS in responding to disasters. This includes state agencies, counties, cities, public school districts, and all other special districts.

THE SEMS SYSTEM

SEMS incorporates proven strategies for emergency management and response:

Incident Command System (ICS): A field-level emergency response system based on management-by-objectives. ICS has been adapted for use in the Emergency Operations Center (EOC) as well. See the Table of Organization below.

Mutual Aid: A statewide system for obtaining additional emergency resources from non-affected jurisdictions.

Multi/Inter-Agency Coordination: Affected agencies work together to coordinate allocations of resources and emergency response activities.

Operational Area Concept: A county and its political subdivisions (cities and special districts); coordinates damage information, resource requests and emergency responses. SEE FAQ.

OASIS: Operational Area Satellite Information System links counties (operational areas) to resources at the Governor's Office of Emergency Services.

California Government Code Section 8607, effective January 1, 1993

FIVE ORGANIZATIONAL LEVELS

1. Field on-scene responders.

2. Local county, city or special districts.

3. Operational Area manages and/or coordinates information, resources and priorities among all local governments within the geographic boundary of a county.

4. Region manages and coordinates information and resources among operational areas.

5. State statewide resource coordination integrated with federal agencies.

FIVE SEMS FUNCTIONS
Table of Organization

Management

Operations   Planning/Intelligence   Logistics   Finance/Administration

- *Management* Provides overall direction and sets priorities for an emergency. Supervises liaison, public information, and other special functions.

- *Operations* Organizes the direct response operations and services, and implements the priorities established by management.

- *Planning/intelligence* Gathers and assesses information and develops operational plans.

- *Logistics* Obtains the resources to support the operations.

- *Finance/Administration* Tracks all costs and approves expenditures related to the operation. Provides other administrative support as required. In the Los Angeles County Emergency Operations Center, this function is called *Finance/Administration/Recovery*. For more information on this function in the County EOC, click here.

The LA County Operational Area includes county government and all cities and other local governments within county borders. This includes special districts like public school districts, sanitation districts, water districts, etc. The Operational Area is part of the Standardized Emergency Management System (SEMS).

SEMS: *the legal authority*

SEMS legislation is contained in Title 19, Division 2 of the California Code of Regulations, which states that local governments within counties should be organized into a single operational area by December 1, 1995. The LA County Board of Supervisors passed a resolution on July 5, 1995 establishing the **LA County Operational Area**.

Does SEMS have other requirements?

Yes. SEMS also includes other requirements that promote effective disaster management and cooperation between governments, such as communications protocols and an emergency management table of organization that's uniform between governments.

Why was the SEMS law passed?

Earthquakes, floods, wildfires, and other major disasters do not respect jurisdictional lines. A single disaster can impact many different governments. SEMS was made law to enhance cooperative relationships in planning for and responding to major disasters.

Is SEMS mandated for local governments?

No, but local governments must adopt SEMS in order to get state funding of emergency response costs. SEMS makes good sense, and almost all local governments have adopted it.

Why is the county the Operational Area leader?

SEMS requires that the county assume the lead unless by formal written agreement a city assumes responsibility to coordinate the operational area. The County of Los Angeles is the leader of the LA County Operational Area.

What does the Operational Area do?

The Operational Area is the communications and coordination link between local governments and state government after a major disaster. This includes establishing Operational Area priorities and managing mutual aid.

What about non-government organizations, businesses, and non-profits?

SEMS is not mandated for non-governmental organizations, but it just makes good sense for government to work cooperatively with businesses and non-profits to help manage the challenges of major disasters.

EXECUTIVE OVERVIEW OF SEMS[3]

Introduction

The days of managing emergencies by "winging it!" are over. Californians have been subjected to a devastating string of major disasters over the last six years. The Loma Prieta Earthquake ... the East Bay (Oakland)Hills Firestorm ... the Los Angeles Riots ... the Northridge Earthquake and the recent Floods of 1995 are evidence that the management of emergencies has become an important part of what government must provide for its citizens. Those involved in emergency response have come to realize that there is a need to standardize the business of emergency management. During an emergency or disaster, various responders must suddenly work in a coordinated fashion to maximize the results of their efforts. No one agency, city or special district can afford to maintain the personnel and equipment levels necessary to handle an earthquake, flood, etc. These entities must work together. The problem has been that response agencies use different management methods, differing terminology and often rely on "old allies" to come to their rescue without any formal methods, agreements or organization to coordinate the overall response.

What is SEMS?

The *Standardized Emergency Management System,* commonly referred to as SEMS, addresses these problems by standardizing the principles and methods of emergency response in California. As a direct result of events during the 1991 East Bay (Oakland) Hills Fire, Senator Dominique Petris authored and introduced Senate Bill 1841. The purpose of this proposed legislation was to standardize the emergency response activities of all agencies within California. Passed and codified as Government Code Section 8607, effective January 1, 1993, this statute directed the Governor's Office of Emergency Services (State O.E.S.), in coordination with all State and local emergency management agencies, to establish by regulation the *Standardized Emergency Management System* (SEMS). Now in effect, SEMS is the system required for managing response to multi-agency and multi-jurisdictional emergencies in California. SEMS establishes the following:

- New organizational levels for managing emergencies
- Standardized emergency management methods
- Standardized training for emergency responders and managers

All local governments, including counties, cities, school districts and special districts, must use SEMS to be eligible for funding of their personnel related costs under state disaster assistance programs.

Purpose of SEMS

SEMS has been established to provide more effective response capabilities to multi-agency and multi-jurisdictional emergencies. By standardizing key elements of the emergency management system, SEMS is intended to:

- Facilitate and improve the flow of information within and between operational levels of the system
- Facilitate and improve coordination among all responding agencies

Use of SEMS will improve the mobilization, deployment, utilization, tracking and demobilization of resources and greatly enhance intelligence gathering and sharing capabilities. Mutual aid requests, damage assessment and situation status information can be shared in a timely coordinated fashion.

Organizational Levels

SEMS consists of five organizational levels which are activated as necessary:

- Field Response
- Local Government
- Operational Area
- Regional
- State

The Regional and State levels are under the jurisdiction of the Governor's Office of Emergency Services, also known as State OES. Formerly, the Operational Area was only required for State of War emergencies and for the most part was non-existent. SEMS requires that all jurisdictions within a geographical county form an Operational Area Organization and that the Operational Area be utilized as needed for any multi-agency or multi-jurisdictional response. The Operational Area is not a political subdivision rather a special purpose organization created to accomplish specific tasks during times of emergency. SEMS regulations require that County governments be the lead agency in the development and operation of each Operational Area. The Local Government level is comprised of all the political subdivisions within the geographical county. Each of these entities is responsible for carrying out their responsibilities within their respective boundaries. The Field Response level is comprised of first line responders representing their respective agencies.

Standardized Management Methods

The intent of SEMS not only deals with the creation of the Operational Area but also requires the standardization of management methods at all levels of operation. The Incident Command System (ICS) is to be used by all field response level personnel. Local governments, Operational Areas, Regional and State levels may use either ICS or some form of Multi Agency Coordination System (MACS) in their emergency management environment which is usually an Emergency Operations Center (EOC). ICS provides standardized procedures and terminology, a unified command structure, a manageable span-of-control, and an action planning process which identifies overall incident response strategies and specific tactical actions. ICS has been used by the fire service for many years, in fact, they developed ICS in the early 1970's. Law enforcement, emergency medical services and other have adopted ICS concepts as well.

In addition to using ICS, SEMS will also facilitate better coordination of mutual aid. Through improved communications within the Operational Areas, resource requests and deployment of personnel can be better coordinated. At the operational area level, collecting, processing and sharing damage assessment situation status and other intelligence information will maximize the efficiency and effectiveness of all response efforts.

Standardized Training

Initially, SEMS approved training has been divided into four courses:

- Introduction to SEMS
- Field response personnel
- EOC personnel
- Executive/Elected personnel

The Introduction and Executive courses are self taught/take home. The Field and EOC courses are more extensive and can be taught in a classroom or in-service environment. State OES has indicated they will provide Train-the-Trainer instruction to prepare persons from each jurisdiction to teach these classes.

Summary

SEMS has four basic requirements: (1) formalization of an *Operational Area Emergency Management Organization* to coordinate response efforts, (2) the use of the *Incident Command System* (ICS) in disaster response, (3) *standardized training* and (4) the *centralized gathering of intelligence and mutual aid requests* into one Emergency Operations Center at the Operational Area level.

Section 8607(e) of the Government Code states "(1) By **December 1, 1996**, each local agency, in order to be eligible for any funding of response-related costs under disaster assistance programs, shall use the Standardized Emergency Management System as adopted pursuant to subdivision (a) to coordinate multiple jurisdiction or multiple agency operations." "(2) Notwithstanding paragraph (1), local agencies shall be eligible for repair, renovation, or any other non-personnel costs resulting from an emergency."

California Code of Regulations Title 19 Section 2409(b) states "All local governments within a county geographic area shall be organized into a single operational area by **December 1, 1995**, and the County board of supervisors shall be responsible for its establishment." Thus, the Kern County Office of Emergency Services has been assigned this responsibility.

Section 2428 states "Emergency response agencies shall ensure that their emergency response personnel maintain minimum training competencies in SEMS pursuant to the approved course of instruction . . ." which has four course components: Introductory, Field, EOC and Executive.

ICS Organization[4]

Every incident or event has certain major management activities or actions that must be performed. Even if the event is small, and only one or two people are involved, these activities will still always apply to some degree.

The organization of the Incident Command System is built around five major management activities. They are:

COMMAND

Sets objectives and priorities

Has overall responsibility at the incident or event

OPERATIONS

Conducts tactical operations to carry out the plan

Develops the tactical objectives

Organization

Directs all resources

PLANNING

Develops the action plan to accomplish the objectives

Collects and evaluates information

Maintains resource status

LOGISTICS

Provides support to meet incident needs

Provides resources and all other services needed to support the incident

FINANCE/ADMINISTRATION

Monitors costs related to incident

Provides accounting Procurement Time recording Cost analyses

These five major management activities are the foundation upon which the ICS organization develops. They apply whether you are handling a routine emergency, organizing for a major event, or managing a major response to a disaster.

On small incidents, these major activities may be managed by one person, the Incident Commander (IC). Large incidents usually require that they be set up as separate Sections

Incident Command-Roles & Responsibilities[5]

INCIDENT COMMANDER

The Incident Commander's responsibility is the overall management of the incident. On most incidents the command activity is carried out by a single Incident Commander. The Incident Commander is selected by qualifications and experience.

The Incident Commander may have a deputy, who may be from the same agency or from an assisting agency. Deputies may also be used at section and branch levels of the ICS organization. Deputies must have the same qualifications as the person for whom they work, as they must be ready to take over that position at any time.

a. Review Common Responsibilities.

b. Assess the situation and/or obtain a briefing from the prior Incident Commander.

c. Determine incident objectives and strategy.

d. Establish the immediate priorities.

e. Establish an Incident Command Post.

f. Establish an appropriate organization.

g. Ensure planning meetings are scheduled as required.

h. Approve and authorize the implementation of an Incident Action Plan.

i. Ensure that adequate safety measures are in place.

j. Coordinate activity for all Command and General Staff.

k. Coordinate with key people and officials.

l. Approve requests for additional resources or for the release of resources.

m. Keep agency administrator informed of incident status.

n. Approve the use of trainees, civilian, and auxiliary personnel.

o. Authorize release of information to the news media.

p. Order the demobilization of the incident when appropriate.

PUBLIC INFORMATION OFFICER

The Public Information Officer is responsible for developing and releasing information about the incident to the news media, to incident personnel, and to other appropriate agencies and organizations.

Only one Public Information Officer will be assigned for each incident, including incidents operating under Unified Command and multi-jurisdiction incidents. The Public Information Officer may have assistants as necessary, and the assistants may also represent assisting agencies or jurisdictions.

Agencies have different policies and procedures relative to the handling of public information. The following are the major responsibilities of the Public Information Officer that would generally apply on any incident:

a. Review Common Responsibilities.

b. Determine from the Incident Commander if there are any limits on information release.

c. Develop material for use in media briefings.

d. Obtain Incident Commander's approval of media releases.

e. Inform media and conduct media briefings.

f. Arrange for tours and other interviews or briefings that may be required.

g. Obtain media information that may be useful to incident planning.

h. Maintain current information summaries and/or displays on the incident and provide information on status of incident to assigned personnel.

i. Maintain log of unit activity.

LIAISON OFFICER

Incidents that are multi-jurisdictional, or have several agencies involved, may require the establishment of the Liaison Officer position on the Command Staff.

Only one Liaison Officer will be assigned for each incident, including incidents operating under Unified Command and multi-jurisdiction incidents. The Liaison Officer may have assistants as necessary, and the assistants may also represent assisting agencies or jurisdictions.

The Liaison Officer is the contact for the personnel assigned to the incident by assisting or cooperating agencies. These are personnel other than those on direct tactical assignments or those involved in a Unified Command.

a. Review Common Responsibilities.

b. Be a contact point for Agency Representatives.

c. Maintain a list of assisting and cooperating agencies and Agency Representatives.

d. Assist in establishing and coordinating interagency contacts.

e. Keep agencies supporting the incident aware of incident status.

f. Monitor incident operations to identify current or potential inter-organizational problems.

g. Participate in planning meetings, providing current resource status including limitations and capability of assisting agency resources.

h. Maintain log of unit activity.

AGENCY REPRESENTATIVES

In many multi-jurisdiction incidents, an agency or jurisdiction will send a representative to assist in coordination efforts.

An Agency Representative is an individual assigned to an incident from an assisting or cooperating agency who has been delegated authority to make decisions on matters affecting that agency's participation at the incident. Agency Representatives report to the Liaison Officer, or to the Incident Commander in the absence of a Liaison Officer.

a. Review Common Responsibilities.

b. Ensure that all agency resources are properly checked-in at the incident.

c. Obtain briefing from the Liaison Officer or Incident Commander.

d. Inform assisting or cooperating agency personnel on the incident that the Agency Representative position for that agency has been filled.

e. Attend briefings and planning meetings as required.

f. Provide input on the use of agency resources unless resource technical specialists are assigned from the agency.

g. Cooperate fully with the Incident Commander and the General Staff on agency involvement at the incident.

h. Ensure the well-being of agency personnel assigned to the incident.

i. Advise the Liaison Officer of any special agency needs or requirements.

j. Report to home agency dispatch or headquarters on a prearranged schedule.

k. Ensure that all agency personnel and equipment are properly accounted for and released prior to departure.

I. Ensure that all required agency forms, reports and documents are complete prior to
departure.

m. Have a debriefing session with the Liaison Officer or Incident Commander prior to
departure.

SAFETY OFFICER

The Safety Officer function is to develop and recommend measures for assuring personnel safety, and to assess and/or anticipate hazardous and unsafe situations.

Only one Safety Officer will be assigned for each incident. The Safety Officer will report to the Incident Commander. The Safety Officer may have assistants as necessary, and the assistants may also represent assisting agencies or jurisdictions. Safety assistants may have specific responsibilities such as air operations, hazardous materials, etc.

a. Review Common Responsibilities.

b. Participate in planning meetings.

c. Identify hazardous situations associated with the incident.

d. Review the Incident Action Plan for safety implications.

e. Exercise emergency authority to stop and prevent unsafe acts.

f. Ensure accountability procedures are in place. (See Section 15.)

g. Size up need for and effectiveness of:

1. Accountability plans/procedures;

2. Rapid Intervention plans/procedures;

3. Protective clothing needs of personnel and assistants;

4. Scene security measures;

5. Safety zones;

6. Avenues of access/egress.

h. Organize, assign and brief assistants as needed.

i. Review and approve the medical plan and ensure that adequate re-hab for all personnel is established.

j. Develop hazardous materials site safety plan as required.

k. Maintain log of unit activity.

Examples of U.S. Government Agency Response Protocols & Multi-Agency Cooperation

The system of government in the United States represents, in many ways, one of the more complex challenges to multi-agency emergency response. The American tradition of separation of powers, divisions of and respect for Federal, state, and local authority, and well-defined agency roles, is an important part of America's political culture. Inherent in this political model is the reality that there will be many agencies with varied, specific responsibilities.

One good example is the U.S. Environmental Protection Agency (EPA). Created in the 1970s, the EPA is tasked with enforcement of federally mandated environmental and natural resource protections. Part of accomplishing that mission requires the EPA to formulate an emergency response program to define its areas of responsibility and prescribe how the EPA resources are to interact with other agencies at the Federal, state, and local levels in the event of an environmental emergency.

Case Study: U.S. Environmental Protection Agency: Inside the Emergency Response Program[6]

THE CARE OF HUMAN LIFE AND HAPPINESS ... IS THE FIRST AND ONLY
LEGITIMATE OBJECT OF GOOD GOVERNMENT.
— *THOMAS JEFFERSON*

Each year, more than 20,000 emergencies involving the release, or threatened release, of oil and hazardous substances are reported in the United States, potentially affecting both large and small communities and the surrounding natural environment. Reports in the local news often portray the swift and effective response of local firefighters and other emergency officials. Behind the scenes, however, a vast *National Response System* (NRS) involving federal, state, and local officials is at work supporting the men and women on the front lines.

The U.S. Environmental Protection Agency plays a leadership role in this national system, chairing the *National Response Team* and directing its own Emergency Response Program. It's goal is the protection of the public and the environment from immediate threats posed by emergencies involving hazardous substances and oil. The program's primary objectives are to take reasonable steps to prevent such emergencies; to prepare emergency response personnel at the federal, state, and local levels for such emergencies; and to respond quickly and decisively to such emergencies wherever and whenever they occur within our national borders.

The Emergency Response Program is a coordinated effort among five key EPA organizations and its 10 *Superfund Regions*. The five headquarters offices are:

The Office of Emergency and Remedial Response which oversees implementation of domestic emergency response including the two major components of the National Response System program, the Superfund Removal Program (*Hazardous Substances*), and the Oil Program, as well as disaster response under the Stafford Act through the Federal Response Plan (FRP);

The *Chemical Emergency Prevention and Preparedness Office*, which has primary responsibility for preparing and planning for chemical emergencies through a network of state and local emergency planning organizations, and provides oversight of EPA International emergency response support and/or assistance and coordination of National Security response issues and key Agency and interagency leadership roles as part of the NRS and the Federal Response Plan (FRP);

Prevention, Pesticides, and Toxic Substances which has primary responsibility for community involvement and community right-to-know; and

Radiation which has primary responsibility for FRERS and radiological expertise under the NRS; and

The *Office of Underground Storage Tanks* which is responsible for preventing petroleum releases from underground storage tanks.

Each of these organizations derives its authority from laws and regulations passed by Congress to specifically address the country's ability to reduce or eliminate the threats to human life and the environment posed by the handling, storage, and use of hazardous substances and oil. While each office plays a different role in EPA's Emergency Response Program, all offices share the common mission of preventing, preparing for, and responding to emergencies involving hazardous substances and oil.

National Response Team

Response planning and coordination is accomplished at the federal level through the *U.S. National Response Team (NRT)*, an interagency group co-chaired by the EPA and the *U.S. Coast Guard* (also see *NRT Member Roles and Responsibilities* for more information on this group). Although the NRT does not respond directly to incidents, it is responsible for three major activities related to managing responses: (1) distributing information; (2) planning for emergencies; and (3) training for emergencies. The NRT also supports the Regional Response Teams.

Distributing Information

The NRT is responsible for distributing technical, financial, and operational information about hazardous substance releases and oil spills to all members of the team. This information is collected primarily by NRT committees whose purpose is to focus attention on specific issues, then collect and disseminate information on those issues to other members of the team. Standing committees of the NRT and the topics that are addressed include:

- **Response Committee**, chaired by EPA, addresses issues such as response operations, technology employment during response, operational safety, and interagency facilitation of response issues (e.g., customs on transboundary issues). Response specific national policy/program coordination and capacity building also reside in this committee.

- **Preparedness Committee**, chaired by the U.S. Coast Guard, addresses issues such as preparedness training, monitoring exercises/drills, planning guidance, planning interoperability, and planning consistency issues. Preparedness specific national policy/program coordination and capacity building also reside in this committee.

- **Science and Technology Committee**, chaired by EPA and the *National Oceanic and Atmospheric Administration* in alternating years, provides national coordination on issues that parallel those addressed by the Scientific Support Coordinator on an incident by incident basis. The focus of this committee is on identifying developed technology and mechanisms for applying those technologies to enhance operational response. The committee monitors research and development of response technologies and provides relevant information to the RRTs and other members of the National Response System to assist in the use of such technologies.

Planning for Emergencies

The NRT ensures that the roles of federal agencies in the NRT for emergency response are clearly outlined in the National Contingency Plan (see the National Contingency Plan Overview). After a major incident, the effectiveness of the response is carefully assessed by the NRT. The NRT may use information gathered from the assessment to make recommendations for improving the National Contingency Plan and the National Response System. The NRT may be asked to help *Regional Response Teams (RRTs)* develop Regional Contingency Plans. The NRT also reviews these plans to determine whether they comply with federal policies on emergency response.

Training for Emergencies

Training is the key to the federal strategy for preparing for oil spills or hazardous substance releases. Although most training is performed by state and local personnel, the NRT develops training courses and programs, coordinates federal training efforts, and provides information to regional, state, and local officials about training needs and courses.

Supporting Regional Response Teams

The NRT supports Regional Response Teams (RRTs) by reviewing Regional or Area Contingency Plans to maintain consistency with national policies on emergency response. The NRT also supports RRTs by monitoring and assessing RRT effectiveness during an incident. The NRT may ask an RRT to focus on specific lessons learned from a particular incident and to share those lessons with other members of the National Response System. In this way, the RRTs can improve their own regional contingency plans while helping to solve problems that might be occurring elsewhere within the National Response System.

Case Study: Safety Management in Disaster and Terrorism Response[7]

Since our nation's beginnings, emergency responders have helped protect the people of the United States from the effects of natural and manmade disasters. From the bucket brigades of colonial times to today's highly complex, multiagency response community, response workers have taken action in emergencies to save lives, preserve property, and protect the public good. The devastating attacks on the World Trade Center and the Pentagon on September 11, 2001, cast a powerful new spotlight on the vital role that responders play in containing and mitigating unexpected crises. Members of the response community disregarded injuries and fatigue, and even gave their lives in their effort to reduce the initial impact of these disasters and bring the situations under control.

The tragic events of September 11 showed that response organizations are a central component of our homeland security system against both natural and manmade threats. This renewed reliance on emergency responders has focused fresh attention on the imperative to protect these individuals from the hazards inherent in their work, not just for the good of the community, but of the nation. While responders should be protected for their own sakes, their safety is also crucial to the effectiveness of the response force as a whole. Injuries to individual members affect their organizations' ability to perform overall, both immediately and in the long term. A responder injured is not only prevented from assisting in today's emergency, but may also be unavailable to respond to an attack tomorrow.

In the military context, this understanding is embodied in the concepts of force protection and force health protection. In applying these concepts, the military aims to preserve its force's fighting strength by protecting individual servicemen and women against the threat of enemy action and by taking steps to minimize the effect of hazards on unit effectiveness, readiness, and morale. The unprecedented potential for multiple terrorist attacks drives home the need for comparable thinking in the response community. Sustainability becomes key: Incidents must be managed with an eye on ensuring the readiness of response organizations to meet future challenges.

Major Disasters Present Special Challenges for Safety Management

Fortunately, disasters of the magnitude of the September 11 events are rare. Usually emergency responders confront incidents on a comparatively small scale that can be handled on a local level and pose more limited safety risks. But a major disaster presents a significant challenge to a locality, a state, a region, and sometimes even a nation. Responding to such an incident tests the capacity of responding organizations and can place large numbers of emergency responders in harm's way. Protecting the safety of responders in those situations is much more difficult.

In contrast to the types of incidents that emergency responders normally face, major disasters share a number of characteristics that create unique difficulties for response organizations.[1]

Large Number of People Affected, Injured, or Killed
While small-scale emergencies involve a few individuals or small groups of people, major disasters severely affect large numbers of citizens across communities, cities, or entire regions. The Northridge earthquake caused more than 60 fatalities and 9,000 injuries and displaced 17,000 to 18,000 people from their homes [Stratton et al. 1996]. The attack on the World Trade Center claimed the lives of more than 2,800 individuals and put many thousands more at risk ["The Numbers" 2002].

Large Geographic Scale
Most emergency incidents involve only a single building or other well-defined site. Major disasters, however, often extend over very large areas. In 1992, Hurricane Andrew left a trail of devastation that extended over 1,000 square miles [Lewis 1993]. Responders to the Oklahoma City bombing confronted a rubble pile more than 35 feet deep made up of approximately one-third of the federal building structure [Oklahoma Department of Civil Emergency Management 2000].

Prolonged Duration
Average emergency response operations are relatively short, lasting only minutes or hours from first responders' arrival on scene to completion of response actions [Study Interviews].[2] In contrast, activities in major disasters can stretch into days, weeks, or even months. Although the total repair and clean up after Hurricane Andrew lasted much longer, the U.S. military relief operation lasted for 50 days [Higham and Donnelly 1992]. In New York City after September 11, 2001, the response was not officially completed until eight months after the attack [Barry 2002].

Multiple, Highly Varied Hazards
Whereas common emergencies usually present emergency responders with a limited number of risks, major disasters involve multiple hazards that can vary widely in nature. The World Trade Center site, for example, exposed response workers to a complex mixture of physical and respiratory perils [Lioy and Gochfeld 2002]. Responders to the Northridge earthquake confronted active fires, collapsing buildings, and hazardous materials [Federal Emergency Management Agency (FEMA) 1994a]. Because of this wide variety, few responders will have experience with everything they might encounter in the aftermath of a major disaster.

Wide Range of Needed Capabilities
Major disasters require supplementary response capabilities not routinely maintained by local response organizations.[3] Many natural disasters and major terrorist incidents require extensive rubble removal and management operations that local response organizations are not equipped to carry out. FEMA-sponsored urban search and rescue teams were needed to respond to both the Northridge earthquake and the September 11 attacks. Such requirements frequently turn the response effort after a major disaster into a multiagency operation that can span all levels of government, nongovernmental organizations, and the private sector.[4]

Influx of Convergent Volunteers and Supplies
In contrast with smaller emergencies generally handled by a local response organization, major disasters often attract large numbers of independent, or "convergent," volunteers. These volunteers may be members of other response organizations that come to a disaster site spontaneously or ordinary citizens who come out of a desire to help [Maniscalco and Christen 2001].[5] Likewise, a

major disaster also frequently prompts individuals and organizations to send large quantities of food and other supplies. Hurricane Andrew was a prominent example, where the influx of people and supplies was so overwhelming that responders referred to it as "the disaster after the disaster" [Study Interviews].

Damage to Infrastructures
While localized disasters leave infrastructures vital to effective emergency response intact, major disasters can damage or destroy them. Hurricane Andrew severely dam-aged the local transportation infrastructure, with road signs destroyed and major roads blocked. The Northridge earthquake caused numerous ruptures in water mains and citywide power outages.

Direct Effects on Responder Organizations
Unlike routine incidents, major disasters can directly affect the operational capacity of response organizations. The emergency responders lost in the World Trade Center collapse are one tragic example. Another occurred in Hurricane Andrew where the homes of at least 128 police officers were damaged or destroyed. Many of the officers reported for duty not knowing what had happened to their families [Taylor 1992].

Responder Safety Management Is Risk Management

The inherently hazardous nature of any emergency situation necessitates that safety be approached from a risk management perspective. Rather than eliminating risk altogether, response managers aim to shield responders from hazards to the greatest extent possible. When making decisions, the level of risk in any given action should be weighed against the potential benefit.[6] This process of safety management can be broken down into three central components, as shown in Figure 1.1:[7]

- gathering information about the nature of the situation, the responders at the scene, and the hazards involved

- analyzing response options and potential protective measures and making decisions

- taking action to implement safety decisions, reduce hazards, or provide health protection to responders.

These three activities take place in a continuous cycle until the response effort comes to a close.[8] As part of this continuous management effort, safety managers constantly reexamine and evaluate their efforts to protect responders as operations at an incident scene continue.[9]

In the course of their routine activities, organizations develop standard approaches for carrying out these functions. But safety management during the response to a major disaster is a far larger and more complex undertaking. Safety management practices that are well developed and effective for standard response activities will very likely be insufficient. In short, the highly demanding and unfamiliar environment after a major disaster makes it difficult, or even impossible, for individual responder organizations to effectively perform the three functions of the safety management cycle.

Major disasters create substantial hurdles on the organizational level as well. For example, the multiagency nature of responses to major disasters makes safety management significantly more complex. In an effort of this magnitude, where many different organizations unfamiliar with each other's operating practices are working side by side, a new set of secondary hazards can arise from the response operation itself. These secondary hazards, such as those generated by fire or law enforcement activities occurring simultaneously with ongoing construction or utility operations, can pose serious risks to all involved responders. In addition, the management problems arising from operations involving many different organizations can also result in communications failures, logistical problems, and other conflicts that can directly or indirectly impact responder safety. These only compound the broad range of primary hazards stemming directly from the disaster.

The Response Community Recognizes a Pressing Need for Improved Safety Management

The events of September 11 brought these safety challenges to the fore with an urgency that the emergency response community, and the nation, had not known before. As one of many initiatives that took place in the disaster's aftermath, the National Institute for Occupational Safety and Health (NIOSH) joined with the Science and Technology Policy Institute (S&TPI), formerly managed by the RAND Corporation, to organize a conference in New York City on protecting emergency workers during responses to conventional and biological terrorist attacks [Jackson et al. 2002]. During the discussions, participants frequently expressed deep concern over safety management practices during major crises in general. The research presented in this report is a direct outgrowth of that concern.

In the following pages, we offer recommendations that response organizations can put in place at both the functional and organizational levels to improve safety management in future response operations. In accordance with an all-hazards perspective, we consider the full range of natural and manmade disasters to ensure that the approach we suggest is flexible and comprehensive.

Tomorrow's Success Depends on Today's Preparations

The emotionally charged, chaotic environment in the immediate aftermath of a major disaster is not the time to start working on procedures or guidelines to improve responder safety. Strategic planning and management well before the event, along with standardized systems and procedures, are key. Preparedness is the crux of effectiveness.

The distinctive characteristics of major disasters make the case for preparedness especially strong. The multiple hazards inherent in situations of this magnitude call for a flexibility from the response community that can only come through preplanning. That major disasters take so many different forms underscores this point. The response community will inevitably be called upon to carry out substantially different activities—that pose highly varied hazards—as different crises arise. Effective safety management requires having the capabilities and resources in place to deal with this variety.

In addition, because major disasters are rare and the safety risks responders face may be unprecedented, response organizations get little to no practice managing them. In this context, scenario-based planning and training assume added value. Similarly, it is also important to build safety management practices that can meet the needs of disasters into organizations' standard operating procedures to the extent possible. While use of safety management practices during smaller-scale events will never be directly analogous to applying them in disasters, the experience will nonetheless make it more likely they can be effectively applied when they are needed most. Although no disaster situation is entirely predictable, the more prepared safety managers are to deal with expected hazards, the more attention and energy they will be able to devote to handling unanticipated issues as they arise.

Finally, the fact that major disasters demand a multiagency response operation makes a common understanding of the needs of different organizations—and the parts different response organizations can play in safety management—a precondition for successfully protecting responders. The recommendations we present in this report focus on the changes organizations can begin making today—both individually and collaboratively—to lay the groundwork for better serving responders' safety needs and managing multiagency safety efforts in the future.

Case Study: US DHS—NATIONAL RESPONSE PLAN (NRP)[8]

What it does for America.

The National Response Plan establishes a comprehensive all-hazards approach to enhance the ability of the United States to manage domestic incidents. The Plan incorporates best practices and procedures from incident management disciplines—homeland security, emergency management, law enforcement, firefighting, public works, public health, responder and recovery worker health and safety, emergency medical services, and the private sector—and integrates them into a unified structure. It forms the basis of how federal departments and agencies will work together and how the federal government will coordinate with state, local, and tribal governments and the private sector during incidents. It establishes protocols to help protect the nation from terrorist attacks and other natural and manmade hazards; save lives; protect public health, safety, property, and the environment; and reduces adverse psychological consequences and disruptions to the American way of life.

Plan Organization

Concept of Operations, Coordinating Structures, Roles and Responsibilities, Definitions, etc.

Base Plan

Glossary, Acronyms, Authorities, and Compendium of National Interagency Plans

Appendixes

Groups capabilities & resources into functions that are most likely needed during an incident (e.g., Transportation, Firefighting, Mass Care, etc.)

Emergency Support Function Annexes

Describes common processes and specific administrative requirements (e.g., Public Affairs, Financial Management, Worker Safety & Health, etc.)

Support Annexes

Outlines core procedures, roles and responsibilities for specific contingencies (e.g., Bio, Radiological, Cyber, HAZMAT Spills)

Incident Annexes

National Response Plan Incident Management Priorities
- Save lives and protect the health and safety of the public, responders, and recovery workers.
- Ensure security of the homeland.
- Prevent an imminent incident, including acts of terrorism, from occurring.
- Protect and restore critical infrastructure and key resources.
- Conduct law enforcement investigations to resolve the incident, apprehend the perpetrators, and collect and preserve evidence for prosecution and/or attribution.
- Protect property and mitigate damages and impacts to individuals, communities, and the environment.
- Facilitate recovery of individuals, families, businesses, governments, and the environment.

Emphasis on Local Response

The Plan identifies police, fire, public health and medical, emergency management, and other personnel as responsible for incident management at the local level.

- The Plan enables incident response to be handled at the lowest possible organizational and jurisdictional level.
- The Plan ensures the seamless integration of the federal government when an incident exceeds local or state capabilities.

Timely Federal Response to Catastrophic Incidents

- The Plan identifies catastrophic incidents as high-impact, low-probability incidents, including natural disasters and terrorist attacks that result in extraordinary levels of mass casualties, damage, or disruption severely affecting the population, infrastructure, environment, economy, national morale, and/or government functions.
- The Plan provides the means to swiftly deliver federal support in response to catastrophic incidents.

Multi-agency Coordination Structure

- The Plan establishes multi-agency coordinating structures at the field, regional and headquarters levels.
- These structures execute the responsibilities of the President through the appropriate federal departments and agencies.
- These structures provide a national capability that addresses both site-specific incident management activities and broader regional or national issues, such as impacts to the rest of the country, immediate regional or national actions required to avert or prepare for potential events, and management of multiple incidents.

New Coordinating Features in the National Response Plan

- **Homeland Security Operations Center (HSOC).**
  The HSOC serves as the primary national-level multiagency hub for domestic situational awareness and operational coordination. The HSOC also includes DHS components, such as the National Infrastructure Coordinating Center (NICC), which has primary responsibility for coordinating communications with the Nation's critical infrastructure during an incident.
- **National Response Coordination Center (NRCC).**
  The NRCC, a functional component of the HSOC, is a multiagency center that provides overall federal response coordination.
- **Regional Response Coordination Center (RRCC).**
  At the regional level, the RRCC coordinates regional response efforts and implements local federal program support until a Joint Field Office is established.
- **Interagency Incident Management Group (IIMG).**
  A tailored group of senior federal interagency experts who provide strategic advice to the Secretary of Homeland Security during an actual or potential Incident of National Significance.

- **Joint Field Office (JFO).**
  A temporary federal facility established locally to provide a central point to coordinate resources in support of state, local, and tribal authorities.

- **Principal Federal Official (PFO).**
  A PFO may be designated by the Secretary of Homeland Security during a potential or actual Incident of National Significance. While individual federal officials retain their authorities pertaining to specific aspects of incident management, the PFO works in conjunction with these officials to coordinate overall federal incident management efforts.

Maintaining the National Response Plan
- The Department of Homeland Security/Emergency Preparedness and Response (EP&R)/Federal Emergency Management Agency (FEMA), in close coordination with the DHS Office of the Secretary, will maintain the National Response Plan.

- The Plan will be updated to incorporate new Presidential directives, legislative changes, and procedural changes based on lessons learned from exercises and actual events.

Endnotes

[1] www.lacity.org/epd/pdf/eompp/**part5**.pdf

[2] http://www.lacoa.org/sems.shtml

[3] http://www.co.kern.ca.us/fire/oes/overview1.htm

[4] http://www.nysemo.state.ny.us/TRAINING/ICS/explain.htm

[5] http://www.northern.vaems.org/Documents/No%20Va%20MCI%20Plan.pdf

[6] http://www.epa.gov/superfund/programs/er/inside.htm

[7] http://www.cdc.gov/niosh/docs/2004-144/chap1.html

[8] http://www.dhs.gov/interweb/assetlibrary/NRP_FactSheet_2005.pdf

FOUR

LOOKOUT, COMMUNICATION, ESCAPE, AND SAFETY (LCES)

Courtesy of Corbis Images.

4

Overview

First responders share a major principle with the Hippocratic Oath taken by physicians: *"First, Do No Harm!"*

Rule number one in any emergency or catastrophic response event is that first and foremost our mission is to begin the rescue, relief, and recovery process. Our own safety is paramount to our ability to heal and repair the scene. If we are inattentive to our safety, we can become victims ourselves, which will only further complicate the catastrophe scene and degrade the effectiveness of our response.

The LCES (or "Laces") concept is a prime example of emergency scene safety protocols. Born out of major wildfire rescue efforts, Laces proscribes necessary precautions needed to ensure the safety and efficacy of a large-scale response operation. This chapter will begin to familiarize the student with a) the Laces concept as a specific example; and b) the overarching concept of emergency planners avoiding further harm to an emergency scene by maintaining their own safety.

LCES[1]

"LCES" stands for Lookouts, Communications, Escape routes and Safety zones. These are the same items stressed in the Fire Orders and "Watch Out Situations". They should be viewed from a "systems" point of view, all interconnected and interdependent. Each should be evaluated independently, but importantly also evaluated as a system. For example, the best safety zone is of no value if your escape route does not offer timely access when needed.

A key concept, the LCES system is identified to **each firefighter prior** to when it must be used. The nature of wildfire suppression dictates continuous evaluation of LCES and when necessary, re-establishment of LCES as time and fire growth progress.

Lookouts or scouts (roving lookouts) need to be in position where **both** the hazard and the firefighter can be seen. Lookouts must be trained to observe the wildland fire environment and to recognize and anticipate fire behavior changes. Each situation determines the number of lookouts that are needed: due to terrain, cover and fire size, one lookout is normally not sufficient. When the hazard becomes a danger, the lookout relays the information to the firefighters so they may reposition to the safety zone. Actually, each firefighter has both the authority and responsibility to warn others when hazards become threats to safety.

Communications is the vehicle which delivers the message to the firefighters, alerting of the approaching hazard. As stated in the current training, communication must be prompt and clear, whether by radio or verbal.

Escape routes are the paths the firefighter takes from his/her current threatened position to an area free from danger. Notice that escape routes is the term used, rather than "escape route". Unlike the other components of the system there **must always** be more than one escape route available. A single escape route may be cut off.

Escape routes are probably the most elusive component of LCES. Their effectiveness changes continuously. As the firefighter works along the perimeter, fatigue and spatial separation increases the time required to reach the safety zone. The most common escape route, or part of one, is the fireline. On indirect or parallel fireline, problem situations become compounded. Unless safety zones have been identified ahead as well as to the rear, firefighter retreat may not be possible.

Safety zones are locations where the threatened firefighter may find adequate refuge from the danger. Improperly and unfortunately, shelter deployment sites have been termed "safety zones". Safety zones should be conceptualized and planned as a location where use of a shelter is unnecessary. If there is a deployment, that location is not a true safety zone. Fireline intensity and topographic location determine a safety zone's effectiveness.

To re-emphasize, the LCES system must be identified **prior** to when it must be used! Each time when firefighters are working around hazards, lookouts must be posted with adequate communications to each firefighter, with a minimum of two escape routes from the work location to an adequate safety zone (NOT A SHELTER DEPLOYMENT SITE).

Safety and tactics should not be considered as separate entities! As with any task, safety and technique necessarily should be integrated. The LCES system should be automatic in any tactical operation where a hazard exist or could be present.

LCES is simply a refocusing of the essential elements of the FIRE ORDERS. This systems approach stresses the importance of the components working together.

LOOKOUTS, COMMUNICATION, ESCAPE ROUTES, SAFETY ZONES[2]

"LCES"

Original Document

By

Paul Gleason

Former Zig Zag Hotshot Superintendent

June, 1991

"LCES and Other Thoughts"

I have been asked to give input on wild land firefighter safety to the Fire and Aviation Staff - Safety and Training, Washington Office. First, let me say I am honored to be able to contribute at this level. The afternoon of June 26, 1990, as I knelt beside a dead Perryville firefighter, I made a promise to the best of my ability to help end the needless fatalities, and alleviate the near misses, by focusing on training and operations pertinent to these goals.

Throughout my career I have dealt with wildland fire suppression, as a Hotshot Crew Supervisor, with only minor injuries occurring to those I have directly supervised. This is primarily because of two reasons, luck (which cannot be ignored) and basic lessons which I learned from the exceptional firefighters I have had the opportunity to work with. Many of the really valuable suppression lessons I learned were prior to fire shelter requirements.

Subject vs. Objective Hazards

A popular mountaineering test divides the alpinists' hazards into two distinct types: subjective, which one has direct control over (e.g., condition of the equipment, the decision to turn back) and objective hazards which are inherent to the alpine environment (e.g., avalanches, rock fall). Objective hazards are a natural part of the environment. They cannot be eliminated and either one must not go into the environment where they exist or adhere to a procedure where safety from the hazard is assured.

Similarly, the wildland firefighter's hazards are either subjective or objective. Examples of subjective hazards would be working below a dozer constructing fireline or the use of improper techniques while felling a tree. The fireline supervisor has direct control over these types of hazards.

The wildland fire environment has four basic objectives hazards; lightning, fire-weakened timber (standing and lying), rolling rocks and entrapment by running fires. When these hazards exist the options are to not enter the environment or to adhere to a safe procedure. I feel the key to this safe procedure is LCES. Although, the following discussion applies to all objective hazards, we will directly address fire entrapments.

LCES

LCES stands for lookout(s), communication(s), escape routes and safety zone(s). These are the same items stressed in the FIRE ORDERS and "Watchout" Situations. I prefer to look at them from a "systems" point of view, that is, as being interconnected and dependent on each other. It is not only important to evaluate each one of these items individually but also together they must be evaluated as a system. For example, the best safety zone is of no value if your escape route does not offer you timely access when needed.

A key concept—the LCES system is identified to each firefighter prior to when it must be used. The nature of wildland fire suppression dictates continuously evaluating and, when necessary, re-establishing LCES as time and fire growth progress. I want to take a minute and briefly review each component and its interconnection with the others.

Lookout(s) or scouts (roving lookouts) need to be in a position where both the objective hazard and the firefighter (s) can be seen. Lookouts must be trained to observe the wildland fire environment and to recognize and anticipate wildland fire behavior changes. Each situation determines the number of lookouts that are needed. Because of terrain, cover and fire size one lookout is normally not sufficient. The whole idea is when the objective hazard becomes a danger the lookout relays the information to the firefighter so they can reposition to the safety zone. Actually, each firefighter has the authority to warn others when they notice an objective hazard which becomes a threat to safety.

Communications(s) is the vehicle which delivers the message to the firefighters, alerting of the approaching hazard. As is stated in current training, communications must be prompt and clear. Radios are limited and at some point the warning is delivered by word of mouth. Although more difficult, it is important to maintain promptness and clearness when communication is by word of Incident intelligence (regarding wildland fire environment, fire behavior and suppression operations) both to and from Incident Management (i.e. Command & General Staff) is of utmost importance. But I don't view this type of communication a normal component of the LCES system. Entrapment occurs on a fairly site-specific level. Incident intelligence is really used to alert of hazards (e.g.. "Watchout" situations) or to select strategical operations. LCES is primarily a Division function: responsibility should be here.

Escape Routes are the path the firefighter takes from their current locations, exposed to the danger, to an area free from danger. Notice that escape routes is used instead of escape route(s). Unlike the other components, there always must be more than one escape route available to the firefighter. Battlement Creek 1976 is a good example of why another route is needed between the firefighter's location and a safety zone.

Escape routes are probably the most elusive component of LCES. Their effectiveness changes continuously. As the firefighter works along the fire perimeter, fatigue and spatial separation increases the time required to reach the safety zone. The most common escape route (or part of an escape route) is the fireline.

LOOKOUT, COMMUNICATION, ESCAPE, AND SAFETY (LCES)

On indirect or parallel fireline, situations become compounded. Unless safety zones have been identified ahead, as well as behind, firefighters retreat may not be possible.

Safety Zone(s) are locations where the threatened firefighter may find refuge from the danger. Unfortunately shelter deployment sites have been incorrectly called safety zones. Safety zones should be conceptualized and planned as a location where no shelter is needed. This does not intend for the firefighter to hesitate to deploy their shelter if needed, just if a shelter is deployed the location is not a tree safety zone. Fireline intensity and safety zone topographic location determine safety zone effectiveness.

Again, a key concept—the LCES system is identified prior to when it must be used. That is lookouts must be posted with communications to each firefighter, and a minimum of two escape routes form the firefighter's work location to a safety zone (not a shelter deployment site) every time the firefighter is working around an objective hazard.

Safety and tactics should not be considered as separate entities. As with any task safety and technique necessarily should be integrated. The LCES system should be automatic in any tactical operation where an objective hazard is or could be present.

LCES is just a re-focusing on the essential elements of the FIRE ORDERS. The systems view stresses the importance of the components working together. The LCES system is a result of analyzing fatalities and near misses for over 20 years of active fireline suppression duties. I believe that all firefighters should be given an interconnecting view of Lookout(s), Communications(s), Escape routes and Safety zone(s).

Division Operations

Establishing a Lookout position in the Operations function has its merits. The Lookout(s) would be assigned directly to the Division Supervisor. They would have only one responsibility, albeit an important one. Lookouts keep one eye on the fire and the other on the Division's firefighters.

Commonly, Weather Watchers, and Field Observers are incorrectly assigned lookout duties. Division Supervisors should solicit input from these sources for their decisions, but these positions are in the Planning sections, not Operations. Lookouts need to be identified prior to tactical deployment of suppression resources and they need to give their undivided attention to the Division's objective hazards and firefighter locations.

Ideally each crew would establish lookouts in potentially hazardous situations. But, this requires the ability to identify these situations and to establish adequate (in amount and location) lookouts for the situation. Additionally, all too often crew supervisors hesitate to remove a crewmember from fireline production and assign them the position of lookout. They do not realize that the assignment of lookouts is not only their authority but also their responsibility.

Incident Management, thru Operations and Planning, would identify the operation's "Watchout" Situations, divisions on which they are (or could) occur and assign qualified lookouts to the Division Supervisor.

Span of Control

Span of control depends directly on the quality of resources and their capabilities. 3-5 subordinates to each supervisor may be sufficient for a static environment where they is direct access to each subordinate; but in the active wildland fire environment experienced leadership is necessary on a tighter ratio. Jerry Monosmith presented solutions via the geographical breakdown of a division into "segments".

Crucial to any solution is the definition of "experienced". How would you define experienced?

Many reasons have been given for the lack of experience including an organization's inability for employee retention and insufficient BASIC supervision skill development.

Downhill/Indirect Firelines

The two situations that firefighters traditionally have found themselves getting into trouble are downhill and indirect fireline operations. The lessons learned on the Loop Fire ('66) developed awareness, and consequential guidelines, for downhill fireline construction. Since then downhill operations have been safer; everyone agrees the only one who works in a chimney is Santa Claus, and he does it in the dead of winter. Unfortunately, we still have a ways to go (i.e., Battlement Creek '76).

Indirect firelines are a different story. In the last half of the 1980's all the entrapments have occurred during indirect operations. Extreme fire behavior with active spotting has put more reliance on indirect strategies. With indirect fireline the firefighter finds themselves removed from the best safety zone, the burn, as well as unable to see the objective hazard.

"Floating Division"

A floating division is the planned division during an indirect operation that exists initially only on paper (a map). It is not anchored. Wildland fire suppression tactics stress the importance of beginning construction at an anchor point (point where there is the least chance of being outflanked). To safely deploy resources on a "floating" division it is extremely important that the division is initially anchored and that the anchor point is also a safety zone. Only then can resources begin work developing the LCES system as they progress.

The success of the operation depends on the safety of personnel and the ability to hold the fireline. It is crucial that indirect fireline location is determined after careful analysis of wildland fire behavior possibilities including that behavior which will results if the fire enters the third-dimension (crowning/spotting from both wind-driven and plume-dominated fires). All too often the full possibilities are not incorporate in location decisions.

Wildland/Urban Interface

Suppression in the wildland/urban interface presents its own unique set of problems. The choice of fireline location is often influenced by the homes which stand between the fire front and a "better" option. Often the standard tactics of anchoring at the rear (or heel) and flanking will leave improvements in the path of the wind-driven fire.

The lack of an ideal fireline location does not in itself constitute unsafe indirect strategy. The "urgency" of the operation causes a break down in solid tactics. During interface suppression operations, maybe more than any operation, the LCES system must be in place.

With the rapid spread rates reached by wind-driven fires only two options are available. The traditional "anchor and flank" strategy or the unorthodox protection of improvements and resources as the wildfire spreads past. The last dictates the necessity for a "defensible space" around each improvement sufficient to serve also as a safety zone, a true safety zone. Unless this precaution has been made the risk to defending the improvement may not be worth the operation.

Judgment Errors

John Dill, head rock climbing rescue ranger in Yosemite NP, recently made an analysis of errors in judgment made preceding an accident. He found three reasons which contribute to the accidents; ignorance, casualness and distraction. After thinking about the firefighter's environment and accidents these same reasons were found to correspond. Allow me to take a moment and help draw the correlations.

Ignorance: Unfortunately, we still have firefighters and fireline supervisors who still end up in wildland fire situations that call for skills and knowledge beyond their level of training. I know it is stressed over and over, but the BASICS, basic wildland fire behavior, basic suppression skills, need to be learned and reviewed. Yet many of the entrapments are the result of no lookouts or an insufficient safety zone, a lack of basics.

Casualness: The rock climber standing at the base of a couple thousand-foot granite walls in Yosemite is reassured in their decision to undertake a challenging ascent because of the helicopter which is poised less than a mile from the proposed ascent. We are doing the same. The situation is viewed more casually because we have an option if the tactic fails – our fire shelter.

Another way casualness enters our environment is through the reinforcement of improper tactics since the fire does not "blowup" while we are working the fireline the first few, or several times. But then we find ourselves entrapped because the familiar situation changes and our reliance on improper tactics just doesn't work this time.

Distraction: Often I have been told that was it not for the on-the-job training that was given by a Division Supervisor the hazard would not have been noticed and tactics would not have been adjusted. Distraction is a very very real problem for firefighters. Fatigue and carbon monoxide do not help with the decision making process either. Fireline personnel should be continually monitoring each other and remain open to communication and others evaluation of the situation at hand.

LCES Checklist[3]

In the wildland fire environment, Lookouts, Communications, Escape Routes, Safety Zones (LCES) is key to safe procedures for firefighters. The elements of LCES form a safety system used by firefighters to PROTECT THEMSELVES AND WORK AS A TEAM WITH OTHERS. This system is put in place before fighting the fire: select a lookout or lookouts, set up a communication system, choose escape routes, and select a safety zone or zones.

LCES IS A SELF-TRIGGERING MECHANISM.

Lookouts assess and reassess the fire environment and communicate threats of safety to firefighters. Firefighters use escape routes to move to safety zones.

LCES is built on two basic guidelines:

1. Before safety is threatened, each firefighter must be informed how the LCES system will be used, and

2. The LCES system must be continuously re-evaluated as conditions change.

Chapter 1 – Firefighting Safety

Lookouts

- Experienced/Competent/Trusted
- Enough lookouts at good vantage points
- Knowledge of crew location
- Knowledge of escape and safety locations
- Map/Weather Kit/Watch/IAP

Communications

- Radio frequencies confirmed
- Backup and check-ins established
- Update on any situation change
- Sound alarm early, not late

Escape Routes

- More than one escape route
- Avoid uphill escape routes
- Scouted: Loose soils/rocks/vegetation

LOOKOUT, COMMUNICATION, ESCAPE, AND SAFETY (LCES)

- Timed: Slowest person/fatigue and temperature factors
- Marked: Flagged for day or night (NFES 0566)
- Evaluate: Escape time vs. rate of spread
- Vehicles parked for escape

Safety Zones

- Survivable without a fire shelter
- Back into clean burn
- Natural Features: Rock areas/water/meadows
- Constructed Sites: Clearcuts/roads/helispots
- Scouted for size and hazards
- Upslope? = more heat impact = larger safety zone
- Downwind? = more heat impact = larger safety zone
- Heavy fuels? = more heat impact = larger safety zone

Escape time and safety zone size requirements will change as fire behavior changes.

Escape Routes and Safety Zones

An **Escape Route** is "a pre-planned and understood route firefighters take to move to a Safety Zone or other low-risk area."

A **Safety Zone** is "a preplanned area of sufficient size and suitable location that is expected to protect fire personnel from known hazards without using fire shelters."

Identification of Escape Routes and Safety Zones is one of the primary responsibilities of any wildland firefighter working on or near the fireline. The following guidelines can be used when selecting Safety Zones:

- Calculations indicate that for most fires, Safety Zones must be wider than 164 feet to ensure firefighter survival.
- The calculation to determine Safety Zone radius is four times the maximum flame height plus 50 square feet per firefighter, or an additional four feet of radius per firefighter. This calculation provides the radius of the Safety Zone, meaning the Safety Zone diameter should be twice the value of the above formula.
- If potential for the fire to burn completely around the Safety Zone exists, the diameter should be twice the values indicated above.
- Factors that will reduce Safety Zone size include reduction in flame height by thinning or burnout operations, shielding the Safety Zone from direct exposure to the flame by locating it on the lee side of ridges or other geographic structures, or reducing flame temperature by applying fire retardant to the area around the Safety Zone.

- All firefighter PPE must be worn.
- Keep in mind that these guidelines do not address convective energy.

Safety Zone Guidelines

- Avoid locations that are downwind from the fire.
- Avoid locations that are in chimneys, saddles, or narrow canyons.
- Avoid locations that require a steep uphill escape route.
- Take advantage of heat barriers such as leeside of ridges, large rocks, or solid structures.
- Burn out safety zones prior to flame front approach.
- For radiant heat only, the distance separation between the firefighter and the flames must be at least 4 times the maximum flame height. This distance must be maintained on all sides, if the fir has ability to burn completely around the safety zone. Convective heat from wind and/or terrain influences will increase this distance requirement. The calculations in the following table assume no slope and no wind.

Flame Height	Distance Separation (firefighters to flame)	Area in Acres
10 ft.	40 ft.	1/10 acre
20 ft.	80 ft.	1/2 acre
50 ft.	200 ft.	3 acres
75 ft.	300 ft.	7 acres
100 ft.	400 ft.	12 acres
200 ft.	800 ft.	50 acres

Distance Separation is the radius from the center of the safety zone to the nearest fuels. When fuels are present that will allow the fire to burn on all sides of the safety zone this distance must be doubled in order to maintain effective separation in front, to the sides, and behind the firefighters.

Area in Acres is calculated to allow for distances separation on all sides for a three person engine crew. One acre is approximately the size of a football field or exactly 208 feet x 208 feet.

Last Resort Survival

Look at your options and immediately act on the best one!

Utilize all Personal Protective Equipment!

Protect your airway!

Escape if you can:

- Drop any gear not needed for fire shelter deployment (keep your fire shelter, hand tool, quart of water, and radio).

LOOKOUT, COMMUNICATION, ESCAPE, AND SAFETY (LCES) 205

- You may be able to use the fire shelter for a heat shield as you move.
- In LIGHT FUELS, you may be able to move back through the flames into the black.
- If you are on the flank of the fire, try to get below the fire.
- Consider vehicles or helicopters for escape.

Find a survivable area:

- Stay out of hazardous terrain features.
- Use bodies of water that are more than 2 feet deep.
- In LIGHT FUELS, you may be able to light an escape fire.
- In other fuels, you may be able to light a backfire.
- Call for helicopter or retardant drops.
- Cut and scatter fuels if there is time.
- Use any available heat barriers (structure, large rocks, dozer berms).
- Consider vehicle traffic hazards on roads.

Pick a fire shelter deployment site:

- Find the lowest point available.
- Maximize distance from nearest aerial fuels or heavy fuels.
- Pick a surface that allows the fire shelter to seal and remove ground fuels.
- Get into the fire shelter before the flame front hits.
- Position your feet toward the fire and hold down the fire shelter.
- Keep your face pressed to the ground.
- Deploy next to each other and keep talking.

Expect:

- Extremely heavy ember showers.
- Superheated air blast to hit before the flame front hits.
- Noise and turbulent powerful winds hitting the fire shelter.
- Pin holes in the fire shelter that allow fire glow inside.
- Heat inside the shelter = Extreme heat outside.
- Deployments have lasted up to 90 minutes.
- When in doubt wait it out.

Downhill Checklist

Downhill fireline construction is hazardous in steep terrain, fast-burning fuels, or rapidly changing weather. Downhill fireline construction should not be attempted unless there is no tactical alternative. When building downhill fireline, the following is required:

- Crew supervisor(s) and fireline overhead will discuss assignments prior to committing crew(s). Responsible overhead individual will stay with job until completed (TFLD or ICT4 qualified or higher).
- Decision will be made after proposed fireline has been scouted by supervisor(s) of involved crew(s).
- LCES will be coordinated for all personnel involved.

Crew supervisor(s) is in direct contact with lookout that can see the fire.

- Communication is established between all crews.
- Rapid access to safety zone(s) in case fire crosses below crew(s).
- Direct attack will be used whenever possible; if not possible, the fireline should be completed between anchor points before being fired out.
- Fireline will not lie in or adjacent to a chute or chimney.
- Starting point will be anchored for crew(s) building fireline down from the top.
- Bottom of the fire will be monitored; if the potential exists for the fire to spread, action will be taken to secure the fire edge.

Common Denominators of Fire Behavior on Tragedy Fires

- Incidents happen on smaller fires or on isolated portions of larger fires.
- Fires look innocent before "flare-ups" or "blow-ups." In some cases, tragedies may occur in the mop-up stage.
- Flare-ups generally occur in deceptively light fuels.
- Fires run uphill surprisingly fast in chimneys, gullies, and on steep slopes.
- Wind direction or wind speed unexpectedly shifts.

Thunderstorm Safety

Approaching thunderstorms may be noted by a sudden reverse in wind direction, a noticeable rise in wind speed, and a sharp drop in temperature. Rain, hail, and lightning occur only in the mature stage of a thunderstorm.

Observe the 30/30 rule: a) If you see lightning and hear thunderclaps within 30 seconds take storm counter-measures identified below. b) Do not resume work in exposed areas until 30 minutes after storm activity has passed.

- Take shelter in a vehicle or building if possible.
- If outdoors, find a low spot away from tall trees, wire fences, utility lines, and other elevated conductive objects. Make sure the place you pick is not subject to flooding.

- If in the woods, move to an area with shorter trees.
- If only isolated trees are nearby, keep your distance twice the tree height.
- If in open country, crouch low minimizing contact with the ground. You can use a pack to sit on, but never lay on the ground.
- If you feel your skin tingle or your hair stand on end, immediately crouch low to the ground. Make yourself the smallest possible target and minimize your contact with the ground.
- Don't group together.
- Don't stay on ridge tops, in wide open areas, near ledges or rock outcroppings.
- Don't operate land line telephones, machinery, or electric motors.
- Don't handle flammable materials in open containers or metal hand tools.
- Handheld radios and cellular telephones can be used.

Clothing and Personal Protective Equipment (PPE)

- All PPE must meet or exceed NFPA 1977 Standard on Protective Clothing and Equipment for Firefighters (current edition).
- Wear hard hat while on the fireline.
- Wear 8-inch laced all-leather boots with slip- and melt-resistant soles and heels.
- Wear flame-resistant clothing while on the fireline and when flying in helicopters. Do not wear clothing, even undergarments, made of synthetic materials which can burn and melt on your skin. Roll down sleeves to the wrist.
- Use leather gloves to protect hands.
- Use eye and face protection whenever there is a danger from material being thrown back in your face.
- Determine and comply with host agency requirements regarding fire shelters on fireline suppression assignments or follow your own agency's requirements if they are more restrictive. The fire shelter is a tool of last resort, not to be used tactically.
- Use hearing protection when working with high noise-level firefighting equipment, such as helicopters, air tankers, chain saws, pumps, etc.
- When operating chain saws, sawyers and swampers will wear additional safety equipment including approved chaps, gloves, hard hat, eye and ear protection.
- Recommend use of an approved dust/smoke mask when in heavy smoke and dusty environments. Use of a dust/smoke mask is not a PPE requirement for all agencies at this time.

- Face and neck protection (Nomex shrouds) are not required PPE. If used, they must meet NFPA 1977. If issued, shrouds should be deployed only in impending flash fuel or high radiant heat situations and not routinely worn throughout the operational period, due to an unacceptable increase in physiological heat stress.

- PPE clothing will be cleaned or replaced whenever soiled, particularly with oils. PPE will be replaced when the fabric is so worn as to reduce fire resistance capability of the garment.

How to Properly Refuse Risk

Every individual has the right and obligation to report safety problems and contribute ideas regarding their safety. Supervisors are expected to give these concerns and ideas serious consideration. When an individual feels an assignment is unsafe they also have the obligation to identify, to the degree possible, safe alternatives for completing that assignment. Turning down an assignment is one possible outcome of managing risk.

A "turn down" is a situation where an individual has determined they cannot undertake an assignment as given and they are unable to negotiate an alternative solution. The turn down of an assignment must be based on an assessment of risks and the ability of the individual or organization to control those risks.

- Individuals may turn down an assignment as unsafe when:
  - There is a violation of safe work practices.
  - Environmental conditions make the work unsafe.
  - They lack the necessary qualifications or experience.
  - Defective equipment is being used.

- Individual will directly inform their supervisor that they are turning down the assignment as given. The most appropriate means to document the turn down is using the criteria (Standard Firefighting Orders, 18 Watch Out Situations, etc.), outlined in the Risk Management Process.

- Supervisor will notify the Safety Officer **IMMEDIATELY** upon being informed of the turn down. If there is no Safety Officer, notification shall go to the appropriate Section Chief or to the Incident Commander. This provides accountability for decisions and initiates communication of safety concerns within the incident organization.

If the supervisor asks another resource to perform the assignment, they are responsible to inform the new resource that the assignment has been turned down and the reasons that it was turned down. If an unresolved safety hazard exists or an unsafe act was committed, the individual should also document the turn down by submitting a SAFENET (ground hazard) or SAFECOM (aviation hazard) form in a timely manner.

These actions do not stop an operation from being carried out. This protocol is integral to the effective management of risk, as it provides timely identification of hazards to the chain of command, raises risk awareness for both leaders and subordinates, and promotes accountability.

After Action Review

What was planned?

- Review the primary objectives and expected action plan.

What actually happened?

- Review the day's actions:
- Identify and discuss effective and non-effective performance.
- Identify barriers that were encountered and how they were handled.
- Discuss all actions that were not standard operating procedure, or those that presented safety problems.

Why did it happen?

- Discuss the reasons for ineffective or unsafe performance. Concentrate on **WHAT**, not **WHO**, is right.

What can we do next time?

- Determine lessons learned and how to apply them in the future.

FIREFIGHTER HEALTH

Fatigue — Work and Rest

- Establish record-keeping systems that track crew work time.
- Plan and strive to provide one hour of sleep or rest for every two hours worked.
- When deviating from work/rest guidelines, the agency administrator or incident commander (IC) must approve in writing.
- Start each operational period with rested crews.
- Provide an adequate sleep environment.
- Monitor individuals for sleep deprivation.

The pulse is a good way to gauge fatigue. The pulse should recover in one minute or less to 110 beats per minute, or, if not, a longer break is needed. A firefighter's wake-up pulse can signal potential problems. If it is 10% or more above normal, it can mean fatigue dehydration, or even a pending illness.

Food and Nutrition

Nutritious food can be a morale booster, but more importantly, it fuels muscles for hard work and internal organs for health and fitness. A firefighter may burn 5,000 to 6,000 calories a day. These calories must be replaced to avoid cramping, fatigue, and impaired judgment. Government-provided food must be low in fats and high in complex carbohydrates.

Drinks provided must replace essential fluids lost from the body during exercise. On a normal fireline assignment, firefighters may replace 12 or more quarts of fluids a day. In some cases, firefighters may need to replace one to two quarts of fluids per hour. Water is an excellent way to replenish fluid loss. Natural juices and sport drinks contain energy-restoring glucose. Avoid caffeinated, carbonated, and "diet" drinks.

Firefighter Rehabilitation

Areas designed for resting, eating, and sleeping should be located in a safe, shady area away from smoke, noise, running fire, falling trees and snags, rolling rocks, moving vehicles, aircraft, and pack stock. Provide reasonable rest periods, especially at high elevations and on hot days.

Driving Limitations

Drivers operating vehicles that require a Commercial Drivers License (CDL) are regulated by the Federal Motor Carriers Safety Regulations Part 393.3 and any applicable State Laws.

All governmental fire agencies are exempted from several requirements of CDL regulation under Department of Transportation 49 CFR but are subject to the NWCG National Incident Operations Driving Standards.

These standards address driving by personnel actively engaged in wildland fire or all risk response activities, including driving while assigned to a specific incident or during initial attack fire response (includes time required to control the fire and travel to a rest location). In the absence of more restrictive agency policy, these guidelines will be followed during mobilization and demobilization as well. Individual agency driving policies shall be consulted for all other non-incident driving.

1. Agency resources assigned to an incident or engaged in initial attack fire response will adhere to the current agency work/rest policy for determining length of duty day.
2. No driver will drive more than 10 hours (behind the wheel) within any duty day.
3. Multiple drivers in a single vehicle may drive up to the duty-day limitation provided no driver exceeds the individual driving (behind the wheel) time limitation of 10 hours.
4. Drivers shall drive only if they have had at least 8 hours off duty before beginning a shift.

Exception to the minimum off-duty hour requirement is allowed when essential to:

    a Accomplish immediate and critical suppression objectives, or

    b. Address immediate and critical firefighter or public safety issues.

5. Documentation of mitigation measures used to reduce fatigue is required for drivers who exceed 16-hour work shifts. This is required regardless of whether the driver was still compliant with the 10-hour individual (behind the wheel) driving limitations.

First Aid

Prompt first aid must be given for all injuries. First aid facilities should be made available in proximity to the fireline and at incident base and camp(s). When activated, the Medical Unit is responsible for all medical emergencies involving assigned incident personnel. Each crew should carry a first aid kit and all supervisory personnel should be trained in basic emergency first aid. While help is on the way, be prepared to move the patient in case of unexpected fire movement.

First Aid Guidelines

Legality:

- Do only what you know how to do and keep records of actions.

Bloodborne Pathogens:

- Personal protective equipment (pocket mask, latex gloves and goggles) should be worn if contact with body fluids is possible.

Treatment Principles:

- Think: prevent further injury; remove from danger. No liquids for the unconscious.
- Fast Exam: airway, breathing and circulation.
- Thorough Exam: head to toe and side to side (symmetry).
- Keep readable records and send a copy with the patient when transporting or evacuating.

Specific Treatments:

- Bleeding: Direct pressure, elevate, and pressure point.
- Shock: Lay patient down, elevate feet, keep warm and replace fluids if conscious.
- Fractures: Splint joints above and below injury and monitor pulse beyond the injury away from the trunk of the body.
- Bee Sting (anaphylaxis): Life-threatening; see if the patient has a sting kit and transport immediately.
- Burns: Remove heat source, cool with water, dry wrap, and replace fluids.
- Diarrhea: Drink fluids in large quantities.
- Eye Injuries: Wash out foreign material, don't open swollen eyes, leave

impaled objects, and pad and bandage both eyes.

- Heat Exhaustion: Skin is gray, cool, and clammy. Rest in cool place and replace electrolytes.
- Heat Stroke: Skin is dry, red, and temperature hot. Cool and transport immediately.

Endnotes

[1] http://www.ci.boulder.co.us/fire/wfire/MOB.htm

[2] http://www.fireleadership.gov/toolbox/documents/lces_gleason.html

[3] www.nwcg.gov/pms/pubs/410-1/410-1.pdf

FIVE

LOGISTICS

Courtesy of Corbis Images.

5

Overview

What do we need, when will we need it, how long will we need it, and how much will it cost? These are the questions of paramount importance to a logistical planner. Logistics is both a noun and a verb—it describes both physical materials and those supplies required for a specific objective as well as the science of managing such provisions.

When managing a catastrophic event, logistics (the noun) may refer to everything from bandages and bottled water to a multi-million dollar helicopters and construction equipment. As a verb, the logistics management of a community-wide catastrophe will be a vast and coordinated effort to identify the correct logistical requirements; source the needed material; ensure replenishment; care, maintenance, and proper operations; and properly secure and return the materials once they are no longer needed.

This chapter will give the student an overview of logistical considerations during a catastrophic event response.

Within a standardized emergency management structure, there are a number of specialized responsibilities. Key among these is the logistics section chief. The logistics chief will be in charge of establishing and securing the necessary equipment and facilities for rescue, response and recovery efforts. The following overview of the logistics chief's responsibilities from FEMA will provide the student with an excellent insight into the basic general logistics responsibilities and considerations for a catastrophic event response.

LOGISTICS SECTION CHIEF CHECKLIST[1]

RESPONSIBILITIES

The Logistics Section Chief, a member of the general staff, is responsible for providing facilities, services, and materials in support of the incident. The

Logistics Section Chief participates in the development of the incident action plan and activates and supervises the branches and units within the Logistics Section.

Instructions: The checklist below presents the minimum requirements for Logistics Section Chiefs. Note that some items are one-time actions, while others are ongoing or repetitive throughout the incident.

COMPLETED/NOT APPLICABLE TASKS

Obtain a briefing from the Incident Commander.

Plan the organization of the Logistics Section.

Assign work locations and preliminary work tasks to section personnel.

Notify the Resources Unit of the Logistics Section units which have been activated, including the names and locations of assigned personnel.

Assemble and brief unit leaders and branch directors.

Participate in the preparation of the incident action plan.

Identify the service and support requirements for planned and expected operations.

Provide input to and review the communications, medical, and traffic plans.

Coordinate and process requests for additional resources.

Review the incident action plan, and estimate section needs for the next operational period.

Ensure that the incident communications plan is prepared.

Advise on current service and support capabilities.

Prepare the service and support elements of the incident action plan.

Estimate future service and support requirements.

Receive the demobilization plan from the Planning Section.

Recommend the release of unit resources in conformity with the demobilization plan.

Ensure the general welfare and safety of Logistics Section personnel.

Maintain the unit log (ICS Form 214 or local form).

MAJOR RESPONSIBILITIES AND TASKS

The major responsibilities of the Logistics Section Chief are stated below. Following each are tasks for implementing the responsibility.

RESPONSIBILITY TASKS

Obtain Briefing from Incident Commander

- Receive an incident briefing, summary of resources dispatched to the incident, and initial instructions concerning work activities.

- Obtain a copy of the incident action plan, if available.

Activate Logistics Section Units

- Determine from the incident briefing what Logistics Section personnel have been ordered.
- Confirm order of appropriate Logistics Section personnel.
- Plan preliminary organization of the Logistics Section.
- Compare the preliminary incident action plan with personnel ordered, as appropriate.
- Identify additional personnel needed.
- Request additional personnel.
- Assign work locations and work tasks to logistics section personnel.
- Notify the Resources Unit of Logistics Section units activated, including names and locations of assigned personnel.

Organize Logistics Section

- Confirm arrival of dispatched Logistics Section personnel.
- Assemble and brief Logistics Section personnel.
- Review initial operations of Logistics Section with section personnel.
- Give instructions for initial operations to section personnel.

Assist in Preparation of the Incident Action Plan

- Attend planning meeting.
- Review suggested strategy and operations for next operational period.
- Advise on current service and support capabilities.
- Estimate logistic capabilities with current capabilities.
- Compare required capabilities with current capabilities.
- Determine additional service and support requirements corresponding to the incident action plan.
- Prepare service and support elements of the incident action plan.
- Identify potential future control operations so as to anticipate logistics requirements.

Request Additional Incident Resources

Note: The Logistics Section Chief performs this function only if the Incident Commander has delegated the corresponding authority.

- Receive requests for resources to be ordered from outside of the incident from members of the general staff or the Resources Unit.
- Coordinate requests for additional resources so as to eliminate duplicate requests.
- Submit the request through the communications center for additional resources from outside the incident. The request goes through normal channels and includes a confirmation/denial of request and ETAs.

Perform Operational Planning for Logistics Section

- Obtain the incident action plan from the Planning Section Chief and review with section personnel as appropriate.
- Identify service and support requirements for planned and expected incident operations.
- Plan organization of the Logistics Section.
- Compare organization plan requirements with dispatched personnel.
- Identify needed or surplus personnel.
- Notify the Resources Unit of names of personnel available for assignment or reassignment.
- Notify personnel being reassigned.
- Request additional personnel needed.
- Request additional support from the Incident Commander if personnel are not available from incident sources.
- Notify the Resources Unit of resources assigned by Logistics Section for support and service needs.
- Assign work locations and specific work tasks to section personnel.

Update Logistics Section Planning

- Review current situation status, resource status, and fire behavior prediction information.
- Obtain information concerning future operations through discussions with incident personnel.
- Estimate future service and support requirements.
- Compare estimated future requirements with expected logistics capabilities.
- Obtain changes to the incident action plan from the Planning Section Chief.
- Obtain the demobilization plan from the Planning Section Chief.

- Identify required modifications to Logistics Section planning. and modify planning as appropriate.
- Inform Logistics Section branch directors, Planning Section Chief, Resources Unit, and others as appropriate of planning modifications.

Direct Operations of Organizational Elements

- Receive reports of significant events.
- Periodically check work progress on assigned tasks of support and service branches and units, as appropriate.
- Coordinate and supervise activities of Logistics Section units.
- Ensure general welfare and safety of logistics personnel.
- Provide input to and review communications, medical, and traffic plans.

Recommend Release of Resources/Supplies

- List resources/supplies recommended for release by type, quantity, location, and time.
- Present recommendations to the Planning Section Chief.
- Coordinate with the Demobilization Unit on the demobilization plan.

Maintain Logs and Records

- Record Logistics Section activities on the unit log (ICS Form 214 or local form).
- Maintain agency records and reports.
- Provide unit logs to the Documentation Unit at the end of each operational period.

LOGISTICS SUPPORT BRANCH DIRECTOR CHECKLIST

RESPONSIBILITIES

The Support Branch Director is responsible for the management of all support activities at the incident.

The Support Branch Director position will be activated only as needed in accordance with incident characteristics, the availability of personnel, and the requirements of the Incident Commander and Logistics Section Chief. The Support Branch Director reports to the Logistics Section Chief.

Instructions: The checklist below presents the minimum requirements for Support Branch Directors. Note that some items are one-time actions, while others are ongoing or repetitive throughout the incident.

COMPLETED/NOT APPLICABLE TASKS

Obtain working materials from the logistics kit.

Identify the Support Branch personnel dispatched to the incident.

Determine initial support operations in coordination with the Logistics Section Chief and Service Branch Director.

Prepare the initial organization and assignments for the initial support operations.

Assemble and brief Support Branch personnel.

Determine if assigned branch resources are sufficient.

Monitor the work progress of units, and keep the Logistics Section Chief informed of activities.

Resolve problems associated with requests from the Operations Section.

Maintain the unit log (ICS Form 214 or local form).

OEM Staff Profiles Case Study: Field Operations and Logistics

One of the most compelling studies of catastrophic event response in recent years will be the terrorist attacks of 9/11. The following information from the Office of Emergency Management in New York City is an excellent illustration of the jobs and responsibilities of operational logistical personnel in a real-world catastrophic event.

Bobby Wilson, Deputy Commissioner of Operations[2]

In the Operations Division, Bobby Wilson was instrumental in organizing vehicle and debris removal following the World Trade Center towers' collapse. With most side streets surrounding the site blocked, emergency services vehicles were unable to access the site. "You couldn't help but be overwhelmed by it when you first got on the scene," Wilson says. "It took a few moments to say in your mind, 'What do we have to do?'" But until emergency personnel could fully access the site, there was little anyone could do. Wilson, working in conjunction with OEM Logistics, NYPD, FDNY Port Authority and USAR Massachusetts Task Force 1, brought in tow trucks, back hoes, front-end loaders and dump trucks to begin clearing the streets for emergency vehicles so the search and rescue effort could get underway.

Michael Berkowitz, Deputy Commissioner of Special Projects

Then working as Planning Supervisor at OEM, Michael Berkowitz was the last to command the EOC at 7 WTC, evacuating with the last group shortly ahead of the South Tower collapse. In the wake of the WTC disaster, Berkowitz oversaw access to the Frozen Area, working closely with DDC and NYPD to establish an orderly re-entry process for affected tenants and businesses.

During this time, he was also involved in peripheral logistics and supervised the establishment of command posts, supply tents, staging areas and respite and feeding locations. In order to better coordinate restricted area clearance,

Berkowitz implemented a credentialing system for quick and efficient identification of authorized personnel. Among his myriad other contributions to the response and recovery effort, Berkowitz also oversaw the last load ceremony out of the site. "In the end, our collaborative efforts helped us finish the job well ahead of schedule," he says. "We really learned the value of proper coordination."

Pete Picarillo, Director of Public-Private Initiatives

Working from various sites around the City, Pete Picarillo, then a senior planner at OEM, undertook the daunting task of coordinating logistics for the rescue and recovery effort. In the days after Sept. 11, Picarillo worked to identify Pier 36 as a suitable storehouse for the scores of supplies needed to support Ground Zero operations. "The real coup d'etat came at Pier 36," Picarillo recalls. "We couldn't find a place to stage our stuff until then." Working with then OEM meteorologist Sean Nolan, Picarillo oversaw supply and equipment distribution and fulfillment for the WTC site through November. At the same time, he worked with the NYPD, FDNY and Port Authority to sort through the glut of donations arriving in the City and channel useful materials to Ground Zero.

In an effort to help the City take control of the WTC work site and Lower Manhattan, Picarillo helped disassemble volunteer operations at various sites around the City in the weeks following the disaster. Among his many other contributions to the rescue and recovery effort, Picarillo also negotiated the Marriott Hotel as a respite center for workers and helped supply the Staten Island landfill site with equipment and food.

Tim Kane, Emergency Preparedness Specialist

The last place Tim Kane thought he'd be on Sept. 13 was on a train bound for New York's Penn Station. After finishing a stint as a summer intern just two weeks earlier, Kane was called back from Salisbury, Maryland, to help with the response and recovery effort. While returning to New York meant he might have to defer his penultimate semester at Salisbury University, Kane decided he couldn't worry about it until he returned. "It was one of those times that you were thankful that making the right decision was so easy," he says. Assisting Pete Picarillo, he spent the next three weeks managing logistics at various sites around the City, helping secure warehouses in and outside the City to divert donated goods, and working to restore facilities to normal operations. Kane returned to Salisbury in early October, finished his semester on time, and came back to work for OEM full-time in June 2002.

Brian Hastings, Citywide Interagency Coordinator

While Brian Hastings was off duty the morning of Sept. 11, he immediately reported to 7 WTC to assist with the response operation on hearing news of the attacks. From 7 WTC, Hastings reported to a field command post at West and Vesey Streets, where he was dispatched to the North Tower. En route to the forward command post at 1 WTC, he was trapped by the South Tower collapse. After emerging from the rubble, Hastings helped evacuate the walking wounded. "I survived the most horrifying attack, but I still had a job to do," he says.

Hastings eventually proceeded north to OEM's staging area at the corner of 6th Avenue and Houston Street, but soon turned back to join efforts to reach trapped rescue personnel following the North Tower's collapse.

For the next nine months, Hastings devoted himself to the rescue and recovery operations at Ground Zero, serving as an interagency coordinator. "From day one, when we started moving debris, until the last day, we did everything possible to bring home everyone that we could have," Hastings reflects. "The job was what we had to do, what we knew how to do, and what we did."

Jeff Armstrong, Director of Response Operations

Among the Sept. 11 disaster's first responders, Jeff Armstrong was busy preparing Pier 92 for an upcoming drill, when the first plane struck 1 WTC. After reporting to the Command Post in the North Tower, Armstrong — an EMS Captain — was immediately assigned to coordinate medical personnel and set up a staging area at West and Vesey Streets.

When the first tower collapsed, Jeff ran to West Street, where he helped direct pedestrians out of harm's way. In the wake of the second tower's collapse, Jeff took shelter in an underground parking garage. When he re-emerged, Armstrong's rescue instincts kicked into gear. Upon hearing that his boss was trapped in debris at the WTC site, Armstrong commandeered an ambulance to reach him, and later in the day, he escorted an ailing colleague to receive medical treatment. During the following weeks and months, Armstrong worked at the Interagency Command Center at IS-89, assisted with credentialing and EOC operations, and ultimately returned to Ground Zero to coordinate operations through May 2002.

John Vigorito, Watch Commander

Like many first responders on Sept. 11, John Vigorito was caught in the debris cloud resulting from 2 WTC's collapse, but managed to take refuge under a Police barricade vehicle. In the disaster's immediate aftermath, Vigorito worked to secure restricted zones, setting up "Checkpoint Charlie" at Canal and West Streets — a location that served to slow the steady stream of vehicles that had descended upon Lower Manhattan. "All of West Street was loaded with dump trucks, as far as the eye could see," John recalls.

Procedures were established to ensure that trucks were staged and available when needed, and a clear route was available for entry and exit. During debris removal, the four primary contractors brought in large numbers of subcontractors, and Vigorito worked closely with the NYPD to ensure that all who arrived at the site had appropriate credentials.

Predesignated Incident Facilities[3]

There are several kinds and types of facilities that can be established in and around the incident area. The determination of kinds of facilities and their locations will be based upon the requirements of the incident and the direction of Incident Command. The following facilities are defined for use with the ICS:

- Command Post—Designated as the CP, the command post will be the location from which all incident operations are directed. There should only be one command post for one incident site. In a unified command structure where several agencies or jurisdictions are involved, the responsible individuals designated by their respective agencies would be co-located at the command post. The planning function is also performed at the command post. Normally the communications center would be established at this location. The command post may be co-located with the incident base if communications requirements can be met.

- Incident Base—The incident base is the location at which primary support activities are performed. The base will house all equipment and personnel support operations, and can support several incident sites. The incident logistics section, which is responsible for ordering all resources and supplies, is also located at the base. There should only be one incident base established; and normally, the base will not be relocated. If possible, incident base locations should always be included in the pre-attack plans. The incident base should be distinguished from a staging area which is a temporary support area.

- Camps—Camps are locations from which resources may be located to better support incident operations. At camps, certain essential support operations (e.g, feeding, sleeping, sanitation) can be maintained. Also at camps, minor maintenance and service of equipment will be done. Camps may be relocated, if necessary, to meet tactical operations requirements.

- Staging Areas—Staging areas are established for temporary location of available resources on three-minute notice. Staging areas will be established by the operations chief at each incident site to locate resources not immediately assigned. A staging area can be anywhere in which mobile equipment can be temporarily parked awaiting assignment.

- Staging areas may include temporary sanitation services and fueling. Feeding of personnel would be provided by mobile kitchens or sack lunches. Staging areas should be highly mobile. The operations chief will assign a Staging Area Manager for each staging area. This manager is responsible for the check-in of all incoming resources; to dispatch resources at the request of the operations chief, and to request logistics section support as necessary for resources located in the staging area.

- Helibases—Helibases are locations in and around the incident area at which helicopters may be parked, maintained, fueled and loaded with retardants, personnel or equipment. More than one helibase may be required on very large incidents. Once established on an incident, a helibase will usually not be relocated.

- Helispots—Helispots are more temporary and less used locations than helibases at which helicopters can land, take off, and in some cases, load patients or supplies. They may be co-located near Casualty Collection Points (CCPs).

Comprehensive Resource Management

Resources may be managed in three different ways, depending upon the needs of the incident.

- Single Resources—These are individual engines, bulldozers, crews, helicopters, plow units, ladders, rescuers or other, that will be assigned as primary tactical units. A single resource will be the equipment plus the required individuals to properly utilize it.

- Task Forces—A task force is any combination of resources that can be assembled for a specific mission. All resource elements within a task force must have common communications and a leader. The leader sometimes will have a separate vehicle. Task forces should be established to meet specific tactical needs and should be demobilized as single resources.

- Strike Teams—Strike teams are a set number of resources of the same kind and type that have an established minimum number of personnel. Strike teams will always have a leader (usually in a separate vehicle) and will have common communications among resource elements. Strike teams can be made up of engines, hand crews, plows, water tankers, or any other kind of resource where common elements become a useful tactical resource.

Disaster Mortuary Requirements[4]

Often times there can be confusion among the public between a mass casualty event and a mass fatality. Mass casualty incidents will require extensive medical care, treatment, and intake facilities to administer to the needs of a situation where multiple persons have been wounded simultaneously. Mass fatalities however, as the name implies, deals with situations where there have been multiple deaths, and where the proper, hygienic, accurate and dignified processing of a large number of fatalities is required.

DMORT (or Disaster Mortuary Operational Response Teams) are an excellent example of the expertise, knowledge, and equipment needed at the scene of a mass fatality. In particular, the Disaster Portable Morgue Unit provides a good insight into the specific and innovative needs necessitated by an event producing a large number of fatal injuries.

Disaster Portable Morgue Unit (DPMU) Team

Incident Morgue Requirements

The DPMU's are caches of highly specialized equipment and supplies pre-staged for deployment to a disaster site. They contain a complete morgue with designated work stations for each processing element and prepackaged equipment and supplies. The DPMU core team travels with this equipment to assist in the set up, operation, packing and restocking of all DPMU equipment. They are known as the "Red Shirts".

There is currently one DPMU stationed in Maryland and a second is now on station in California.

The DPMU can be deployed to the incident site by rail, truck, plane or military transportation.

The DPMU contains over 10,000 individual items including:

Pathology equipment including forceps, scalpels, hemostats etc.

Anthropology Equipment including Measuring devices, instruments etc.

Radiology equipment including a Dental X-ray, 2 Full Body X-ray machines and Developers etc.

Photography/Personal Effects including Camera, film, Ladder etc.

Information Resources including Computers, Fax machines, copiers, forms, WinId2, VIP etc.

Wheeled Exam tables

Support equipment including Partitions and Supports; Electrical Distribution; Plumbing/ Hot water heaters; Personal Protective Gear etc.

Incident Morgue Requirements but we will always make it work...

- Convenient to scene
- Adequate capacity
- Completely secure
- Easy access for vehicles
- Ventilation
- Hot/cold water
- Drainage
- Non-porous floors
- Sufficient electrical capacity
- Refrigerated Trucks
- Forklift(s)
- Fuel - diesel, propane etc.
- Communications
- Office Space
- Rest/debriefing area
- Refreshment area

- Restrooms
- >=8000 sq. ft.

Email the DPMU Team with other questions

NOTE: People have to be a current member of a regional team and must go through their Regional Team Leaders to request consideration to a core group including the DPMU.

Evidence Gathering—FBI "Disaster Squad"

Friendly rivalries among public safety agencies are not unknown phenomena. As we have discussed throughout this text, often times a community-wide catastrophic event will require multiple responding agencies with different expertise and capabilities. Unfortunately, in the overall objective of response, rescue, relief and recovery, the different tactical missions of responding agencies sometimes may come into conflict with one another.

Forensic investigation is an example of one such specialty that may find itself in conflict with other efforts. As one crime scene investigator once put it jokingly, he often thinks of the fire department as the "Evidence Elimination Squad" because a firefighter's immediate and necessary goal of containing a emergency and rescuing victims sometimes results in the loss of important evidence—evident that law enforcement may need to determine the cause of the event, and the identities of whom may be responsible for criminal misconduct.

Nevertheless, proper investigation must occur and is a vital element of any event response. If a catastrophic incident is the result of terrorism, sabotage, arson, negligence, or design error, it is critical that these facts are determined and that those responsible are apprehended; otherwise a profound risk remains that such an event will occur again.

Disaster Squad Marks a Milestone[5]

Sixty years ago, a small plane carrying twenty-five people — including two FBI employees — crashed in a cornfield in Virginia, killing all on board. FBI representatives were sent to the crash site to claim FBI property and to offer whatever help they could. Upon their arrival the representatives found total chaos — no one seemed to know what to do. One major problem? Identifying the victims. So the FBI offered its fingerprinting expertise and was able to positively identify eight of the twenty-five victims.

The Virginia plane crash clearly demonstrated a need for a national "disaster squad" — experts ready to travel to the scene of a disaster at a moment's notice to assist local authorities in identifying victims. Thus, the FBI Disaster Squad was born.

Today, as it has been for the past sixty years, this humanitarian service provided by the FBI is free of charge. During the 1990s, this elite group of experts was involved in many high-profile incidents, both in the United States and abroad, from Desert Storm; Waco, Texas; the bombing of the Oklahoma City federal building; the Value Jet plane crash into the Florida Everglades; the Khobar Towers bombing in Saudi Arabia; TWA Flight 800; and Egypt Air Flight 990.

The squad's area of expertise is varied and includes thorough knowledge of fingerprint identification methods, forensic dentistry, forensic anthropology, and the proper operational procedures to follow after a disaster. Requests for Disaster Squad assistance come from many different sources, including a ranking law enforcement official at a disaster scene, a medical examiner or coroner in charge of victim identification, a ranking official of a public transportation carrier, the National Transportation Safety Board (NTSB), the Federal Aviation Administration (FAA), or a foreign government through the State Department (in foreign disasters involving U.S. citizens). During its on-site work, the squad works with various agencies, including the FAA, NTSB, the Armed Forces Institute of Pathology (AFIP), local Medical Examiners Offices, and the local FBI field office.

Why is it so important to positively identify a victim? Beyond the most obvious reason — so that the family of the victim can claim the remains and give them a proper burial — there are also economic and legal reasons to do so. Payment of insurance, settlement of estates, and dissolution of business partnerships all depend on the person in question being positively identified.

The Human Factor

The Disaster Squad is currently made up of approximately forty people - four Agents and the rest Latent Fingerprint Specialists — all from the Forensic Analysis Branch in the **FBI Laboratory**. Two of the four Agents are licensed dentists and work as dental pathologists when on the scene of a disaster.

When they are called on to assist, squad members act in appropriate roles — "lead supervisors" have multiple disaster scene experience; "experienced" members have worked on two more scenes; and "limited" members have one disaster under their belt. Squad members are sent out on a rotating basis. The number of people sent and the time spent working on a disaster scene (from two days up to three weeks) vary depending on the severity of the incident.

All squad members have to keep up with necessary immunizations to travel overseas, and they have physicals every other year. Because of the grisly nature their work, Disaster Squad members receive critical stress debriefings when they return from a site. Depending upon the severity and size of a disaster, the critical stress counselor can travel to the site of a disaster to be immediately available to squad members if necessary. The squad works closely with the Employee Assistance Program (EAP) at Headquarters as well.

Working on disasters, though, is only a small part of their duties. While at Headquarters in Washington, DC, these specialists are kept busy with a continuous stream of latent fingerprint work submitted by law enforcement agencies from around the country.

Modern Take on a Traditional Job

For sixty years, the work of the Disaster Squad has varied only slightly. Victims are still recovered by human hands, and fingerprints, footprints, and dental impressions are still taken.

However, advances in automation during the 1990s have sped up the actual identification process. Searching FBI databases — like the Integrated Automated Fingerpring Identification System (IAFIS) — and state and local databases now takes a matter of hours instead of days, weeks, or even months.

It is interesting to note that the states of California and Texas fingerprint everyone who applies for a driver's license in those states. Databases like these could be invaluable in identifying disaster victims — it means law enforcement would not be limited to searching databases containing prints of criminals or of individuals who have submitted civil fingerprint cards at one time or another. (No word on whether other states may be inclined to follow the leads of California and Texas.)

Disaster Squad Accomplishments

To date, the FBI Disaster Squad has assisted in 207 disasters involving some 8,235 victims. It has positively identified 4,490 victims by fingerprint, palm print, or foot print.

On January 31, 2000, members of the squad traveled to the California coast to assist in the identification of victims from the Alaskan Airlines crash. The squad examined 115 body parts and made positive identification of 24 individuals by finger or foot prints.

On April 12, 2000, squad members traveled to Dover Air Force Base in Delaware to assist in identifying victims from the April 2000 crash of a military aircraft. The crash happened during a training mission in Arizona, killing all 19 individuals on board. Remains were transported to Dover, and the squad was able to obtain fingerprints from 10 bodies and positively identify all 10 based on the prints. The remaining victims were identified by dental records, DNA, or other means.

TRIAGE AND DISASTER PLANNING[6]

Triage is the somewhat grim but highly important task of organizing a catastrophic event and prioritizing the administration of medical care. Generally in a triage event, those with the proper training and expertise will assess victims along some variation of these basic criteria:

- Those that can be saved, if they receive immediate medical care
- Those that will need first aid and medical care but aren't in immediate jeopardy
- Those that can walk out on their own
- Those already dead

Understanding triage and having an existing plan to classify and prioritize emergency medical treatment on the front can save valuable time, better focus energies and resources, and maximize the efficacy of response and rescue operations.

Triage Plan:

I. The first arriving ambulance, or highest trained health professional in emergency care at the scene after assessing the scene and making proper notifications, will be responsible for scene triage. In cases of multiple scene locations, the first unit or highest trained health professional in emergency care at each scene will perform the triage function

II. The triage team may be assisted by other agencies, however, any health professional performing triage should be licensed at a minimum as an Emergency Medical Technician

III. The triage team members will wear triage vests, obtained from the disaster kits on their own and other arriving ambulances

IV. Patients will be triaged in a four (4) color system, following EMS protocol guidelines. These guidelines are summarized as follows

   A. **BLACK**- Priority 4 - Expired, or no chance of surviving before delivery to a hospital facility.

   B. **RED** - Priority 1 - Critically ill or injured patient who needs immediate treatment.

   C. **YELLOW** - Priority 2 - Not currently life threatening illness or injury, but potentially life threatening if care is unduly delayed.

   D. **GREEN** - Priority 3 - Non-urgent condition which will require medical attention, but not immediate treatment.

V. Triage assessment information should only include enough information to classify the patient. Additional assessment information may be added to the card at a later time

VI. Triage information should be placed on the triage card obtained from the ambulance disaster kit. Additional cards should be obtained from other EMS units arriving at the scene. The cards should be torn to indicate the correct patient color code. It should then be attached to a limb of the patient in plain view to other workers

VII. In cases of major disasters (20 or more ill or injured) patients should be immobilized (where necessary) and moved to a patient holding area. The holding area will be designated by a sign or a flag indicating the condition of patients (by triage color)

VIII. Upon completion of the initial triage process, the triage team will report to the Medical Commander for assignment

TRIAGE AND DISASTER PLANNING

IX. Patient Injury/Illness Classifications - All persons will be assigned a priority rating according to the following procedure

A. <u>Priority 1</u> - Critically ill or injured patient who needs immediate treatment:
   1. Head Injury with GLASGOW Coma Scale of <10.
   2. Penetrating trauma to the head, neck, chest or abdomen.
   3. Significant blunt trauma with any of the following:
      a. Respiratory distress.
      b. Altered level of consciousness.
   4. Cardiac arrest or post cardiac arrest.
   5. Stridor or other respiratory distress.
   6. Any patient who is intubated or with assisted respirations.
   7. Status Epilepticus.
   8. Hypotension with clinical signs of shock.
   9. Unresponsive patient.
   10. Unstable vital signs.
   11. Unrelieved cardiac specific chest pain

B. <u>Priority 2</u> - Not currently life threatening illness or injury, but potentially life threatening if care is unduly delayed:
   1. GLASGOW Coma scale of 11 - 14 (acute change).
   2. Tender distended abdomen.
   3. Pelvic instability.
   4. Bilateral femur fractures.
   5. Postictal state following a seizure.
   6. Minimal aberrations in vital signs.
   7. ALS standing orders have been initiated.
   8. Significant mechanism of injury with presently stable vital signs.

C. <u>Priority 3</u> - Non-urgent condition which will require medical attention, but not immediate treatment:
   1. GLASGOW Coma scale of 15 or patient's normal mental status.
   2. Stable vital signs.

D. <u>Priority 4</u> - Hopelessly injured, non-salvageable in mass causality incident or obviously dead.

Note: The above guidelines will be used for establishing a priority status of each patient. Common sense MUST be utilized in prioritizing each and every patient.

PREPARING FOR THE MEDIA MEGA-EVENT[7]

By Ken Hilte

Oklahoma City, North Hollywood Bank Robbery, Columbine, Paducah, Waco, Heaven's Gate, the Texas Seven. Public Information Officers (PIO's) employed by law enforcement agencies across the country recognize those places as being synonymous with the media mega-event; that event which brings media attention not only from your local media community, but that of the neighboring communities, the state, the nation and in some cases, international attention. At best, you may anticipate 16-20 hour days, for multiple days, weeks, or even longer.

How does one possibly prepare for the mega-event? What arrangements can the law enforcement professional make ahead of time in preparation for the dreaded day? What can the PIO expect from the media? Through discussions with those who have weathered those storms, we have learned that there are common strategies that are effective and which can be employed by virtually any law enforcement agency.

Build Relationships Now!

The time to build trusting relationships with your local media and other area PIO's is now. For the price of a cup of coffee, an occasional lunch, and a few minutes at the local radio, television or newspaper offices, you will reap benefits twenty-fold. There is no substitute for the trust you will earn with your local reporters.

This trust, of course, is not purchased with coffee alone but earned through mutual respect. It is earned by developing a reputation for being honest and dependable. Always return your calls as promptly as possible. If you can't answer the reporter's question, tell them you "can't answer," and when possible, tell them why you can't answer. It goes without saying that you never, ever lie. Recognize that the business of reporting the news today is very competitive. Cultivate and guard your reputation, as well as that of your organization, as being fair and not playing favorites.

Next, develop relationships with the PIO's from your neighboring agencies. Certainly I'm talking about those other police agencies, but I'm also speaking of the PIO's from your area hospitals, fire departments, health departments, military bases, and other public service agencies such as utilities, the telephone company, etc. Simple and informal agreements regarding "*who* will speak to *what*" are beneficial.

Consult your Health Department and Coroner's Office and come to an agreement on who will release the number of deaths in the event of mass casualty. Generally, the numbers grow as time passes. Agreeing, for example, that the Coroner's Office will be the soul source of those numbers will prevent a case of competing figures.

My own contacts with my hospital PIO's taught me to not identify the specific hospital to which a victim was taken. Instead, they preferred I use the general term "area hospital." Furthermore, we agreed that they would restrict their comments to "medical condition," and let me speak to "what happened." Likewise, they agreed to teach me the meaning of the terms they use, i.e., "Good, Stable, and Critical Condition." I agreed to "parrot" those terms with no other comment.

One common factor in all the mega-events is that you will become overwhelmed by events. Hundreds of calls will flood to your pager and voice mail system. Your Dispatch Center and officers from throughout the operation will be calling for your assistance to deal with media requests.

You will desperately need and appreciate the help of other professional PIO's. They can help gather information, disseminate information when appropriate, return calls, arrange interviews, and "manage" your time, i.e., moving you from one interview to the next. They can help set-up the media center, prepare a podium and an advantageous speaking position, prevent media "creep" (the tendency of the media throng to surround the PIO and to move within two inches of him or her). Having someone available to take care of your personal needs like food, drink, breaks, and personal errands is extremely helpful.

In some regions there are professional organizations comprised of local PIO's who meet regularly to discuss recent events and share their experiences with one another. These forums can be used to provide training and to learn from each other's success and shortcomings. It is also an opportunity to invite media personnel to meet the PIO's in a nonstressful setting and teach them to see news from their perspective. More importantly, however, it affords the local PIO's a chance to get to know one another and build those relationships that will result in mutual aid when needed.

The Jefferson County, Colorado, Sheriff's Office spokesman during the media frenzy that followed the tragedy at Columbine High School has publicly stated that he never made a single call for help to his fellow Denver area PIO's. They simply showed up. He added that their help was indispensable.

There are, of course, professional media organizations also. I encourage you to join them and get to know the media in your area. Boston Police and their area media were among the first in the nation to agree to voluntary guidelines on what their respective responsibilities are during high profile media events. Among other things, they agreed to no live shots of SWAT teams, escaping victims, and agreed not to attempt to contact the hostages or criminals while the incident is on going. Law enforcement agreed to provide frequent and regular briefings, and better locations for pool cameras. The Radio and Television News Directors Association (RTNDA) later adopted the agreement and similar agreements have sprung up around the nation.

Set-Up a Media Center

When choosing a location for the media center, several issues must be considered. Your first consideration should, of course, be for the safety of others. The center, however, should be in the general vicinity of the incident, and when possible, should allow the media some type of appropriate visual backdrop. The media will not tolerate you establishing your media center miles away in an unrelated location. They are going to go to the scene of the action. It is also important that it be close enough to the incident to allow you to easily travel back and forth to the command post to receive updates.

Often times, it seems that the location of the media center has already been selected by the media prior to your arrival. If that location meets your needs then leave it. If not, move it as soon as possible. It is difficult for television camera crews to rearrange their lighting, cable and equipment. The sooner this is done, the better.

Pick a location that doesn't allow you to be surrounded and that provides you with an "escape route." Allowing yourself to be surrounded prevents you from leaving and receiving timely updates. Each news station wants its own interview with you and will attempt to prevent you from leaving in order to get it.

Consider a podium and / or an elevated speaking position that will allow all to see and hear you. This is to your mutual benefit. If they get their footage the first time you say it, they won't have to keep asking the same question for their own cameras.

Use barrier tape or barricades to keep the media throng about one to two feet away from you. Otherwise they will creep within inches of you and prevent others from being able to see you. You must, however, be close enough for them the hear you and reach you with their microphones.

The Media Briefings

The media briefing is different from a media conference in that the briefing is held on scene, often outdoors and usually on short notice. The media conference, on the other hand, is usually conducted indoors and offers you the luxury of more time to prepare.

As with any situation, know what your message is prior to making contact with the media. Be sure to contact the on-scene commander and discuss what will or will not be released to the media at this time.

Take questions after your statement. Be sure to repeat the question prior to answering it. This provides everyone the opportunity to hear what it is you're responding to. Repeated and redundant questions can be frustrating but are inevitable. If there is a large media gathering, and you turned your head to the left to answer a question, you can expect the question to be repeated from the right. They do this so you will face their cameras for that answer as well.

Do not hesitate to answer "I don't know," "I hadn't heard that," or "I can't tell you that at this time." Write the questions down to which you don't know the answers and offer to find out in time for the next briefing. Do not, however, burden yourself with chasing down rumors. The information, for the most part, will find its way to you through the command post.

Knowing when to end the briefing is largely a matter of personal experience. Once all the questions seem to be ones you've answered before or start to border on minutia, it's probably time to close. If the briefing has evolved from them asking questions to asking for personal interviews, it's time to move on.

When ending the briefing, do so decisively and tell them when you'll be back. Do not allow yourself to be held up by "just one more question."

The length and frequency of these briefings is dictated by the nature of the events. In the beginning of the incident, briefings will be more detailed and should probably be held no more than 30-45 minutes apart. As the situation draws out and there are less breaking developments, updates will be shorter and farther apart.

The Interviews

Personal interviews should be done as your schedule allows. The press briefings, however, take precedence. The rules for interviews are the same as they are for any media event, but there are differences when dealing with the mega-event.

It is tempting to cater to the national media, however now is not the time to ignore your local media. It is very important to meet their needs. Long after the satellite trucks, network affiliates, and morning show anchors are gone, your local media will remain.

When dealing with the national networks, recognize they are working with different time deadlines. For instance, if you live in the Rocky Mountain Time Zone, a live interview with a network morning show from New York means a 5:00 a.m., appointment.

Furthermore, the national networks do not necessarily cooperate with the other shows in their own network. For example, the evening news, the morning show (Good Morning America, Today, The Early Show), the news magazine show (60 Minutes, 60 Minutes II, 20/20, Dateline, 48-Hours, Night Line, etc.), and the Sunday morning talk show (Meet The Press, Face The Nation, This Week, etc.), although all from the same national network, will want to do their own interviews. Multiplied by the three networks, Fox News, CNN and others, and it's easy to see how one can overextend oneself in trying to accommodate everyone.

The hectic pace can often mean consecutive interviews at 4:45 a.m., 5:00 a.m., and 5:15 a.m. Your fellow PIO's can be of tremendous help in scheduling interviews and keeping you on schedule.

There are advantages to limiting your spokesperson/s to one person during the mega-event. Your message stays focused and on-target. You're aware of what has and has not been said. It is however, exhausting, and there are certain tasks and announcements that others can make for you. Any announcements made by others should be at the direction of the primary PIO and limited to the message given. Play it safe. Anything else should be referred back to the primary PIO.

In many events, it is entirely fitting and proper for the sheriff or police chief to respond to the scene and address the media. It is important, however, that the chief or sheriff does not replace the PIO in his / her role at the scene. The smart law enforcement PIO will have made arrangements with his chief or sheriff long in advance as to how they are going to handle those scenes. When the need for the chief or sheriff to address the media arises, the PIO should carefully brief the chief or sheriff on the latest developments and discuss the material that the chief or sheriff will be covering. The chief or sheriff should limit his comments to that material and should refer inquiries beyond that material to his / her PIO.

Media Helicopters

In the event your tactical operation should become compromised by the presence of media helicopters, you may contact the air traffic control center for your area. Note that the air traffic control center is not necessarily the same as your local air traffic control tower. In accordance with Federal Aviation Regulation (FAR) 91.137a(2), the center may issue a Temporary Flight Restriction (TFR) for the effected area for reasons related to maintaining a "Safe Environment." You must, however, be prepared to provide the following information via telephone or fax:

1. Name, telephone number and organization of the person recommending or requesting the TFR.
2. A brief description of the situation.
3. Estimated duration of restrictions.
4. Name of the agency responsible for on-scene emergency activities.
    a. Ground Unit Frequency
    b. Code Name/Call Sign
    c. Telephone Number
5. Description of affected area (latitude/longitude) and the altitudes. You may use a GPS to identify the longitude and latitude.
6. Description of activity posing a hazard to persons and property in the area.
7. Description of the hazard that would be magnified, spread, or compounded by flying aircraft or rotor wash.
8. Any other airborne operations present (i.e., police helicopter, medical evacuation helicopter).

Again, you should take the time to learn the location, telephone number and procedures of your air traffic control center before you need them. When you call, ask for a copy of the FAA Form (ZDV Form 7210-34[6/98]) which is used to request a Temporary Flight Restriction. The information above can be penned-in and simply read off or faxed to the air traffic control center.

If your jurisdiction is within a large enough media market that you are already subject to media helicopters, cultivate those relationships with your news directors and let them know about your concerns. Chances are you'll be able to avert having to go through the Federal Aviation Administration.

Competition, Fairness & Respect

As stated earlier, the news media business is extremely competitive. From their perspective, it is one thing for you not to release the information, but it is quite another for you to deny them equal access to it. You are not only the "Golden Goose" insofar as information, you also serve as the moderator and arbitrator of what is fair.

If, for example, you end a media briefing saying that you'll return in three hours and you'll not have any more information until then, it is improper for you to release that information in one-on-one interviews before then.

During the "Question & Answer" portion of your media briefing, do not let one reporter dominate the session. Insure that many, if not all, are given the opportunity to ask a question. Follow-up questions, although reasonable, can become burdensome and prevent others the chance to pose a question.

Furthermore, be prompt and on time. Do not let one reporter's interview drag on to the point it keeps you from being on time for the next.

Lastly, if you know of interesting footage, the media will appreciate you referring them to it. Likewise, if you, for example, see them setting up cameras along a route where they think the suspect will be taken, and you "know" they've already missed the shot, or it's not coming, tell them so. Your professional courtesy will long be remembered and appreciated.

Learn From Others

Take advantage of every opportunity to learn from other PIO's who have experienced the mega-event. The International Association of Chiefs of Police (IACP) sponsors an annual Public Information Officer's Conference, and presenters are often PIO's who have weathered the storm of a high profile incident. These conferences allow me to learn from their experiences, and I have personally found them to be very valuable.

Closing

The concept of "community oriented policing" demands that we communicate with our community. The media is our conduit to the community and can be valuable partners in that effort. The relationship between law enforcement and media can, and should be, mutually beneficial.

The law enforcement media mega-event is dynamic and fast paced. There is no foolproof blueprint for success. There are, however, certain actions a PIO can take to prepare oneself. The time spent in preparation is well worth it.

The demands upon your time and patience will be insatiable. Don't complain to the reporters or expect sympathy. They are operating in the same high-stress environment you are and in many cases are working under tighter deadlines. Maintain your professional demeanor no matter what the circumstances. As one veteran PIO stated, "People will long remember the cool head in the storm."

KRH/kh
2/15/01

Lieutenant Ken Hilte, a 17 year veteran of the El Paso County Sheriff's Office in Colorado Springs, is currently serving as a Patrol Watch Commander. He recently completed a two year assignment as the agency's Public Information Officer. Lieutenant Hilte is also a part-time instructor at Pikes Peak Community College. He has a Masters Degree in Management from Troy State University and has been previously published in Sheriff, American Jails, and Wisconsin Counties, Inc.

On March 20, 1995, members of the Aum Shinrikyo (or "Supreme Truth") cult carried several small plastic bags filled with liquid Sarin, a toxic nerve agent, onto several Tokyo subway cars. At a pre-determined time, cult operatives places the bags down on the floor, ruptured them with umbrella tips, and exited the trains as the liquid Sarin began to vaporize with the air. By the end of the day, 3,800 people were injured, 1,000 hospitalized, and 12 were dead. In addition to those in need of hospitalization, there were thousands more who rushed to Tokyo hospitals with hysterical symptoms—that is, they believed that they may have been exposed to Sarin when in fact they had not.

One can only imagine the stress and strain that was placed upon Tokyo's public health apparatus. Hospitals had to not only receive thousands of terrified patients, but then triage the critically ill from those to treat and those to release to those not in need of treatment at all.

A community-wide disaster can test the limits of any public health system. To that end, planners must be aware of what the exact limits of their hospital and emergency medicine system. These limits must be known and, as much as possible, extended and modified to prepare for adequate health treatment of dozens, hundreds, or even thousands in the aftermath of a catastrophe.

National Preparedness and a National Health Information Infrastructure[8]

Statement of the
**American Hospital Association before the
National Committee on Vital and Health Statistics Panel on
National Preparedness and a National Health Information Infrastructure**

February 26, 2002

Good morning. I am Roslyne Schulman, senior associate director for health policy at the American Hospital Association (AHA). I am a member of the AHA's hospital disaster readiness staff team and chair of the team's resources subgroup. On behalf of the AHA's nearly 5,000 hospitals, health systems, networks, and other providers of care, I appreciate this opportunity to present our views on how national preparedness could be enhanced through improvements to the national health information infrastructure.

The terrorist attacks of September 11, 2001 and the subsequent anthrax attacks have changed how Americans view safety and security. Over the past five months, the nation has focused on strengthening our national security and emergency readiness. As part of America's vital health care infrastructure, hospitals play a central role in that effort – a role that is sure to be enhanced as we move forward. The attacks redefined the meaning of disaster readiness for hospitals. Hospitals are now compelled to plan for what was previously unthinkable — disasters that are intentionally inflicted; involving large numbers of casualties; and involving the use chemical, biological or radiological agents.

Mass casualty incidents, by definition, overwhelm the resources of individual hospitals. They may overwhelm the resources of a community's entire health care system. Therefore, the response to mass casualty incidents is likely to require a broad array of community resources to supplement the health care system and requires coordination between these components. The minimum components of an effective response will involve public health, hospitals, physicians, community emergency management officials and "traditional" first responder organizations (fire, police and emergency medical services (EMS)). State and Federal government resources will be tapped, depending on the scale of the disaster.

Mass casualty incidents that result from an infectious agent, as would occur in a bioterrorist attack, differ from other types of disasters in many ways, including:

- The onset of the incident may remain unknown for several days before symptoms appear;
- Even when symptoms appear, they may be distributed throughout the community's health system and not recognized immediately by any one provider;
- Once identified, the initial symptoms are likely to mirror those of the flu or other common illness, so that the health care system will have to care for both those infected and the "worried well".

In order to increase readiness to respond to mass casualty incidents, particularly those involving biological agents, the AHA believes that hospitals must adopt a community-wide perspective and broaden the scale and scope of their disaster plans to link with and involve community partners. For instance, hospitals should establish an open and ongoing relationship with the local health department and its leadership. Biological incidents, in particular, require community-wide surveillance and control effort to assemble apparently isolated symptoms into a recognizable pattern that alerts the community's health care and public health system about the potential for an epidemic and initiates appropriate public health interventions, such as immunizations and prophylactic antibiotics. In addition, hospitals have an opportunity to use their existing EMS, trauma coordination, and other relationships as a framework upon which to build expanded relationships for mass casualty readiness. These existing programs also provide a framework for communications linkages, and data collection and sharing.

Establishing these community-wide relationships can serve readiness by facilitating the creation or linkage of data reporting systems to provide a community-wide assessment of health needs and health care resources. Because large-scale disasters increase the demand on all of the community's health resources simultaneously, there will not be enough time or available staff to survey hospitals and other facilities in order to inventory capabilities after the incident starts. Systems that are designed to share a common architecture and that integrate real-time data from institutional operations will provide the best means to matching community needs to available resources. However, there are many challenges to making these community linkages work.

Improvements needed in the communications infrastructure

First, hospitals and others in the frontline responder community depend on effective communications to provide emergency medical care, rescue accident victims, and respond to disasters. One of the key lessons learned from the September 11th terrorist attacks and the subsequent anthrax attacks is that we must enhance our ability to gather information and to communicate it efficiently to all relevant parties. In disasters, particularly those involving large numbers of casualties, it is critical that hospitals have pre-established communications linkages with other frontline responders that are reliable and interoperable. However, in disasters, most organizations experience problems with interoperability. Communications often degrade as a result of saturated cellular phone systems, and wireless communications systems that interfere with public safety communications. Public health services must be linked using secure connections to the Internet. High speed, dedicated access to the Internet should be available for all public and private health care facilities and related organizations. There is a critical need for funding to upgrade, modernize and link frontline responder communications systems and to address interoperability problems.

Need for real-time assessment of health care capacity

In the event of a disaster, many communities are not able to assess, in a rapid and accurate way, what health care resources are available for response. Readiness could be enhanced if all communities had a real-time system in place to assess hospital capacity. This would ideally include frequently updated information on the number, type, and location of available hospital beds; available stocks of drugs, supplies and equipment; and the number and location of trained staff.

Appropriate staffing poses a special concern in mass casualty incidents. For example, most hospital disaster plans provide for staff augmentation by extending the working hours of present staff or by calling in supplemental staff. If all of the disaster plans in a community are collected, they appear to provide for a substantial increase of staff. This includes medical staff, nursing staff, technicians and technologists, and support services staff. However, it is common for each hospital's disaster plan to be prepared individually. Thus, there is a real potential for double counting of potential staff. That is, two or more hospitals may envision using the same resources for staff augmentation. In a mass casualty incident, where the full human resources of the community are stressed, hospitals improve their preparedness by working together to develop an unduplicated estimate of the number and sources of additional staff.

In addition, disaster readiness would be enhanced by the development of a community-wide concept of "reserve staff" — identifying physicians, nurses, and hospital workers who are retired; have changed careers to work outside of healthcare services; or now work in areas other than direct patient care. However, this concept of reserve staff will only be a viable alternative if adequate funds are made available to regularly train and update reserves so that they can immediately step into roles in the hospital.

Need for improvements in disease surveillance and disease reporting systems

An effective public health and medical response to a covert bioterrorist attack will also depend upon the ability of individual clinicians, field providers, and public health departments to quickly detect, accurately diagnose, rapidly contain, and effectively treat an uncommon disease or illness. Improving the capacity of hospitals, public health departments, laboratories, and clinicians to engage in disease surveillance and disease reporting will be critical in determining that a cluster of disease may be related to the intentional release of a biological or chemical terrorism agent and in expediting an effective response. The monitoring of sudden changes in syndromic information gathered by emergency departments, EMS communication centers, health departments and telephone nurse triage call centers can also provide advance warning of community health threats.

While disease reporting and syndromic surveillance systems are critical in responding to biological and chemical terrorist threats, they also could serve to improve the health of the public in other ways in the future, such as tracking population health status and health service utilization. In addition, data captured from surveillance systems, once analyzed, can generate appropriate follow-up actions such as the provision of "just in time" educational materials to providers to assist in the medical management of patients.

To facilitate this level of readiness, hospitals and public health departments will need adequate resources and significantly upgraded surveillance systems to detect and respond to unusual diseases or patterns of symptoms. Public health laboratories will need to upgrade their capacity to carry out their essential analytic and reporting functions. And all public health, laboratory, and medical partners will require enhanced electronic information and communications systems to assure rapid and secure reporting and information exchange.

Furthermore, a successful surveillance system will, to the maximum extent possible, utilize and build upon sources of information already collected by hospitals and emergency departments. Automated retrieval of existing data from clinical databases in hospitals is preferable to systems that require manual entry of data, and may represent the best solution for the rapid provision of surveillance data to public health departments. Such a solution should also be less burdensome and costly for providers. In an environment in which every hour of patient care provided in a hospital emergency department results in one additional hour of paperwork, it would be difficult to justify adding to this burden through new and manual data collection.

Challenge for hospitals: HIPAA restrictions on information sharing

The AHA would also like to raise with NCVHS a serious conflict between the Health Insurance Portability and Accountability Act of 1996 (HIPAA) privacy regulations and efforts to improve hospital disease surveillance capabilities. The HIPAA privacy regulations place unnecessary roadblocks in the path of state hospital associations' efforts to share important health and demographic information with the hospitals in their states. The ability to continue to share such information could be critical to identifying an unusual pattern of symptoms that could indicate that a bioterrorist attack has occurred.

While the privacy rules permit state hospital associations to aggregate and analyze medical data from their member hospitals, they would not allow hospital associations to share this "protected health information" from one hospital to another hospital. Further, while the regulations do include an exception that would allow public health agencies to collect "protected health information" without consent, it is not clear that state hospital associations would fall under this exception.

As a consequence, once the regulations go into effect in April 2003, state hospital associations would be barred from sharing critical disease surveillance data with contributing hospitals. Among the data that hospital associations would be prohibited from sharing are: county or neighborhood by zip code; specific age of the patient; and the date on which the hospital treated the injury or illness. These are data elements that are integral to disease surveillance activities. At a minimum, HHS should either reform or clarify the rules to allow state hospital associations to share the critical elements of data with contributing hospitals and health researchers. This could be done most effectively by carving out those data from the list of identifiable data included in the rule. In addition, HHS should permit the use of a master business associate agreement under which all contributing hospitals could share such data with their state association.

Thank you for the opportunity to testify. I would be happy to answer any questions.

National Disaster Medical System & DMAT[9]

Disaster Medical Assistance Team (DMAT)

The Department of Homeland Security (DHS), through the National Disaster Medical System (NDMS) fosters the development of Disaster Medical Assistance Teams (DMATs). A DMAT is a group of professional and para-professional medical personnel (supported by a cadre of logistical and administrative staff) designed to provide medical care during a disaster or other event. Each team has a sponsoring organization, such as a major medical center, public health or safety agency, non-profit, public or private organization that signs a Memorandum of Agreement (MOA) with the DHS. The DMAT sponsor organizes the team and recruits members, arranges training, and coordinates the dispatch of the team.

- To supplement the standard DMATs, there are highly specialized DMATs that deal with specific medical conditions such as crush injury, burn, and mental health emergencies.

- Other teams within the NDMS Section include Disaster Mortuary Operational Response Teams (DMORTs) that provide mortuary services, Veterinary Medical Assistance Teams (VMATs) that provide veterinary services, National Nursing Response Teams (NNRTs) that will be available for situations specifically requiring nurses – and not full DMATs. Such a scenario might include assisting with mass chemoprophylaxis (a mass vaccination program,) or a scenario that overwhelms the nation's supply of nurses in responding to a weapon of mass destruction event. Others teams are the National Pharmacy Response Teams (NPRTs) that will be used in situations such as those described for the NNRTs but where pharmacists, not nurses or DMATs, are needed, and the National Medical Response Teams (NMRTs) that are equipped and trained to provide medical care for potentially contaminated victims of weapons of mass destruction.

- DMATs deploy to disaster sites with sufficient supplies and equipment to sustain themselves for a period of 72 hours while providing medical care at a fixed or temporary medical care site. In mass casualty incidents, their responsibilities may include triaging patients, providing high-quality medical care despite the adverse and austere environment often found at a disaster site, and preparing patients for evacuation. In other types of situations, DMATs may provide primary medical care and/or may serve to augment overloaded local health care staffs. Under the rare circumstance that disaster victims are evacuated to a different locale to receive definitive medical care, DMATs may be activated to support patient reception and disposition of patients to hospitals. DMATs are designed to be a rapid-response element to supplement local medical care until other Federal or contract resources can be mobilized, or the situation is resolved.

- DMAT members are required to maintain appropriate certifications and licensure within their discipline. When members are activated as Federal employees, licensure and certification is recognized by all States. Additionally, DMAT members are paid while serving as part-time federal employees and have the protection of the Federal Tort Claims Act in which the Federal Government becomes the defendant in the to provide interstate aid. event of a malpractice claim.

- DMATs are principally a community resource available to support local, regional, and State requirements. However, as a National resource they can be federalized

Communications Challenges

"Interoperability" has become an important concept in the emergence of homeland security technology. Having a large number of agencies from different communities and using different equipment can present great challenges to successful emergency communications. Since 9/11 and some of the tragedies stemming from that day it has become know that responders were unable to always successfully communicate to one another. This has placed a new emphasis and focus on the need for effective, universal, or interoperable communications solutions in the face of a multi-agency catastrophic event response.

The following pieces will provide an examination of methods and considerations for communities addressing the problem of managing emergency communications in a multi-agency event.

Chicago Office of Emergency Management and Communications[10]

(Communications) Technology

The Office of Emergency Management and Communications (OEMC) has a number of key technological and infrastructure elements that enhances its operations:

Joint Operations Center
Chicago's Joint Operations Center (JOC), designed for use during major emergencies, provides complete communications capabilities for all police, fire, medical and city, state, and federal command personnel. The JOC has recently undergone a complete renovation, including new technology, new meeting and strategy rooms, and a complete redesign of the room's layout, for optimal functionality. Each workstation is equipped with a telephone and computer with online capabilities, thus, allowing those staffing the JOC to connect directly to their office and networks. Three large-screen televisions enable the JOC to monitor local, national and international events. The JOC has also improved video technology, which will now link the OEMC, with City Hall and the Chicago Police Department Headquarters, with live video conferencing.

Emergency Telephone Notification System
If an incident occurs, the OEMC has the ability to pinpoint the location, designate a radius around the location, and call all land-line phones in this predetermined area. This system's technology is capable of calling up to 1,000 phones per minute. A message may be chosen from a library of options, or a custom message can be created for a specific incident.

Regional Alert and Information Monitoring Center
The OEMC has expanded the existing notification system to include key regional and state officials. In addition, this technology is capable of tracking every event in the region on a daily basis. Moreover, the City is able to track information regarding the location of vital assets in use at all times.

Field Communications Van

Since June 2000, the OEMC has been operating a mobile communications vehicle capable of interconnecting radios using disparate radio bands for multiple first responders. The vehicle is designed to support not only Chicago's first responders, but also county, state and federal agencies operating around the City. The operational plan calls for the vehicle to respond within 75 miles of the City to support the communications needs of a multi-jurisdictional response.

The vehicle is equipped with a JPS Inc, ACU-1000 transportable communications suite. This suite allows the operator to connect multiple radios across numerous frequencies, as well as cellular and land line telephones. With just a few mouse clicks on a laptop controller, the operator can connect up to four (4) separate conversations at the same time.

Security Camera Initiative
The City of Chicago and the OEMC have spearheaded a citywide initiative that encompasses the design and installation of up to two hundred remote security cameras at various city locations. These cameras are connected to the OEMC, allowing emergency management personnel to view and record real-time video at multiple sites across the city.

High Rise Building and Evacuation Information
The City of Chicago instituted the Chicago High Rise Evacuation Ordinance in May 2002. This ordinance not only mandates workable evacuation plans for many of Chicago's high rise buildings, but also provides critical information, such as floor plans, evacuation routes, utility locations and tenant rosters that can be used by first responders on the scene of a major incident.

Concepts of EMS Communications[11]

Emergency communications begins with the detection of an emergency incident and ends only with the full resolution of the emergency.

An Emergency Medical Services and Trauma Care Communications System must provide the means to use, mobilize, manage and coordinate emergency medical resources during normal and adverse situations. An EMS communications system must integrate sufficient communications paths and operational capabilities to provide access to the emergency services or public safety networks.

The state EMS Communications System will use the Washington State Patrol microwave network as its framework. Statewide coverage is obtained by locating repeater sites near selected medical trauma facilities. These sites will not be located with local communications systems, where possible. Selecting different locations for state and local communications systems should expand local communications coverage and provided redundancy in most cases. The EMS Communications System will use VHF, UHF, satellite and other technologies to maximize its efficiency.

As a two-component system, the EMS Communications System consists of the statewide component that uses the WSP microwave network and local/ regional EMS communications system. In the planning of the statewide EMS

Communications system, the State (first-component) identifies the goals and factors that need to be coordinated statewide. These goals are used as guidelines in the development of local/regional EMS communications systems.

The Office of Emergency Medical Services does not operate EMS communications systems. It acts in the role of coordinator and facilitator for local/regional communications systems. OEMS focus is on the process and its results. The state will not be involved with the daily operations of individual communications systems. It is concerned with the interfaces and interactions between communications systems. Items of interest include the degree to which the communications systems provide public access, medical communications for basic and advanced life support, radio coverage and EMS communications training standards.

As an integral part of this concept, local/regional (second-component) EMS communications plans will be prepared according to state guidelines. The local/regional EMS communications plans are tailored to satisfy local/regional emergency medical service system needs. They will be compatible and interoperable with other emergency medical services throughout the state. Technical and daily operations of local/regional EMS communications systems are the responsibilities of the local/ regional EMS agencies. OEMS will provide assistance when requested.

EMS communications are the exchanges of information necessary for the functioning of the Emergency Medical System. Emergency communications begins with the detection of an emergency incident and continues through the dispatching of manpower and equipment necessary to respond to the emergency scene. It extends through the treatment of the patient at the scene and during the transport of the patient to the hospital. EMS communications ends only with the full resolution of the emergency. ...

Logistical Planning—Lessons from 9/11

Nearly Two Years After 9/11, the United States is Still Dangerously Unprepared and Underfunded for a Catastrophic Terrorist Attack, Warns New Council Task Force[12]

Overall Expenditures Must Be as Much as Tripled to Prepare Emergency Responders Across the Country...

June 29, 2003 - Nearly two years after 9/11, the United States is drastically underfunding local emergency responders and remains dangerously unprepared to handle a catastrophic attack on American soil, particularly one involving chemical, biological, radiological, nuclear, or high-impact conventional weapons. If the nation does not take immediate steps to better identify and address the urgent needs of emergency responders, the next terrorist incident could be even more devastating than 9/11.

These are the central findings of the Council-sponsored Independent Task Force on Emergency Responders, a blue-ribbon panel of Nobel laureates, U.S. military leaders, former high-level government officials, and other senior experts, led by former Senator **Warren B. Rudman** and advised by former White House terrorism and cyber-security chief **Richard A. Clarke**. This report marks the first time that data from emergency responder communities has been brought together to estimate national needs.

The Task Force met with emergency responder organizations across the country and asked them what additional programs they truly need—not a wish list—to establish a minimum effective response to a catastrophic terrorist attack. These presently unbudgeted needs total $98.4 billion, according to the emergency responder community and budget experts (See attached budget chart.)

Currently the federal budget to fund emergency responders is $27 billion for five years beginning in 2004. Because record keeping and categorization of state and local spending varies greatly across states and localities, the experts could not estimate a single total five-year expenditure by state and local governments. Their best judgment is that state and local spending over the same period could be as low as $26 billion and as high as $76 billion. Therefore, total estimated spending for emergency responders by federal, state and local governments combined would be between $53 and $103 billion for the five years beginning in FY04.

Because the $98.4 billion unmet needs budget covers areas not adequately addressed at current funding levels, the total necessary overall expenditure for emergency responders would be $151.4 billion over five years if we are currently spending $53 billion, and $201.4 billion if we are currently spending $103 billion. Estimated combined federal state, and local expenditures therefore would need to be as much as tripled over the next five years to address this unmet need. Covering this funding shortfall using federal funds alone would require a five-fold increase from the current level of $5.4 billion per year to an annual federal expenditure of $25.1 billion.

"While we have put forth the best estimates so far on emergency responder needs, the nation must urgently develop a better framework and procedures to generate guidelines on national preparedness," said Rudman, who served as Task Force chair. "And the government cannot wait to increase desperately needed funding to emergency responders until it has these standards in place," he said.

The Task Force credits the Bush administration, Congress, governors and mayors for taking important steps since 9/11 to respond to the risk of catastrophic terrorism, and does not seek to apportion blame about what has not been done or not done quickly enough. The report is aimed, rather, at closing the gap between current levels of emergency preparedness and minimum essential preparedness levels across the United States.

"This report is an important preliminary step in a process of developing national standards and determining national needs for emergency responders," said Council President **Leslie H. Gelb**, "but the report also highlights the need for much more work to be done in this area."

The Independent Task Force, *Emergency Responders: Drastically Underfunded, Dangerously Unprepared*, based its analysis on data provided by front-line emergency responders—firemen, policemen, emergency medical workers, public health providers and others—whose lives depend upon the adequacy of their preparedness for a potential terrorist attack.

The study was carried out in partnership with the Concord Coalition and the Center for Strategic and Budgetary Assessment, two of the nation's leading budget analysis organizations.

Jamie Metzl, Council Senior Fellow and a former National Security Council and Senate Foreign Relations Committee official, directed the effort. The Task Force drew upon the expertise of more than twenty leading emergency responder professional associations and leading officials across the United States. (A list of participating associations is attached below.)

The Task Force identified two major obstacles hampering America's emergency preparedness efforts. First, because we lack preparedness standards, it is difficult to know what we need and how much it will cost. Second, funding for emergency responders has been sidetracked and stalled due to a politicized appropriations process, slowness in the distribution of the funds by federal agencies, and bureaucratic red tape at all levels of government.

To address the lack of standards and good numbers, the Task Force makes the following recommendations:

- Congress should require that the Department of Homeland Security (DHS) work with state and local agencies and officials and emergency responder professional associations to establish clearly defined standards and guidelines for emergency preparedness. These standards must be sufficiently flexible to allow local officials to set priorities based on their needs, provided that they reach nationally-determined preparedness levels within a fixed time period.

- Congress should require that the DHS and the Department of Health and Human Services submit a coordinated plan for meeting identified national preparedness standards by the end of FY07.

- Congress should establish a system for allocating scarce resources based less on dividing the spoils and more on addressing identified threats and vulnerabilities. To do this, the Federal government should consider such factors as population, population density, vulnerability assessment, and presence of critical infrastructure within each state. State governments should be required to use the same criteria for distributing funds within each state.

- Congress should establish within DHS a National Institute for Best Practices in Emergency Preparedness to work with state and local governments, emergency preparedness professional associations, and other partners to share best practices and lessons learned.

- Congress should make emergency responder grants in FY04 and thereafter on a multi-year basis to facilitate long-term planning and training.

To deal with the problem of appropriated funds being sidetracked and stalled on their way to Emergency Responders, the Task Force recommends:

- The U.S. House of Representatives should transform the House Select Committee on Homeland Security into a standing committee and give it a formal, leading role in the authorization of all emergency responder expenditures in order to streamline the federal budgetary process.

- The U.S. Senate should consolidate emergency preparedness and response oversight into the Senate Government Affairs Committee.

- Congress should require the Department of Homeland Security to work with other federal agencies to streamline homeland security grants to reduce unnecessary duplication and to establish coordinated "one-stop shopping" for state and local authorities seeking grants.

- States should develop a prioritized list of requirements in order to ensure that federal funding is allocated to achieve the best possible return on investments.

- Congress should ensure that all future appropriations bills for emergency responders include strict distribution timelines.

- The Department of Homeland Security should move the Office of Domestic Preparedness from the Bureau of Border and Transportation Security to the Office of State and Local Government Coordination in order to consolidate oversight of grants to emergency responders within the office of the Secretary.

The Task Force on Emergency Responders is a follow on to the Council's highly acclaimed *Hart-Rudman Homeland Security Task Force*, which made concrete recommendations last October on defending the country against a terrorist attack.

Established in 1921, the **Council on Foreign Relations** is a nonpartisan membership organization, publisher, and think tank, dedicated to increasing America's understanding of the world and contributing ideas to U.S. foreign policy. The Council accomplishes this mainly by promoting constructive debates, clarifying world issues, producing reports, and publishing *Foreign Affairs*, the leading journal on global issues.

Full Text and the Executive Summary of the Council-sponsored Independent

Task Force *Emergency Responders: Drastically Underfunded, Dangerously Unprepared.*

TASK FORCE MEMBERS

Warren B. Rudman (Chair)
Partner, Paul, Weiss, Rifkind, Wharton and Garrison
Former Senator, New Hampshire

Charles Graham Boyd
Chief Executive Officer and President, Business Executives for National Security
Former Deputy Commander in Chief, U.S. European Command

Richard A. Clarke (Senior Adviser)
Senior Adviser, Council on Foreign Relations
Chairman of Good Harbor Consulting, LLC
Former Senior White House Adviser

William J. Crowe
Senior Advisor, Global Options
Former Chairman, Joint Chiefs of Staff

James Kallstrom
Senior Executive Vice President, MBNA America
Former Director, Office of Public Security for the State of New York

Joshua Lederberg
President-Emeritus and Sackler Foundation Scholar, Rockefeller University
Nobel Laureate

Donald Marron
Chairman, UBS America and Chairman, Lightyear Capital

Jamie Metzl (Project Director)
Senior Fellow and Coordinator for Homeland Security Programs, Council on Foreign Relations
Former National Security Council aide
Former Senate Foreign Relations Committee official

Philip A. Odeen
Former Chairman, TRW, Inc.

Norman J. Ornstein
Resident Scholar, American Enterprise Institute for Public Policy Research

Dennis Reimer
Director, Oklahoma City National Memorial Institute for the Prevention of Terrorism
Former Chief of Staff, USA

George P. Shultz
Thomas W. and Susan B. Ford Distinguished Fellow, the Hoover Institution, Stanford University; Former Secretary of State, Secretary of the Treasury,

Secretary of Labor, and Director, Office of Management and Budget

Anne-Marie Slaughter
Dean, the Woodrow Wilson School of Public and International Affairs, Princeton University

David Stern
Commissioner, National Basketball Association

Paul Tagliabue
Commissioner, National Football League

Harold E. Varmus
President and Chief Executive Officer, Memorial Sloan-Kettering Cancer Center
Nobel Laureate

John W. Vessey
Former Chairman, Joint Chiefs of Staff

William H. Webster
Partner, Milbank, Tweed, Hadley & McCloy
Former Director, Central Intelligence Agency
Former Director, Federal Bureau of Investigation

Steven Weinberg
Director of the Theory Group, University of Texas
Nobel Laureate

Mary Jo White
Partner and Chair of the Litigation Department, Debevoise & Plimpton
Former U.S. Attorney for the Southern District of New York

Emergency Responders Five-Year Unmet Needs Budget (FY04-FY08)*

Response Area	Need	Estimated Five-Year Cost
Fire Services	Strengthen hazardous materials preparation and response, and EMS, including equipment and training.	$36.8 billion
Urban Search and Rescue	Prepare fire departments and EMS for technical rescue and enhance FEMA's national search and rescue teams.	$15.2 billion
Hospital Preparedness	Upgrade communications, personnel protective equipment, mental health services, decontamination and training for hospitals.	$29.6 billion
Public Health	Enhance CDC and epidemiological services;	

	upgrade state and local public health department capacities to respond to terrorism.	$6.7 billion
Emergency 911 Systems	Implement a national emergency telephone number system with effective first responder deployment capacity.	$10.4 billion
Interoperable Communications	Ensure dependable, interoperable communications for first responders.	$6.8 billion
Emergency Operations Centers	Provide physical and technical improvements in emergency operations centers.	$3.3 billion
Animal/Agriculture Emergency Response	Develop regional and state teams to respond to emergencies and enhance lab support capacity.	$2.1 billion
Emergency Medical Services Systems	Improve state and local EMS infrastructure including mutual aid, planning, and training.	$1.4 billion
Emergency Management Planning and Coordination	Enhance basic emergency coordination and planning capabilities at state/local levels.	$1 billion
Emergency Response Regional Exercises	Fund annual regional exercises.	$0.3 billion
SUBTOTAL		$113.6 billion
Undesignated offsets from federal grants**		($15.2 billion)
TOTAL		$98.4 billion

* *These budgetary figures are based on estimates provided by the Emergency Responders Action Group. Where possible these figures have already been reduced to account for anticipated federal spending in relevant response areas.*

***This assumes a thirty percent match by state and local governments.*

Emergency Responders Action Group Participating Organizations

American College of Emergency Physicians
American Hospitals Association

American Veterinary Medical Association
Century Foundation
Council of State Governments
County Executives of America
International Association of Chiefs of Police
International Association of Emergency Managers
International Association of Fire Chiefs
International Association of Fire Fighters
International City County Management Association
Joint Commission on the Accreditation of Health Care Organization
National Association of Counties
National Association of County and City Health Officials
National Association of Emergency Medical Technicians
National Association of Public Hospitals and Health Systems
National Emergency Numbers Association
National Fire Protection Association
National League of Cities
National Memorial Institute for the Prevention of Terrorism
National Sheriffs' Association
National Volunteer Fire Council
Trust for America's Health
United States Conference of Mayors

Community "Arks"[13]

Contact: Lisa Shields, Vice President, Communications, (212)434-9888

Project ARK

Project ARK, a disaster shelter program, is a cooperative effort between the City of Sunnyvale, the American Red Cross, and four school districts in Sunnyvale—Sunnyvale Unified, Santa Clara Unified, Cupertino Unified, and Fremont Union.

Project ARK provides for twelve emergency supply containers, known as ARKS, strategically placed at school sites around the City. Each ARK can support up to 300 people. The containers are 40 feet long, 8 feet wide, and 8 1/2 feet high. They contain emergency supplies including cots, blankets, water, portable lanterns, flashlights, portable toilets, personal care kits, and other essential items needed to open an emergency shelter.

Schools were chosen as sites for disaster shelters for several reasons. Schools are built to a stronger seismic code than most buildings, have gyms or other areas for sleeping and feature kitchen and rest room facilities for large numbers of people.

In the event of a disaster, the ARK's and the supplies within, may be used to activate an American Red Cross Mass Care Shelter.

ARKS are strategically located throughout the city at the following school sites:

- Bishop Elementary School
- Columbia Elementary School
- Cupertino Middle School
- Fremont Middle School
- Lakewood Elementary School
- Peterson Middle School
- Ponderosa Elementary School
- Sunnyvale Middle School

Project ARK is a preparedness component of the Santa Clara Chapter's disaster shelter program. This program involves the use of supply containers stocked with emergency supplies and placed at strategic locations throughout the Chapter's jurisdiction.

Neighborhood groups are recruited, organized, and trained to provide mass care services in the event of a disaster. In the event of a disaster or other local emergency, the supplies in the ARKs are used to open Red Cross shelters.

Emergency Operations Centers

One of the most critical and complex need for a coordinated event planning is the creation of an Emergency Operations Center or EOC. The EOC enables Incident Commanders and their staffs to command, control, communicate, and coordinate the response, rescue, and recovery operations. The following excerpt is from the City of Los Angeles' emergency plan and provides and excellent insight into the logistical needs of a major urban EOC.

Los Angeles Emergency Plan[14]

The City operates two EOCs. The primary center is located on the P-4 level of City Hall East, the alternate center is a Mobile Emergency Operations Center (MEOC). While the facilities and their capabilities are considerably different, they both operate in accordance with SEMS.

4.1 Primary Emergency Operations Center

The City's primary EOC occupies approximately 2,600 square feet....

The internal arrangement of workstations within the EOC has been developed around the five primary SEMS functions. The configuration provides for a staffing level of 70 persons. The arrangement of functional sections and workstations provides for close interaction between persons who have to communicate with each other frequently and for ease of movement within the facility.

EOC equipment and support systems are covered in Section 4.7.

4.2 Organization of the EOC

The EOC is organized into seven sections, one for each of the five SEMS functions (Management, Operations, Planning/Intelligence, Logistics and Finance/Administration) and two additional sections for Public Information and Liaison. Each section can be divided into functional divisions and/or units which are activated as required. Space is provided for support staff and representatives from other agencies....

4.3 Duties and Responsibilities of EOC Sections

EOC section responsibilities are briefly described. Detailed responsibilities and checklists for section and unit positions are documented in a separate EOC Procedures Manual. Attachment C contains a functional checklist for each position.

4.3.1 EOC Management (Director/Deputy Director)

The EOC management function is performed by the EOC Director who will initially represent either the Fire or Police Department, depending on the nature of the emergency. If the emergency is primarily people related, e.g., civil disorder, other criminal behavior, major public event, etc., the Police Department will be the lead agency. For all other events, the Fire Department will serve as the initial lead agency.

All sections within the EOC organization report to the EOC Director. The Director provides overall coordination and direction of EOC operations, and ensures that all functional activities within the EOC are appropriately activated, staffed, and operating effectively. The Director reports to the EOB.

4.3.2 Liaison Section

The Liaison Section provides coordination for City and non-City agencies that may have representatives temporarily assigned to the EOC. Representatives to this section will vary based on the nature of the emergency. The CAO EPD coordinates the Liaison Section.

4.3.3 Information and Public Affairs Section

The EOC Information and Public Affairs Section is responsible for developing information about the emergency, responding to media inquiries and communicating to the public through the broadcast and print media. The section will be the principal point for the development of City-wide public service announcements and emergency broadcast coordination during a declared local emergency. All media requests or inquiries submitted to the EOC for emergency-related information will be handled by the section. The Mayor's Office coordinates the Information and Public Affairs Section. Staffing for the section will be drawn from other EOO divisions and City departments as necessary.

4.3.4 Operations Section

The Operations Section ensures that all essential emergency-related information and resource requests are received, processed and internally coordinated within the EOC. Functional workstations have been established in the Operations Section for ten of the EOO divisions, with each division providing one or more representatives. Division representatives are responsible for providing incoming situation information and resource requests to the EOC, and ensuring that essential information and results of internal EOC coordination efforts are passed on to DOCs and Incident Command Posts (ICPs) as appropriate. Depending on the nature of the emergency, the Operations Section will be coordinated by either the Fire or Police Department.

4.3.5 Planning and Intelligence Section

The Planning and Intelligence Section is responsible for collecting, evaluating, processing and distributing information about the emergency to all functional elements and agencies in the EOC. The section will maintain all internal wall displays, maintain current information in the automated EOC Information Management System (EOCIMS) and prepare situation summaries and EOC action plans. In most cases, either the Fire or Police Department coordinate the Planning and Intelligence Section.

4.3.6 Logistics Section

The Logistics Section provides resources support and services to City-wide emergency operations. Logistics obtains and provides essential city personnel, facilities, equipment, supplies and services not found within those EOO divisions and departments represented in the EOC Operations Section and maintains an inventory of EOC-designated critical city resources. The Department of General Services coordinates the Logistics Section.

4.3.7 Finance/Administration Section

The Finance/Administration Section provides general administrative, finance and legal support related to EOC activities. With the support of the Operations and Planning and Intelligence Sections, the Finance/Administration Section compiles and processes damage assessment information. The CAO Disaster Grants Group coordinates the section.

4.4 Activation of the EOC

Depending on the nature of the emergency, the EOC can be activated at three different levels (level one, two or three) in order to provide appropriate staffing. This insures a standardized method for EOC activation and eliminates the need for always having full scale activations.

The EOC may be activated by the Mayor, City Council President, any member of the EOB, any EOOzdivision or any City department. The EOC may be activated without declaration of a local emergency whenever an event or pending event requires resources beyond those normally available from one City department.

The EOC will be opened and initially staffed by members of the Fire

Department's Operations Control Division (OCD).

Steps involved in the activation.

- The request to activate the EOC shall be made by notifying OCD. All requests to activate the EOC shall include the reason for activation and the required level of activation. Agencies shall respond in accordance with Figures 4-4, 4-5 and 4-6. The requesting department or the EOC Director shall identify any additional EOO divisions, departments or outside agencies required to respond based on the needs of the specific event.

- OCD will immediately direct the City Hall Operator to notify concerned entities of the EOC Activation. The activation order to the City Hall Operator will include the reason for activation designation of the lead agency (Fire or Police), requested level of activation (one, two or three) and identification of any additional EOO divisions, departments, or outside agencies required to respond.

- The City Hall Operator will use pre-established lists to contact designated representatives of City departments and outside agencies, informing them of the reason for activation, lead agency and level of activation in accordance with Figure 4-3. The City Hall Operator shall first notify those City departments and outside agencies which are required to respond in accordance with Figures 4-4, 4-5 and 4-6. All other EOO divisions, departments and appropriate outside agencies shall be subsequently advised of the EOC activation by the City Hall Operator for information only.

- Upon completion of the notification process, the City Hall Operator shall submit a copy of the "EOC Activation Notification Record" to the EOC Planning/Intelligence Section Coordinator.

- Departments are responsible for further internal notifications of EOC activations.

- The CAO EPD shall notify all EOB members of EOC activations and deactivations....

4.5 Overview of EOC Operations

The steps listed provide a summary overview of EOC operations.

1. The EOC is initially activated at one of three levels by the department with lead responsibility for the emergency (Fire/Police).

2. Upon activation, designated staff will proceed to the EOC, sign in and activate their assigned work stations.

3. The EOC staff is organized by functional sections: management, operations, planning, logistics, finance/administration, liaison and information and public affairs. Sections may be further divided into divisions or units. Section coordinators are assigned under all activation levels, and are a primary point of contact for personnel assigned to the EOC.

4. Each functional position in the EOC has a job description and a checklist to be followed. The checklist contains the basic guidance for operation of the workstation. Job descriptions and checklists are found in the EOC Information Binders at each workstation.

5. For most EOC positions, the checklist requires: contacting the primary DOC; obtaining and maintaining information needed within the EOC; and ensuring that all appropriate EOC-generated material is made available to DOCs and/or ICPs.

6. The primary role for divisional and liaison functional coordinators in the EOC is the passing of information from and to EOO divisions, departments and City and other agencies, and the coordination of resource and support requests. The EOC is designed so that information exchange and coordination may take place freely between sections, divisions and units. All requests for resources or support are processed through designated channels in order to maintain accountability.

7. Information obtained by division and unit coordinators should be made available to the Section Coordinator and, as appropriate, provided to the Situation Assessment Unit.

8. The EOC Director and General Staff (Section Coordinators) will hold periodic planning meetings. A written EOC action plan may result from these meetings, which may change priorities related to division and department operations. The Director will provide and coordinate periodic briefings for all EOC personnel.

9. Deactivation of functional positions is accomplished by the Director or Section Coordinator.

10. All EOC personnel are responsible for: maintaining current status information regarding their section, division, unit, department or agency; maintaining a duty log; briefing relief personnel; and completing deactivation procedures when instructed.

4.6 Action Planning in the EOC

In the first hours of an emergency, EOC action plans will be verbal statements of actions to be taken as stated by the EOC Director. Within the first four hours of an activation, the EOC Director should convene a meeting of the General Staff (Section Coordinators) and any others he/she selects to attend. The purpose of this meeting will be to document EOC related objectives and actions to be taken within the next operational period. EOC action plans should be concise, actions planning meetings brief. Prior to the meeting, the EOC

Director, General Staff and other attendees should receive an updated situation report and be clear on overall resources availability. General Staff should determine in advance specific section-level objectives to be presented at the meeting.

A recommended format for an EOC action planning meeting agenda is shown in Figure 4.7 [below]

EOC Action Planning Meeting Agenda
EOC ACTION PLAN

DATE: _____ OPERATIONAL PERIOD FROM _____ TO _____

Action Plan Steps	Guideline for Content	Responsibility
1. Review prior Op. Period Objectives as Appropriate	Determine status of each prior objective. Completed, or % complete. Decide objectives to be carried forward to next Op. Period.	EOC Director General Staff Participate
2. State objectives	List one to five near-term primary objectives to be achieved at the EOC level. Be specific.	EOC Director General Staff contributes
3. Establish priorities related to objectives	Discuss objectives and put them in priority order.	EOC Director General Staff contributes
4. As required adopt strategies to achieve objectives	Some objectives may allow for different strategies. Also, there may be cost, legal or political policy implications to be considered in how to achieve an objective. (Strategies will be influenced by resources availability)	EOC Director General Staff con tributes
5. Make Assignments to implement the strategy for each objective.	Be specific. This is the step that will be used to see if the objectives are being met. What assignments Who does them What resource are needed What additional resources are required?	Operations Section Coordinator Planning Section Coordinator Logistics Section Coordinator

6. Review/Establish length of next Operational Period.	If the assignments and actions needed to meet the objectives will take four hours, then that will be the length of the Operational Period. Operational Periods tend to be short at the beginning of an emergency and longer as time goes on.	Planning Section Coordinator
7. Establish organizational elements as required.	Review staffing needs, and complete an EOC Organization Chart for the next Operational Period.	Planning Section Coordinator and remainder of General Staff
8. Logistical or other technical support required	Describe what is needed and develop a resource order if necessary.	Logistics Section Coordinator
9. Attachments	Determine what may be needed to help explain or support the plan. E.g., policy constraints, communications plan, weather forecast, etc.	Planning Section Coordinator

4.7 EOC Equipment and Support Systems

The following is a summary of primary equipment located within the EOC. Detailed operating procedures are located at each work station.

4.7.1 Television

Ceiling-mounted TV monitors are located throughout the EOC. In addition, ten-inch television sets are located at principal workstations. Monitoring of television audio is done through headsets.

4.7.2 Commercial Radio

Three commercial radio stations may be monitored at each work station. Selection and volume controls are located on the rear panel of each work station. Monitoring of commercial radio at each work station should be done using the headsets.

4.7.3 Government Radio

City department radio frequencies can be assigned to work stations. Departments may receive and send on these frequencies depending upon department policy.

4.7.4 Telephone

Three telephone systems operate within the EOC:

1. Each work station is equipped with one or more Meridian telephones which are a part of the Pacific Bell CENTREX system. These phones have

extensive capabilities which are described in EOC Information Binders at each workstation.

2. Each work station has one or more 3-digit COMLINE telephones, which operate on a

City-owned switched telephone system. The COMLINE telephone system is used for internal EOC calls and has limited outside calling capabilities to other city locations.

3. Designated workstations are equipped with direct hard-wired ring-down telephone links to designated points. These phones are direct point-to-point and have no dial or switching capability. Lifting of the handset causes the phone to ring at the designated location.

4.7.5 Other Support Equipment

The EOC is equipped with eight automatic fax machines. A copy machine is located in the Planning/Intelligence area.

The EOC is equipped with an internal (Microphone) public address system and a P/A system accessible by telephone number access. Each system covers both the EOC and the Police DOC without the ability to de-select one. The microphone system has plug-in jacks at several locations in the EOC only.

4.7.6 EOC Information Management System

EOCIMS consists of a local area computer network within the EOC and a wide area network which connects the EOC to various DOCs. The primary function of EOCIMS is a messaging system which allows users to compose, send, and respond to messages from all stations on the network. Other functions include:

- Resource Request Form;
- Situation Report;
- Maps;
- Procedures and Checklists;
- Station Log; and
- Telephone Directory

Detailed procedures for using EOCIMS are found in the EOC Procedures Manual.

4.8 Information Flow within the EOC

EOC operations have been designed to encourage the free-flow transfer of information between all organizational elements. This simply means that all personnel are free to contact any other person for purposes of obtaining or transferring information.

Incoming and outgoing EOC information is normally routed first to division representatives in the Operations Section.

The Situation Assessment Unit in the Planning/Intelligence Section is the primary point of contact for information exchange within the EOC. This unit collects, processes, displays and distributes information through EOCIMS, wall displays, mapping and situation summaries. This information is available to all EOC personnel.

The Planning/Intelligence Section also produces action plans which contain information of use to all EOC organizational elements.

Face-to-face contact and communication between EOC staff is encouraged. Telephones and EOCIMS are also available for internal EOC communications.

The resource request function in EOCIMS is the primary means for making resource requests. All requests for services or resources must be made in writing (via computer) to the appropriate person. This requirement ensures resource accountability. Requests for service or resources will normally flow along organizational channels, or as described by current EOC procedures.

Written hard copy messages using a paper EOC message form may be used as a backup to EOCIMS.

4.9 Multi-agency Coordination within the EOC

SEMS regulations encourage all jurisdictions to use a multi-agency coordination (MACS) process within EOCs. Multi-agency coordination in the City EOC is accomplished through three separate processes.

4.9.1 EOC General Staff

The EOC General Staff (Section Coordinators) ensure that necessary coordination takes place between agencies and departments. Ad-hoc task forces made up of individuals from different departments or agencies may be established to work on specific issues, as required.

4.9.2 Planning Meetings

The EOC Director will periodically call planning meetings. The EOC Director and General Staff will participate in the meetings, with other personnel and outside agency representatives attending as needed.

4.9.3 Emergency Operations Board

The EOB provides management-level multi-agency coordination for the City during an emergency.

Board meetings are open to the public, unless in executive session.

4.10 Mobile Emergency Operations Center (MEOC)

4.10.1 Background

The MEOC is the City's mobile, alternative EOC. It consists of two 35-foot motor coaches, communications, generator and supply trucks, all of which are self powered, and a restroom trailer (STANS). MEOC work space can be expanded through the use of adjoining canopies and tents.

The MEOC is intended for use as:

- A back-up for the City's primary EOC; and
- A supplemental facility to support City response operations during an emergency.

The MEOC can be deployed to any location large enough to allow access and accommodate vehicle placement. A pre-designated operating location for the MEOC is in Parking Lot # 32 at Dodger Stadium. This site has adequate radio coverage throughout a good portion of the City, a clear microwave path to Mt. Lee, and a pre-positioned interconnection to the Pacific Bell Telephone System. The MEOC can also be activated from its storage facility at the North Central Animal Care and Control Center.

4.10.2 MEOC Operations

When set up at a remote site, the two motor home units are parked in parallel, approximately 25-feet apart. The communications and generator trucks are parked nearby and are connected to the motor coaches by cable. Canopies can be extended to partially cover the area between the coaches. Organized according to SEMS, this area will be used as operational space with tables, chairs and telephones. Electrical power is supplied to the coaches from the generator truck, although each coach has an on-board back-up generator.

Each coach has 16 single-person work stations. These are essentially communication positions used to receive and transmit messages, and have limited work space. All positions are configured in the same manner, with a console to control radio, telephone and TV audio. The MEOC communications equipment includes a broad variety of radios which cover most, if not all, of the systems in use by the City. A patch bay in the communications truck enables radios to be assigned to specific work stations in the MEOC vans.

The City is presently replicating EOCIMS for use in the MEOC.

Personnel from each EOO division and assisting agencies are assigned to the work stations in the MEOC. During operations, personnel assigned to division work stations can communicate by telephone and/or radio with their respective DOCs, dispatch centers and ICPs as well as to any other off-site location accessible through the telephone or radio systems.

4.10.3 MEOC Operating Instructions

Detailed operating instructions for the communications equipment in the MEOC are contained in a separate MEOC Operations Manual.

4.10.4 Use and Activation of the MEOC

The MEOC is an EOO resource which may be used during actual emergencies and by any EOO division or City department for planned events, training and emergency exercises when not in use as the alternate EOC.

The CAO EPD coordinates use and activation of the MEOC. During regular business hours, requests for MEOC use or activation shall be directed to the

EPD. During non-business hours, requests for use or activation shall be directed to the Police Department's Detective Headquarters Division ([213] 485-3261).

Deployment of MEOC vehicles is coordinated by the Police Department's Uniformed Support Division. Deployment of the MEOC shall be made by personnel from either Uniformed Support Division or the Department of General Services (GSD), in that order.

MEOC maintenance and storage is coordinated by the Uniformed Support Division with support from the following agencies:

1. The Department of General Services, Fleet Services Division provides staff and materials for maintenance and repair of the vehicles.

2. Information Technology Agency (ITA) provides staff and materials for maintenance of communications and computer equipment.

3. The Fire Department provides staff and materials for preventive and emergency maintenance of all vehicle equipment.

4. The Police Department provides staff and materials to deploy, clean the interior and exterior of the MEOC and maintains the drivers log.

5. When the MEOC is deployed, General Services and ITA will ensure that the appropriate technician(s) accompany the vehicles to set up operations.

Mobile Operations Capability Guide for Emergency Managers and Planners[15]

Introduction

Disasters may require resources beyond the capabilities of the local or State authorities.

In response to Regional requests for support, the Federal Emergency Management Agency (FEMA) provides mobile telecommunications, operational support, life support, and power generation assets for the on-site management of disaster and all-hazard activities. This support is managed by the Response and Recovery Directorate's Mobile Operations Division (RR-MO).

The Mobile Operations Division has a small headquarters staff and five geographically dispersed Mobile Emergency Response Support (MERS) Detachments and the Mobile Air Transportable Telecommunications System (MATTS) to:

- Meet the needs of the government emergency managers in their efforts to save lives, protect property, and coordinate disaster and all-hazard operations.

- Provide prompt and rapid multi-media communications, information processing, logistics, and operational support to Federal, State, and Local agencies during catastrophic emergencies and disasters for government response and recovery operations.

The MERS and MATTS support the Disaster Field Facilities. They support the Federal, State, and Local responders—not the disaster victims.

This guide describes what telecommunications, logistics, and operational support are available, where they are located, how to obtain them, and what factors need to be considered.

Endnotes

[1] http://www.fema.gov/txt/onp/toolkit_unit_11.txt

[2] http://www.nyc.gov/html/oem/html/other/sub_news_pages/911/profiles_oplog.html

[3] http://www.vdfp.state.va.us/components.htm

[4] http://www.dmort.org/DNPages/DMORTDPMU.htm

[5] http://www.fbi.gov/hq/lab/disaster/disaster.htm

[6] http://www.gfn.org/emsregb/shiawasee/TRIAGE.htm

[7] http://shr.elpasoco.com/media_event.asp

[8] http://www.ncvhs.hhs.gov/020226p1.htm

[9] http://ndms.dhhs.gov/dmat.html

[10] http://egov.cityofchicago.org/city/webportal/

[11] http://www.doh.wa.gov/hsqa/emstrauma/communic.htm

[12] http://www.cfr.org/pub6086/press_release/nearly_two_years_after_911_the_united_states_is_still_dangerously_unprepared_and_underfunded_for_a_catastrophic_terrorist_attack_warns_new_council_task_force.php

Full Text and the Executive Summary of the Council-sponsored Independent Task Force *Emergency Responders: Drastically Underfunded, Dangerously Unprepared*.

[13] http://sunnyvale.ca.gov/Departments/Public+Safety/Emergency+Preparedness/Volunteer+Opportunities/Project+ARK.htm

[14] http://www.fema.gov/rrr/mers01.shtm

SIX

FIRST RESPONDER RESPONSIBILITIES

6

Overview

In the post-planning phase we have identified four major elements of catastrophic event response: Response, Rescue, Relief, and Recovery. Among these four phases, response and recovery tend to get the most attention—as they tend to be the most dramatic and time-sensitive in terms of stemming further losses. As a result of this attention, particularly in the post-9/11 world, the term *First Responder* has entered the public lexicon. First responders can be law enforcement, fire, emergency-medical and private-industry personnel. In this chapter we will examine the specific roles and responsibilities expected of those who are first on the scene at a catastrophic event.

Job Description: Certified First Responder[1]

Responsibilities

Certified **First** Responders (CFR) may function in the context of a broader role, i.e., law enforcement, fire rescue or industrial response. With a limited amount of equipment, the CFR answers emergency calls to provide efficient and immediate care to ill and injured patients. After receiving notification of an emergency, the CFR safely responds to the address or location given, using the most expeditious route, depending on traffic and weather conditions. The CFR must observe traffic ordinances and regulations concerning emergency vehicle operation. The CFR:

- functions in uncommon situations;
- has a basic understanding of stress response and methods to ensure personal well-being;
- has an understanding of body substance isolation;
- understands basic medical-legal principles;
- functions within the scope of care as defined by state, regional and local regulatory agencies;

- complies with regulations on the handling of the deceased, protection of property and evidence at the scene, while awaiting additional EMS resources.

Before initiating patient care, the CFR will "size-up" the scene to determine if the scene is safe, to identify the mechanism of injury or nature of illness, the total number of patients and to request additional help, if necessary. In the absence of law enforcement, the CFR creates a safe traffic environment, such as the placement of road flares, removal of debris and redirection of traffic for the protection of the injured and those assisting in the care of injured patients. Using a limited amount of equipment, the CFR renders emergency medical care to adults, children and infants based on assessment findings. Duties include but are not limited to:

- opening and maintaining an airway;
- ventilating patients;
- administering cardiopulmonary resuscitation;
- providing emergency medical care of simple and multiple system trauma such as:
  - controlling hemorrhage,
  - bandaging wounds,
  - manually stabilizing injured extremities.
- providing emergency medical care to:
  - assist in childbirth
  - manage general medical complaints, altered mental status, seizures, environmental emergencies, behavioral emergencies and psychological crises.
  - searching for medical identification emblems as a guide to appropriate emergency medical care.
  - reassuring patients and bystanders by working in a confident, efficient manner.
  - Avoiding mishandling and undue haste while working expeditiously to accomplish the task.

Where a patient must be extricated from entrapment, the CFR:

- assesses the extent of injury and assists other EMS providers rendering emergency medical care and protection to the entrapped patient
- performs emergency moves and assists other EMS providers in the use of prescribed techniques and appliances for safely removing the patient

- assists other EMS providers, in lifting the stretcher, placing the stretcher in the ambulance, and seeing that the patient and stretcher are secured
- if needed, radios the dispatcher for additional help or special rescue and/or utility services
- in case of multiple patients, performs basic triage.
- reports directly to the responding EMS unit, emergency department or the communications center the nature and extent of injuries, the number of patients, and the condition of each patient.
- identifies assessment findings that may require communicating with medical control for advice.
- constantly assesses patient and administers additional care while awaiting additional EMS resources and while enroute to the emergency facility.

For purposes of records and diagnostics the CFR reports verbally and in writing, observations and emergency medical care of the patient at the emergency scene and in transit, to the responding EMS unit or the receiving medical facility staff. Upon request, the CFR provides assistance to the transporting unit staff or the receiving medical facility staff.

After each call, the CFR:

- restocks and replaces used supplies,
- cleans all equipment following appropriate disinfecting procedures,
- carefully checks all equipment to ensure the availability for the next response
- maintains emergency vehicle in efficient operating condition
- ensures that the emergency vehicle is clean and washed and kept in a neat orderly condition
- in accordance with local, state or federal regulations, decontaminates the interior of any vehicle used to transport patients after transport of a patient with contagious infection or hazardous materials exposure.

Additionally the CFR:

- determines that the emergency vehicle is in proper mechanical condition by checking items required by service management
- maintains familiarity with specialized equipment used by the service
- attends continuing education and refresher education programs as required by employers, medical control and licensing or certifying agencies.

First Responder Responsibilities[2]

The functions of the MFB firefighter First Responder's are:

- To provide an emergency medical response when so dispatched by MAS according to protocols agreed to between MFB and MAS under arrangements approved by the Minister for Health
- At the scene of medical emergencies, to provide emergency care as follows:
  - to assess dangers and to control these, if appropriate, in order to prevent additional injury to the patient and to minimise risks to emergency personnel to gain access to the patient
  - to rapidly assess whether there is any immediate life treat to the patient (i.e.. dangers/response/airway/breathing/circulation) and, if so to provide any immediate emergency life support required, including use of a shock advisory defibrillator, within the limitations of the officer's training and equipment and as circumstances allow
  - where there is no immediate life-treat to the patient, to assess for other major clinical problems and initiate first aid, as appropriate, while waiting for the MAS response
  - to relay information about the scene and the patient to MAS and the oncoming ambulance, when appropriate
  - to remain with the patient and provide ongoing care as circumstances allow
  - to hand patient care over to MAS paramedics as soon as paramedics arrive on scene and to remain to assist where requested if practicable
  - to maintain the MFB equipment necessary for this emergency medical response
  - to gather and record operational, patient care and implementation pilot information according to the requirements of the MFB and the pilot program
  - to maintain and ethical approach to patient care and to respect patient confidentiality at all times
  - to maintain and further develop emergency medical First Responder skills through active participation in the MFB EMS quality assurance and continuing education programs, and to undertake emergency medical responder re-accreditation and re-certification, as scheduled by the MFB.

Roles of Emergency Services at Inter-Agency Scenes[3]

Chapter 8 - Roles of the other Emergency Services

The roles detailed below reflect national guidance and are generally accepted by most services. There may be some local variations and where this occurs all services should be aware of them. E.g. in Scotland, police do not formally use the Gold, Silver and Bronze system of command and control. For further notes on inter-service command and liaison matters refer to para 1.2.7 in chapter 1.

8.1 Role of the Police at an Incident

The Police have their own policies and procedures for operational command.

The primary areas of Police responsibility, when attending an incident, can be summarised as follows:

- The saving of life in conjunction with other emergency services.
- Co-ordination of the emergency services and other subsidiary organisations.
- The protection and preservation of the scene.
- The investigation of the incident, in conjunction with other investigative bodies where applicable.
- The collation and dissemination of casualty information.
- Identification of victims on behalf of the Coroner who is the principal investigator when fatalities are involved.
- The restoration of normality at the earliest opportunity.

While the Police have the overall co-ordinating role, it is important to note that no single organisation has the sole responsibility for 'Command' at a large or major incident. The Fire Service will be expected to exercise some control over other emergency services within the hazard zone at incidents involving fire, rescue and hazardous materials. At incidents other than fire, the police role is primarily one of co-ordination and facilitation of the overall situation. In such circumstances, the police will co-ordinate the response of the emergency services and facilitate the mobilisation and access of resources to the incident site. Where the Fire Service responds to a hazardous area to undertake its rescue role, they will normally set up and maintain an inner cordon and take charge of operations inside it.

Where it is appropriate and safe to do so, the police may take over control of access. This should be achieved by mutual agreement at the time, with a clear briefing and acknowledged handover of responsibility. The police may establish an inner cordon with different boundaries for a different purpose, eg. for investigative purposes. The police will normally be responsible for the setting up and maintaining an outer cordon.

8.2 Police Incident Command Structure

Dependent on the size and location of the incident, three levels of police command may be implemented.

8.3 Police Forward Control Point (Bronze or Operations Command)

Normally the first control to be established, or the nearest to the scene of the incident and responsible for immediate deployment and security. Initially under the command of the Police Incident Officer, the functions of the Forward Control Point may vary considerably dependent upon the type of incident, setting up arrangements and location of the Incident Control Post. Initially, the first police vehicle at the scene will serve as the Forward Control Point/Incident Control Post with the first officer on the scene acting as Incident Officer. His/her initial responsibility is to assume interim command, assess the situation and inform police control; but should not get involved in rescue work.

Where possible, all emergency service forward controls should be sited adjacent to one another but the fire service may influence their location in the interests of safety.

8.4 Police Incident Control Post (Silver or Tactical Command)

If the size and/or nature of the incident requires it, a separate Police Incident Control Post will be set up to, co-ordinate and manage the response to the incident at the tactical level, providing a central point of contact for all emergency and specialist services. The Police Incident Control Post will be the responsibility of a co-ordinator and also under the command of the Police Incident Officer who will be senior in rank to the police officer initially having assumed command.

8.5 Police Major Incident Control Room (Gold or Strategic Command)

The need for such a control is very much dependent on the size and scope of the incident. In some cases, even though there may be a number of casualties, all aspects of the operation can be co-ordinated through the Incident Control Post. However, with large scale/protracted incidents, a Major Incident Room (Gold) may be established to co-ordinate the multi-agency response at the strategic level.

In summary:

- Police Gold is the overall incident commander located at the major incident control Room.
- Police Silver is the incident officer, located at the incident control post.
- Police Bronze is the sector commander(s) located at the forward control point(s).

8.6 Role of the Ambulance Service at an Incident

The primary areas of Ambulance Service responsibility, when attending an incident, can be summarised as follows:

- Provide a focal point at the incident, through an Ambulance Control Point, for all Medical resources.

- The saving of life, in conjunction with other Emergency Services.

- The treatment and care of those injured at the scene, either directly or in conjunction with other medical personnel.

- Either directly, or in conjunction with medical personnel, determine the priority evacuation needs of those injured. (Triage)

- Determine the main Receiving and Supporting hospitals for the receipt of those injured.

- Arrange and ensure the most appropriate means of transporting those injured to the Receiving or Supporting hospitals.

- Ensuring that adequate medical staff and support equipment resources are available at the scene.

- The provision of communications facilities for National Health Service resources at the scene.

- The restoration to normality at the earliest possible opportunity.

8.7 Ambulance Service Incident Command Structure

The Ambulance Service, like the Police, employ a three tier approach to Incident Command; these tiers are known as Gold, Silver and Bronze, although the role of Gold Command differs slightly from that of the Police.

8.8 Ambulance Forward Control Point (Bronze Command)

Normally the first control to be established, or the nearest to the scene, where the Incident Officer/Forward Incident Officer can direct the operation with mobile communications. The Forward Control will also act as a focal point for the NHS/Medical resources at the initial point of patient contact on the scene. There may be a requirement for more than one Forward Control, which will be sited outside the Inner Cordon. The access of Ambulance staff to the Inner Cordon will be controlled by the Fire Service.

8.9 Ambulance Control Point (Silver Command)

An emergency control vehicle, readily identified by a green flashing light, providing an on-site communications facility, which may be distant from the incident. It is to this location that all NHS/Medical resources should report and from where the Incident Officer will operate. Ideally this point should be in close proximity to the Police and Fire Service Control vehicles, subject to radio interference constraints.

8.10 Ambulance Control Management (Gold Command)

The Ambulance Control Management Officer should not be involved directly with the controlling of the Ambulance Service resources but rather have a listening brief, providing an overview of how the incident is progressing. Through this monitoring, he/she will provide a valuable backup to the Ambulance Incident Officer, highlighting any likely problem areas and taking account of the implications for normal day-to-day operations. He/she should be responsive to the needs of the Ambulance Incident Officer at the scene.

In summary:

- Ambulance Gold is the ambulance control management officer.
- Ambulance Silver is the incident officer located at the ambulance control point.
- Ambulance Bronze is the forward incident officer located at the forward control point.

Endnotes

[1] www.health.state.ny.us/nysdoh/ems/srgcfr.pdf

[2] http://www.mfbb.vic.gov.au/default.asp?casid=154

[3] http://www.upton.ma.us/html/inter-agency.html

SEVEN

EMERGENCY OPERATIONS PLANNING (EOP) VS. INCIDENT MANAGEMENT SYSTEM

Courtesy of Corbis Images.

7

Overview

This chapter will examine two distinct concepts in modern catastrophic event planning by contrasting incident management principles with emergency operations planning. In the simplest terms, incident management—a concept we have already discussed at great lengths in this text—involves preparation, planning, and steps for tactical execution during the rescue and response phase of a catastrophic event. Emergency operations planning on the other hand advocates what is called an "all-hazards" approach which uses a comprehensive planning strategy to address an overall prevention and mitigation posture for a broad variety of potential threat scenarios.

In this chapter, we will briefly examine the Department of Homeland Security's National Incident Management System as a refresher of ICS concepts, which have already been discussed in this text. The remainder of the chapter will be allocated to FEMA's Emergency Operations Plan for a Weapons of Mass Destruction (WMD) Incident. Throughout your reading, compare and contrast what you have already learned about the ICS system with the layout, detail, and structure of an all-hazards EOP approach. In particular, notice the emphasis on pre-event planning and preparation and identification of resources for a variety of different possibilities evident in an EOP plan.

National Incident Management System[1]

Developed by the Secretary of Homeland Security at the request of the President, the National Incident Management System (NIMS) integrates effective practices in emergency preparedness and response into a comprehensive national framework for incident management. The NIMS will enable responders at all levels to work together more effectively to manage domestic incidents no matter what the cause, size or complexity.

The benefits of the NIMS system will be significant:

- Standardized organizational structures, processes and procedures;
- Standards for planning, training and exercising, and personnel qualification standards;
- Equipment acquisition and certification standards;
- Interoperable communications processes, procedures and systems;
- Information management systems; and
- Supporting technologies — voice and data communications systems, information systems, data display systems and specialized technologies.

Emergency Operations Planning[2]

Guide for All-Hazard Emergency Operations Planning

State and Local Guide (101)

Chapter 6

Attachment G — Terrorism

Federal Emergency Management Agency

April 2001

TABLE OF CONTENTS

	Page
A. PURPOSE	6-G-3
B. THE HAZARD	6-G-3
1. Nature of the Hazard	6-G-3
2. Hazard Agents	6-G-5
3. Potential Targets	6-G-8
4. Release Area	6-G-8
C. SITUATION AND ASSUMPTIONS	6-G-9
1. Situation	6-G-9
2. Assumptions	6-G-9
D. CONCEPT OF OPERATIONS	6-G-11
1. Direction and Control	6-G-11
2. Communications	6-G-14
3. Warning	6-G-14
4. Emergency Public Information	6-G-14

5. Protective Actions	6-G-15
6. Mass Care	6-G-15
7. Health and Medical	6-G-16
8. Resources Management	6-G-16
E. ORGANIZATION AND ASSIGNMENT OF RESPONSIBILITIES	6-G-16
1. Local Emergency Responders	6-G-16
2. Interjurisdictional Responsibilities	6-G-17
3. State Emergency Responders	6-G-17
4. Local Emergency Planning Committees, State Emergency Response Commissions, and Tribal Emergency Response Commissions	6-G-17
5. Federal Emergency Responders	6-G-17
F. ADMINISTRATION AND LOGISTICS	6-G-18

FIGURE

1. Coordination Relationships in Terrorism Incident Response	6-G-13

TABLES

1. General Indicators of Possible Chemical Agent Use	6-G-5
2. General Indicators of Possible Biological Agent Use	6-G-6
3. General Indicators of Possible Nuclear Weapon/ Radiological Agent Use	6-G-7
4. Suggested Emergency Operations Plan Elements	6-G-10
5. Responses to a WMD Incident and the Participants Involved	6-G-12
A. Suggested Format for a Terrorist Incident Appendix to a Basic All-Hazards Emergency Plan	6-G-A-1
B. Federal Departments and Agencies: Counterterrorism-Specific Roles	6-G-B-1
C. Hotlines and Online Resources	6-G-C-1
D. Incident Indications and First Responder Concerns	6-G-D-1
E. Potential Areas of Vulnerability	6-G-E-1
F. Definitions	6-G-F-1
G. Acronyms	6-G-G-1

CHAPTER 6

HAZARD-UNIQUE PLANNING CONSIDERATIONS

ATTACHMENT G – TERRORISM

A. PURPOSE

The purpose of Attachment G is to aid State and local emergency planners in developing and maintaining a Terrorist Incident Appendix (TIA) to an Emergency Operations Plan (EOP) for incidents involving terrorist-initiated weapons of mass destruction (WMD).[3] The planning guidance in this Attachment was prepared with the assistance of the Departments of Defense, Energy, Agriculture, Health and Human Services, Justice, and Veterans Affairs; the Environmental Protection Agency; the Nuclear Regulatory Commission; the National Emergency Management Association; and the International Association of Emergency Managers.

State and local governments have primary responsibility in planning for and managing the consequences of a terrorist incident using available resources in the critical hours before Federal assistance can arrive. The information presented in this Attachment should help planners develop a TIA that integrates the Federal, State, and local responses. The TIA resulting from this guidance should supplement existing State and local EOPs. A suggested format for a TIA is shown in Tab A.

Federal departments and agencies have developed plans and capabilities for an integrated Federal response to a WMD incident. This Attachment summarizes that response for State and local planners. The Federal Response Plan (FRP), including its Terrorism Incident Annex, provides additional information.

While primarily intended for the use of planners, this Attachment contains information that may be of value to first responders. Planners should consider whether, and how best, to incorporate such information into their plans, procedures, and training materials for first responders.

B. THE HAZARD

The TIA should identify and discuss the nature of the WMD hazard(s), the hazard agents, potential targets, and release areas, as described below.

1. **Nature of the Hazard.** The hazard may be chemical, biological, nuclear/radiological, and/or explosive.

    a. **Initial Warning.** While specific events may vary, the emergency response and the protocol followed should remain consistent. When an overt WMD incident has occurred, the initial call for help will likely come through the local 911 center. This caller probably will not identify the incident as a terrorist incident, but rather state that there was an explosion, a major "accident," or a mass casualty event. Information relayed through the dispatcher prior to arrival of first responders on scene, as well as the initial assessment, will provide first responders with the basic data to begin responding to the incident. With increased awareness and training about WMD

EMERGENCY OPERATIONS PLANNING (EOP) VS. INCIDENT MANAGEMENT SYSTEM

incidents, first responders should recognize that a WMD incident has occurred. The information provided in this Attachment applies where it becomes obvious or strongly suspected that an incident has been intentionally perpetrated to harm people, compromise the public's safety and well-being, disrupt essential government services, or damage the area's economy or environment.

   b. **Initial Detection.** The initial detection of a WMD terrorist attack will likely occur at the local level by either first responders or private entities (e.g., hospitals, corporations, etc.). Consequently, first responders and members of the medical community—both public and private—should be trained to identify hazardous agents and take appropriate actions. State and local health departments, as well as local emergency first responders, will be relied upon to identify unusual symptoms, patterns of symptom occurrence, and any additional cases of symptoms as the effects spread throughout the community and beyond. First responders must be protected from the hazard prior to treating victims. Tab D contains an overview of first responder concerns and indicators related to chemical, biological, and nuclear/radiological WMDs.

The detection of a terrorism incident involving covert biological agents (as well as some chemical agents) will most likely occur through the recognition of similar symptoms or syndromes by clinicians in hospital or clinical settings. Detection of biological agents could occur days or weeks after exposed individuals have left the site of the release. Instead, the "scene" will shift to public health facilities receiving unusual numbers of patients, the majority of whom will self-transport.

   c. **Investigation and Containment of Hazards.** Local first responders will provide initial assessment or scene surveillance of a hazard caused by an act of WMD terrorism. The proper local, State, and Federal authorities capable of dealing with and containing the hazard should be alerted to a suspected WMD attack after State/local health departments recognize the occurrence of symptoms that are highly unusual or of an unknown cause. Consequently, State and local emergency responders must be able to assess the situation and request assistance as quickly as possible. For a list of Federal departments and agencies with counterterrorism-specific roles, see Tab B; for telephone and online resources from selected organizations, see Tab C.

2. **Hazard Agents**

   a. **Chemical.** Chemical agents are intended to kill, seriously injure, or incapacitate people through physiological effects. A terrorist incident involving a chemical agent will demand immediate reaction from emergency responders—fire departments, police, hazardous materials (HazMat) teams, emergency medical services (EMS), and emergency room staff—who will need adequate training and equipment. Hazardous chemicals, including industrial chemicals and agents, can be introduced via aerosol devices (e.g., munitions, sprayers, or aerosol generators), breaking containers, or covert dissemination. Such an attack might involve the release of a chemical warfare agent, such as a nerve or blister agent or an industrial chemical, which may have serious consequences. Some indicators of the possible use of chemical agents are listed in Table 1. Early in an investigation, it may not be obvious whether an outbreak was caused by an infectious agent or a hazardous chemical; however, most chemical attacks will

Table 1. General Indicators of Possible Chemical Agent Use

Stated Threat to Release a Chemical Agent

Unusual Occurrence of Dead or Dying Animals

- For example, lack of insects, dead birds

Unexplained Casualties

- Multiple victims
- Surge of similar 911 calls
- Serious illnesses
- Nausea, disorientation, difficulty breathing, or convulsions
- Definite casualty patterns

Unusual Liquid, Spray, or Vapor

- Droplets, oily film
- Unexplained odor
- Low-lying clouds/fog unrelated to weather

Suspicious Devices or Packages

- Unusual metal debris
- Abandoned spray devices
- Unexplained munitions

be localized, and their effects will be evident within a few minutes. There are both persistent and nonpersistent chemical agents. Persistent agents remain in the affected area for hours, days, or weeks. Nonpersistent agents have high evaporation rates, are lighter than air, and disperse rapidly, thereby losing their ability to cause casualties after 10 to 15 minutes, although they may be more persistent in small, unventilated areas.

    b. **Biological.** Recognition of a biological hazard can occur through several methods, including identification of a credible threat, discovery of bioterrorism evidence (devices, agent, clandestine lab), diagnosis (identification of a disease caused by an agent identified as a possible bioterrorism agent), and detection (gathering and interpretation of public health surveillance data).

When people are exposed to a pathogen such as anthrax or smallpox, they may not know that they have been exposed, and those who are infected, or subsequently become infected, may not feel sick for some time. This delay between exposure and onset of illness, or incubation period, is characteristic of infectious diseases. The incubation period may range from several hours to a few weeks, depending on the exposure and pathogen. Unlike acute incidents involving explosives or some hazardous chemicals, the initial response to a biological attack on civilians is likely to be made by direct patient care providers and the public health community.

Terrorists could also employ a biological agent that would affect agricultural commodities over a large area (e.g., wheat rust or a virus affecting livestock), potentially devastating the local or even national economy. The response to agricultural bioterrorism should also be considered during the planning process.

Responders should be familiar with the characteristics of the biological agents of greatest concern for use in a bioterrorism event (see Tab C for resources). Unlike victims of exposure to chemical or radiological agents, victims of biological agent attack may serve as carriers of the disease with the capability of infecting others (e.g., smallpox, plague). Some indicators of biological attack are given in Table 2.

Table 2. General Indicators of Possible Biological Agent Use

Stated Threat to Release a Biological Agent

Unusual Occurrence of Dead or Dying Animals

Unusual Casualties

- Unusual illness for region/area
- Definite pattern inconsistent with natural disease

Unusual Liquid, Spray, or Vapor

- Spraying and suspicious devices or packages

  c. **Nuclear/Radiological.** The difficulty of responding to a nuclear or radiological incident is compounded by the nature of radiation itself. In an explosion, the fact that radioactive material was involved may or may not be obvious, depending upon the nature of the explosive device used. Unless confirmed by radiological detection equipment, the presence of a radiation hazard is difficult to ascertain. Although many detection devices exist, most are designed to detect specific types and levels of radiation and may not be appropriate for measuring or ruling out the presence of radiological hazards. Table 3 lists some indicators of a radiological release.

Table 3. General Indicators of Possible Nuclear Weapon/Radiological Agent Use

- A stated threat to deploy a nuclear or radiological device
- The presence of nuclear or radiological equipment (e.g., spent fuel canisters or nuclear transport vehicles)
- Nuclear placards or warning materials along with otherwise unexplained casualties

The scenarios constituting an intentional nuclear/radiological emergency include the following:

(1) Use of an **Improvised Nuclear Device (IND)** includes any explosive device designed to cause a nuclear yield. Depending on the type of trigger device used, either uranium or plutonium isotopes can fuel these devices. While "weapons-grade" material increases the efficiency of a given device, materials of less than weapons grade can still be used.

(2) Use of a **Radiological Dispersal Device (RDD)** includes any explosive device utilized to spread radioactive material upon detonation. Any improvised explosive device could be used by placing it in close proximity to radioactive material.

(3) Use of a **Simple RDD** that spreads radiological material without the use of an explosive. Any nuclear material (including medical isotopes or waste) can be used in this manner.

 d. **Conventional Explosive Devices.** The easiest to obtain and use of all weapons is still a conventional explosive device, or improvised bomb, which may be used to cause massive local destruction or to disperse chemical, biological, or radiological agents. The components are readily available, as are detailed instructions to construct such a device. Improvised explosive devices are categorized as being explosive or incendiary, employing high or low filler explosive materials to explode and/or cause fires. Bombs and firebombs are cheap and easily constructed, involve low technology, and are the terrorist weapon most likely to be encountered. Large, powerful devices can be outfitted with timed or remotely triggered detonators and can be designed to be activated by light, pressure, movement, or radio transmission. The potential exists for single or multiple bombing incidents in single or multiple municipalities. Historically, less than five percent of actual or attempted bombings were preceded by a threat. Explosive materials can be employed covertly with little signature, and are not readily detectable. Secondary devices may be targeted against responders.

 e. **Combined Hazards.** WMD agents can be combined to achieve a synergistic effect—greater in total effect than the sum of their individual effects. They may be combined to achieve both immediate and delayed consequences. Mixed infections or intoxications may occur, thereby complicating or delaying diagnosis. Casualties of multiple agents may exist; casualties may also suffer from multiple effects, such as trauma and burns from an explosion, which exacerbate the likelihood of agent contamination. Attacks may be planned and executed so as to take advantage of the reduced effectiveness of protective measures produced by employment of an initial WMD agent. Finally, the potential exists for multiple incidents in single or multiple municipalities.

3. **Potential Targets.** In determining the risk areas within a jurisdiction (and in multiple jurisdiction areas participating in an emergency response), the vulnerabilities of potential targets should be identified, and the targets themselves should be prepared to respond to a WMD incident. In-depth vulnerability assessments are needed for determining a response to such an incident. For examples of vulnerability areas to be considered, see Tab E. In addition, reference Risk Management Plans and Emergency Planning and Community Right-to-Know Act (EPCRA) Plans, which include potential target areas and information on industrial chemical facilities, can be obtained from the Local Emergency Planning Committee (LEPC) in your area.

4. **Release Area.** Standard models are available for estimating the effects of a nuclear, chemical, or biological release, including the area affected and consequences to population, resources, and infrastructure. Some of these models include databases on infrastructure that can be useful in preparing the TIA. A good source of information on available Federal government models is the *Directory of Atmospheric Transport and Diffusion Consequence Assessment Models*, published by the Office of the Federal Coordinator for Meteorology (OFCM). The directory is available both in print and online on OFCM's web page, http://www.ofcm.gov (select "Publications," then "Publications Available Online," then the directory). The directory includes information on the capabilities and limitations of each model, technical requirements, and points of contact.

C. SITUATION AND ASSUMPTIONS

1. **Situation.** The situation section of a TIA should discuss what constitutes a potential or actual WMD incident. It should present a concise, clear, and accurate overview of potential events and discuss a general concept of operations for response. Any information already included in the EOP need not be duplicated in the TIA. The situation overview should include as much information as possible that is unique to WMD response actions, including the suggested elements listed in Table 4.

WMD situation planning should include provisions for working with Federal crisis and consequence management agencies. The key to successful emergency response involves smooth coordination with multiple agencies and officials from various jurisdictions regarding all aspects of the response.

2. **Assumptions.** Although situations may vary, planning assumptions remain the same.

    a. The first responder (e.g., local emergency or law enforcement personnel) or health and medical personnel will in most cases initially detect and evaluate the potential or actual incident, assess casualties (if any), and determine whether assistance is required. If so, State support will be requested and provided. This assessment will be based on warning or notification of a WMD incident that may be received from law enforcement, emergency response agencies, or the public.

b. The incident may require Federal support. To ensure that there is one overall Lead Federal Agency (LFA), the Federal Emergency Management Agency (FEMA) is authorized to support the Department of Justice (DOJ) (as delegated to the Federal Bureau of Investigation [FBI]) until the Attorney General transfers the overall LFA role to FEMA. (Source: FRP, Terrorism Incident Annex) In addition, FEMA is designated as the lead agency for consequence management within the United States and its territories. FEMA retains authority and responsibility to act as the lead agency for consequence management throughout the Federal response. In this capacity, FEMA will coordinate Federal assistance requested through State authorities using normal FRP mechanisms.

c. Federal response will include experts in the identification, containment, and recovery of WMD (chemical, biological, or nuclear/radiological).

d. Federal consequence management response will entail the involvement of FEMA, additional FRP departments and agencies, and the American Red Cross as required.

Table 4. Suggested Emergency Operations Plan Elements

- **Maps**   Use detailed, current maps and charts.
  - Include demographic information.
  - Use natural and manmade boundaries and structures to identify risk areas.
  - Annotate evacuation routes and alternatives.
  - Annotate in-place sheltering locations.
- **Environment**[a]   Determine response routes and times.
  - Include bodies of water with dams or levees (these could become contaminated).
  - Specify special weather and climate features that could alter the effects of a WMD (e.g., strong winds, heavy rains, etc.).
- **Population**[b] Identify those most susceptible to WMD effects or otherwise hindered or unable to care for themselves.
  - Identify areas where large concentrations of the population might be located, such as sports arenas and major transportation centers.
  - List areas that may include retirement communities.

- Note location of correctional facilities.

- Note locations of hospitals/medical centers/schools/day care centers where multiple evacuees may need assistance.

- Identify non-English-speaking populations.

- **Metropolitan**   Identify multi-jurisdictional perimeters and boundaries.

- Identify potentially overlapping areas for response.

- Identify rural, urban, suburban, and city (e.g., city-sprawl/surroundings) mutual risk areas.

- Identify specific or unique characteristics such as interchanges, choke points, traffic lights, traffic schemes and patterns, access roads, tunnels, bridges, railroad crossings, and overpasses and/or cloverleafs.

[a] *The Environmental Protection Agency (EPA) will work with local and State officials on environmental planning issues.*

[b] *The Department of Veterans Affairs (VA), in close cooperation with the Department of Health and Human Services (HHS), will work with State and local officials on these issues.*

e. Jurisdictional areas of responsibility and working perimeters defined by local, State, and Federal departments and agencies may overlap. Perimeters may be used to control access to the affected area, target public information messages, assign operational sectors among responding organizations, and assess potential effects on the population and the environment. Control of these perimeters may be enforced by different authorities, which will impede the overall response if adequate coordination is not established.

D. CONCEPT OF OPERATIONS

The TIA should include a concept of operations section to explain the jurisdiction's overall concept for responding to a WMD incident. Topics should include division of local, State, Federal, and any intermediate interjurisdictional responsibilities; activation of the EOP; and the other elements set forth in Chapter 4 (Basic Plan Content) of State and Local Guide (SLG) 101. A suggested format for a TIA is given in Tab A.

1. **Direction and Control.** Local government emergency response organizations will respond to the incident scene(s) and make appropriate and rapid notifications to local and State authorities (Table 5).[4] Control of the incident scene(s) most likely will be established by local first responders from either fire or police. The Incident Command System (ICS) that was initially established likely will transition into a Unified Command System (UCS) as mutual-aid partners and State and Federal responders arrive to augment the local responders. It is recommended that local, State, and Federal regional law enforcement officials develop consensus "rules of engagement" early in the planning process to smooth the transition from ICS to UCS. This UC structure will facilitate both crisis management and consequence management

activities. The UC structure used at the scene will expand as support units and agency representatives arrive to support crisis and consequence management operations. The site of a terrorist incident is a crime scene as well as a disaster scene, although the protection of lives, health, and safety remains the top priority.

Figure 1 summarizes the coordination relationships between the UC and other response entities. It is assumed that normal disaster coordination accomplished at State and local emergency operations centers (EOCs) and other locations away from the scene would be addressed in the basic EOP. Any special concerns relating to State and local coordination with Federal organizations should be addressed in the TIA.

Local, State, and Federal interface with the FBI On-Scene Commander (OSC) is coordinated through the Joint Operations Center (JOC). FEMA (represented in the command group) will recommend joint operational priorities to the FBI based on consultation with the FEMA-led consequence management group in the JOC. The FBI, working with local and State officials in the command group at the JOC, will establish operational priorities.

Response to any terrorist event requires direction and control. The planner must consider the unique characteristics of the event, identify the likely stage at which coordinated resources will be required, and tailor the direction and control process to merge into the ongoing public health response.

Table 5. Responses to a WMD Incident and the Participants Involved

Events	Participants
1. Incident occurs.	
2. 911 center receives calls, elicits information, dispatches first responders, relays information to first responders prior to their arrival on scene, makes notifications, and consults existing databases of chemical hazards in the community, as required.	911 Center, first responders.
3. First responders arrive on scene and make initial assessment. Establish Incident Command. Determine potential weapon of mass destruction (WMD) incident and possible terrorist involvement; warn additional responders to scene of potential secondary hazards/devices. Perform any obvious rescues as incident permits. Establish security perimeter. Determine needs for additional assistance. Begin triage and treatment of victims. Begin hazard agent identification.	Incident Command: Fire, Law Enforcement, Emergency Medical Services (EMS), and HazMat unit(s).
4. Incident Command manages incident response; notifies medical facility, emergency management (EM), and other local organizations outlined in	

EMERGENCY OPERATIONS PLANNING (EOP) VS. INCIDENT MANAGEMENT SYSTEM

Emergency Operations Plan; requests notification of Federal Bureau of Investigation (FBI) Field Office.	Incident Command.
5. Special Agent in Charge (SAC) assesses information, supports local law enforcement, and determines WMD terrorist incident has occurred. Notifies Strategic Information and Operations Center (SIOC), activates Joint Operations Center (JOC), coordinates the crisis management aspects of WMD incident, and acts as the Federal on-scene manager for the U.S. government while FBI is Lead Federal Agency (LFA).	FBI Field Office: SAC.
6. Local Emergency Operations Center (EOC) activated. Supports Incident Command, as required by Incident Commander (IC). Coordinates consequence management activities (e.g., mass care). Local authorities declare state of emergency. Coordinates with State EOC and State and Federal agencies, as required. Requests State and Federal assistance, as necessary.	Local EOC: Local agencies, as identified in basic Emergency Operations Plan (EOP).
7. Strategic local coordination of crisis management activities. Brief President, National Security Council (NSC), and Attorney General. Provide Headquarters support to JOC. Domestic Emergency Support Team (DEST) may be deployed. Notification of FEMA by FBI/SIOC triggers FEMA actions.[a]	SIOC: FBI, Department of Justice (DOJ), Department of Energy (DOE), Federal Emergency Management Agency (FEMA), Department of Defense (DoD), Department of Health and Human Services (HHS), and Environmental Protection Agency (EPA).
8. Manage criminal investigation. Establish Joint Information Center (JIC). State and local agencies and FEMA ensure coordination of consequence management activities.	FBI; other Federal, State, and local law enforcement agencies. Local Emergency Management (EM) representatives. FEMA, DoD, DOE, HHS, EPA, and other Federal Response Plan (FRP)

a FEMA may initiate FRP response prior to any FBI/SIOC notification.

9. State EMS support local consequence management. Brief Governor. Declare state of emergency. Develop/coordinate requests for Federal assistance through FEMA Regional Operations Center (ROC). Coordinate State request for Federal consequence management assistance.

    State EOC: State EMS and State agencies, as outlined in EOP.

10. DEST provides assistance to FBI SAC. Merges into JOC, as appropriate.

    DEST: DoD, DOJ, HHS, FEMA, EPA, and DOE.

11. FEMA representative coordinates Consequence Management Group. Expedites Federal consequence management activities and monitors crisis management response to advise on areas of decision that could impact consequence management response.

    FBI, FEMA, EPA, DoD, DOE, HHS, and other FRP agencies.

12. Crisis management response activities to incident may continue. Operations, Hazardous Materials Response Unit (HMRU), Joint Technical Operations Team, Joint Inter-Agency Intelligence Support, and additional authorities, as needed.

    FBI, Incident Command System (ICS), Special

13. Federal response efforts coordinated and mission assignments determined. A consequence management support team deploys to incident site. All EOCs coordinate.

    ROC and regional-level agencies.

14. An Emergency Response Team - Advance Element (ERT-A) deploys to State EOC and incident site, as needed. Base installation sites identified for mobilization centers. Liaisons from WMD-related agencies requested for Emergency Support Team (EST) and ROC. Disaster Field Office (DFO) liaisons as needed (may be after extended response phase).

    ERT-A: Regional-level FEMA and FRP primary support agencies, as needed.

15. A consequence management support team provides operational technical assistance to Unified Command.

    FEMA, DOE, DoD, HHS, EPA, and FBI.

16. Recovery operations. Transition of LFA from FBI to FEMA.

2. **Communications.** In the event of a WMD incident, rapid and secure communication is crucial to ensure a prompt and coordinated response. Strengthening communications among first responders, clinicians, emergency rooms, hospitals, mass care providers, and emergency management personnel must be given top priority in planning.

3. **Warning.** Every incident is different. There may or may not be warning of a potential WMD incident. Factors involved range from intelligence gathered from various law enforcement or intelligence agency sources to an actual notification from the terrorist organization or individual. The EOP should have HazMat facilities and transportation routes already mapped, along with emergency procedures necessary to respond.

   a. The warning or notification of a potential WMD terrorist incident could come from many sources; therefore, open communication among local, State, and Federal law enforcement agencies and emergency response officials is critical. The local FBI Field Office must be notified of any suspected terrorist threats or incidents.

   b. **Threat Level.** The FBI operates with a four-tier threat level system:

(1) **Level Four (Minimal Threat).** Received threats do not warrant actions beyond normal liaison notifications or placing assets or resources on a heightened alert.

(2) **Level Three (Potential Threat).** Intelligence or an articulated threat indicates the potential for a terrorist incident; however, this threat has not yet been assessed as credible.

(3) **Level Two (Credible Threat).** A threat assessment indicates that a potential threat is credible and confirms the involvement of WMD in a developing terrorist incident. The threat increases in significance when the presence of an explosive device or WMD capable of causing a significant destructive event, prior or actual injury or loss is confirmed or when intelligence and circumstances indicate a high probability that a device exists.

(4) **Level One (WMD Incident).** A WMD terrorism incident has occurred resulting in mass casualties that requires immediate Federal planning and preparation to provide support to State and local authorities. The Federal response is primarily directed toward the safety and welfare of the public and the preservation of human life.

4. **Emergency Public Information.** Accurate and expedited dissemination of information is critical when a WMD incident has occurred. Preservation of life and property may hinge on instructions and directions given by authorized officials. In the event of a terrorist attack, the public and the media must be provided with accurate and timely information on emergency operations. Establishing and maintaining an effective rumor control mechanism will help clarify emergency information for the public. Initial interaction with the

media is likely to be implemented by an information officer, as directed by the Incident Commander. To facilitate the release of information, the FBI may establish a Joint Information Center (JIC) comprised of representatives from Federal, State, and local authorities for the purpose of managing the dissemination of information to the public, media, and businesses potentially affected by the incident. An act of terrorism is likely to cause widespread panic, and ongoing communication of accurate and up-to-date information will help calm fears and limit collateral effects of the attack.

5. **Protective Actions.** Evacuation may be required from inside the perimeter of the scene to guard against further casualties, either from contamination by an agent released or the possibility that additional WMD or secondary devices targeting emergency responders are present. "In-place sheltering" may be required if the area must be contained because of the need for quarantine or if it is determined to be safer for individuals to remain in place. The TIA should be flexible enough to accommodate either contingency. As with any emergency, State and local officials must be involved in making protective action decisions. Multi-jurisdictional issues regarding mass care, sheltering, and evacuation should be pre-coordinated and included in the TIA.

6. **Mass Care.** The location of mass care facilities will be based partly on the hazard agent involved. Decontamination, if it is necessary, may need to precede sheltering and other needs of the victims to prevent further damage from the hazard agent, either to the victims themselves or to the care providers. The American Red Cross (the primary agency for mass care), the Department of Health and Human Services, and the Department of Veterans Affairs should be actively involved with the planning process to determine both in-place and mobile mass care systems for the TIA. A "mid-point" or intermediary station may be needed to move victims out of the way of immediate harm. This would allow responders to provide critical attention (e.g., decontamination and medical services) and general lifesaving support, then evacuate victims to a mass care location for further attention. General issues to consider for inclusion in the TIA are:

    a. Location, setup, and equipment for decontamination stations, if any.

    b. Mobile triage support and qualified personnel.

    c. Supplies and personnel to support in-place sheltering.

    d. Evacuation to an intermediary location to provide decontamination and medical attention.

    e. Determination of safety perimeters (based on agent).

7. **Health and Medical.** The basic EOP should already contain a Health and Medical Annex. Issues that may be different during a WMD incident and that should be addressed in the TIA include decontamination, safety of victims and responders, in-place sheltering versus evacuation, and multihazard/multiagent triage. Planning should anticipate the need to handle large numbers

of people who may or may not be contaminated but who are fearful about their medical well-being.

The response to a bioterrorism incident will require the active collaboration of the clinicians and local public health authorities responsible for disease monitoring and outbreak investigation. Their activities should be factored into the overall response process.

8. **Resources Management.** The following considerations are highly relevant to WMD incidents and should be addressed, if appropriate, in one or more appendixes to a resource management annex:

   a. Nuclear, biological, and chemical response resources that are available through interjurisdictional agreements (e.g., interstate pacts).

   b. Unique resources that are available through State authorities (e.g., National Guard units).

   c. Unique resources that are available to State and local jurisdictions through Federal authorities (e.g., the National Pharmaceutical Stockpile, a national asset providing delivery of antibiotics, antidotes, and medical supplies to the scene of a WMD incident).

   d. Unique expertise that may be available through academic, research, or private organizations.

E. ORGANIZATION AND ASSIGNMENT OF RESPONSIBILITIES

As with any hazard-specific emergency, the organization for management of local response may vary for a WMD incident and should therefore be defined in the TIA. The effects of a terrorist act involving a WMD have the potential to overwhelm local resources, which may require assistance from State or Federal governments. The following response roles and responsibilities should be articulated in the TIA.

1. **Local Emergency Responders.** Local fire departments, law enforcement personnel, HazMat teams, and EMS will be among the first to respond to a WMD incident. As response efforts escalate, the local emergency management agency and health department will help coordinate needed services.

Primary Duties. The duties of local departments, such as fire, law enforcement, and EMS, along with those of the local emergency management agency and health department should be addressed in their respective EOPs. Any special duties necessary to respond to a suspected terrorist WMD incident should be set forth in the local TIA.

2. **Interjurisdictional Responsibilities.** The formal arrangements and agreements for emergency response to a WMD incident among neighboring jurisdictions, State, Tribal, local, and neighboring States (and those jurisdictions physically located in those States) should be made **prior** to an incident. When coordinating and planning, the Risk Assessment and Risk Area sections of the TIA (areas where potential multiple jurisdictions could overlap and inter-

play) will be readily identifiable. Federal response is already predisposed for interagency and interdepartmental coordination.

3. **State Emergency Responders.** If requested by local officials, the State emergency management agency has capabilities to support local emergency management authorities and the Incident Commander (IC).

Primary Duties. The duties of all responding State agencies should be addressed in the State EOP. Any special duties necessary to respond to a WMD incident should be set forth in the State's TIA.

4. **Local Emergency Planning Committees (LEPCs), State Emergency Response Commissions (SERCs), and Tribal Emergency Response Commissions (TERCs).** These entities are established under the Superfund Amendments and Reauthorization Act of 1986 (SARA) Title III and the implementing regulations of the Environmental Protection Agency (EPA). LEPCs develop and maintain local hazardous material emergency plans and receive notifications of releases of hazardous substances. SERCs and TERCs supervise the operation of the LEPCs and administer the community right-to-know provisions of SARA Title III, including collection and distribution of information about facility inventories of hazardous substances, chemicals, and toxins. LEPCs will have detailed information about industrial chemicals within the community. It may be advisable for LEPCs, SERCs, and TERCs to establish Memoranda of Agreement (MOAs) with agencies and organizations to provide specialized resources and capabilities for response to WMD incidents.

Primary Duties. Any responsibilities germane to terrorism preparedness or response should be outlined in local, State and Tribal hazardous materials emergency response plans or the hazardous materials annex to the local emergency plan.

5. **Federal Emergency Responders.** Upon determination of a credible WMD threat, or if such an incident actually occurs, the Federal government may respond through the appropriate departments and agencies. These departments and agencies may include FEMA, the Department of Justice (DOJ) and FBI, the Department of Defense (DoD), the Department of Energy (DOE), the Department of Health and Human Services (HHS), the EPA, the Department of Agriculture (USDA), the Nuclear Regulatory Commission (NRC), and possibly the American Red Cross and Department of Veterans Affairs. The roles and responsibilities for Federal departments and agencies participating in both crisis management and consequence management are discussed in more detail in Tab B. See the United States Government Interagency Domestic Terrorism Concept of Operations Plan and the Terrorism Incident Annex to the Federal Response Plan for information on the roles and responsibilities of Federal departments and agencies responding to terrorism incidents involving WMD.

Primary Duties. Upon determining that a WMD terrorist incident is credible, the FBI Special Agent in Charge (SAC), through the FBI Headquarters, will initiate liaison with other Federal agencies to activate their operations centers. The responsible FEMA region(s) may activate a Regional Operations Center (ROC) and deploy a representative(s) to the affected State(s). When the responsible FEMA region(s) activates a ROC, the region(s) will notify the responsible FBI Field Office(s) to request a liaison. If the FBI activates the Strategic Information and Operations Center (SIOC) at FBI Headquarters, then other Federal agencies, including FEMA, will deploy a representative(s) to the SIOC, as required. Once the FBI has determined the need to activate a Joint Operations Center (JOC) to support the incident site, Federal, State, and local agencies may be requested by FEMA to support the Consequence Management Group located at the JOC.

F. ADMINISTRATION AND LOGISTICS

There are many factors that make response to a WMD terrorist incident unique. Unlike some natural disasters (e.g., hurricanes, floods, winter storms, drought, etc.), the adminstration and logistics for response to a WMD incident require special considerations. For example, there may be little or no forewarning, immediately obvious indicators, or WMD knowledge (lead time) available to officials and citizens. Because the release of a WMD may not be immediately apparent, caregivers, emergency response personnel, and first responders are in imminent danger themselves of becoming casualties before the actual identification of the crime can be made. Incidents could escalate quickly from one scene to multiple locations and jurisdictions.

TAB A

SUGGESTED FORMAT FOR A TERRORIST INCIDENT APPENDIX TO A BASIC ALL-HAZARDS EMERGENCY PLAN

Supplement to a State or Local Basic Emergency Operations Plan

A. PROMULGATION DOCUMENT

B. SIGNATURE PAGE

C. AUTHORITIES AND REFERENCES

D. TABLE OF CONTENTS

E. PURPOSE

The purpose of the Terrorist Incident Appendix (TIA) is to develop a consequence management plan for responding to and recovering from a terrorist-initiated weapon of mass destruction (WMD) incident. The TIA supplements the Emergency Operations Plan (EOP) already in effect.

F. THE HAZARD

1. **Nature of the Hazard** {Identify WMD hazards that could potentially affect the jurisdiction.}

2. **Incident** {Statement of the situations that would cause the consequence management plan for a WMD incident to go into operation.}

3. **Hazard Agents** {Separate sections for each of the following hazards may be

used, as risk area, treatment, etc., are unique to each incident. The plan for identification of the hazard agent may be included here, as well as an assessment of the risk and definition of the risk area.}

   a. **Chemical** {Statement on chemical terrorism. A Tab with the names of chemicals, composition, reference materials (activation, lethality, treatment, handling, mixture, etc.) may be created and included in the TIA.}

(1) Assessment of risk

(2) Risk area

   b. **Biological** {Statement on biological terrorism. Reference material (identification, handling, treatment, lethality, etc.,) may be created and included in the TIA in a Tab.}

(1) Assessment of risk

Risk area

   c . **Nuclear/Radiological** {Statement on nuclear terrorism. Reference material can be listed in a Tab and may include lethality, handling, treatment, etc.}

(1) Assessment of risk

(2) Risk area

   d. **Explosives** {Statement on explosives terrorism. A Tab with the names of explosives, composition, reference materials (activation, lethality, treatment, handling, mixture, etc.) may be created and included in the TIA.}

(1) Assessment of risk

(2) Risk area

G. SITUATION AND ASSUMPTIONS

1. **Situation:** Basic information on the terrorist incident threat or potential threat. A description of the locale for which the plan is being written. Any information listed below that is already included in the EOP need not be duplicated here. A general description of the area may be given, with the following information in a Tab. Consideration should be given to maintaining information in a secure place.

   a. **Environment**

(1) Geographic conditions (terrain).

(2) Weather (climate).

   b. **Population:** General and special needs individuals, retirement communities and nursing homes, schools, day care centers, correctional facilities, non-English-speaking communities, etc.

   c. **Metropolitan:** Rural/urban/suburban/city (city-sprawl/surroundings).

   d. **Critical Infrastructure/Transportation:** Major highways, secondary roads, tertiary roadways, dirt/gravel roads. Details may include interchanges, choke points, traffic lights, traffic schemes and patterns, access roads, tunnels, bridges, railroad crossings, overpasses/cloverleafs.

e. **Trucking/Transport Activity:** Cargo loading/unloading facilities (type of cargo), waterways (ports, docks, harbors, rivers, streams, lakes, ocean, bays, reservoirs, pipelines, process/treatment facilities, dams, international roll-on/roll-off container shipments, HazMat [oil] flagged registry).

f. **Airports:** Carriers, flight paths, airport layout (air traffic control tower, runways, passenger terminal, parking).

g. **Trains/Subways:** Physical rails, interchanges, terminals, tunnels, cargo/passengers.

h. **Government Facilities:** Post office, law enforcement, fire/rescue, town/city hall, local mayor/governor's residences, Federal buildings, judicial personnel (i.e., judges, prosecutors, residences, offices).

i. **Recreation Facilities:** Sports arenas, theaters, malls, theme parks.

j. **Other Facilities:** Financial institutions (banking facilities/loan institutions), universities, colleges, hospitals, and research institutes (nuclear, biological, chemical, medical clinics).

k. **Military Installations**

l. **HazMat Facilities:** Emergency Planning and Community Right-to-Know Act (EPCRA) sites with Risk Management Plan requirements, Comprehensive Environmental Response, Compensation, and Liability Act (CERCLA) sites, nonreporting Resource Conservation and Recovery Act (RCRA) facilities (i.e., combustion sites, generating sites, and treatment, storage, and disposal [TSD] sites), facilities inventoried by the Toxic Release Inventory System (TRIS), utilities and nuclear facilities, chemical stockpile and/or manufacturing sites.

2. **Assumptions:** This plan will go into effect when a WMD incident has occurred or a credible threat has been identified.

H. CONCEPT OF OPERATIONS

1. **Direction and Control** {Based on the above assessments, provide wiring diagram/flow chart showing the chain of command and control. These diagrams/charts may be specific to WMD or more generally pertinent to any incident.}

2. **Communications** {May elaborate on communications described in the basic EOP.}

   a. Security of communications among responding organizations.

   b. Coordination of communications with Federal responders.

3. **Warning**

4. **Emergency Public Information** {The plan should identify specific methods (channels) to notify the public that an incident has occurred, direct their actions, and keep them informed as the situation progresses. Evacuation and sheltering in place are key actions that may need to be communicated to the public, and continuous updating will be required.}

5. **Protective Actions**

    a. In-place sheltering.

    b. Evacuation routes/means of conveyance should be predetermined based on area and type of agent.

    c. Evacuation support.

6. **Mass Care**

    a. Safe location of mass care facilities

    b. Structural safety

    c. Health and medical services

    d. Provisions for food and water

    e. Policy and procedures for pet care

7. **Health and Medical**

8. **Resources Management**

9. **Recovery Operations**

I. ORGANIZATION AND ASSIGNMENT OF RESPONSIBILITIES

In concert with guidance already in existence, supplementing the EOP, the roles and responsibilities are outlined here for all jurisdictions and entities.

1. Local

2. Interjurisdictional Responsibilities

3. State

4. Tribal

5. Federal

J. ADMINISTRATION AND LOGISTICS

The administrative framework for WMD response operations is outlined here.

1. General support requirements

2. Availability of services

3. Mutual aid agreements

4. Emergency Management Assistance Compacts

5. Administrative policies and procedures (e.g., financial record keeping)

K. TABS

1. Acronyms.

2. Key definitions.

3. Points of contact.

4. Each of the WMD hazard agents may have a separate Tab with subcategories and subsets of information specific to each, including the identification of departments and agencies that have authority and expertise relevant to incidents involving specific agents.

   a. Index of chemical agents.

   b. Index of biological agents.

   c. Index of nuclear/radiological materials.

TAB B

FEDERAL DEPARTMENTS AND AGENCIES: COUNTERTERRORISM-SPECIFIC ROLES

A. FEDERAL EMERGENCY MANAGEMENT AGENCY

FEMA is the lead agency for consequence management and acts in support of the FBI in Washington, DC, and on the scene of the crisis until the U.S. Attorney General transfers the Lead Federal Agency (LFA) role to FEMA. Though State and local officials bear primary responsibility for consequence management, FEMA coordinates the Federal aspects of consequence management in the event of a terrorist act. Under Presidential Decision Directive 39, FEMA supports the overall LFA by operating as the lead agency for consequence management until the overall LFA role is transferred to FEMA and in this capacity determines when consequences are "imminent" for purposes of the Stafford Act. (Source: Federal Response Plan Terrorism Incident Annex, April 1999) Consequence management includes protecting the public health and safety and providing emergency relief to State governments, businesses, and individuals. Additional information on Federal response is given in the United States Government Interagency Domestic Terrorism Concept of Operations Plan (http://www.fema.gov/r-n-r/conplan/).

Web site: **www.fema.gov**

1. **Office of the Director/Senior Advisor to the Director for Terrorism Preparedness.** The Senior Advisor (1) keeps the FEMA Director informed of terrorism-related activities, (2) develops and implements strategies for FEMA involvement in terrorism-related activities, and (3) coordinates overall relationships with other Federal departments and agencies involved in the consequence management of terrorism-related activities.

2. **Preparedness, Training, and Exercises Directorate (PT).** This office provides planning guidance for State and local government. It also trains emergency managers, firefighters, and elected officials in consequence management through the Emergency Management Institute (EMI), National Fire Academy (NFA), and the National Emergency Training Center (NETC) in Emmitsburg, Maryland. EMI offers courses for first responders dealing with the consequences of a terrorist incident. PT conducts exercises in WMD terrorism consequence management through the Comprehensive Exercise Program. These exercises provide the opportunity to investigate the effectiveness of the Federal Response Plan (FRP) to deal with consequence management and test the ability of different levels of response to interact. PT also manages FEMA's Terrorism Consequence Management Preparedness Assistance used by State and local governments for terrorism preparedness planning, training, and exercising.

3. **Mitigation Directorate.** This office has been assigned the responsibility of providing the verified and validated airborne and waterborne hazardous material models. The office also is responsible for developing new, technologically advanced, remote sensing capabilities needed to assess the release and dispersion of hazardous materials, both in air and water, for guiding consequence management response activities.

4. **Response and Recovery Directorate.** This office manages Federal consequence management operations in response to terrorist events. In addition, it manages the Rapid Response Information System, which inventories physical assets and equipment available to State and local officials, and provides a database of chemical and biological agents and safety precautions.

5. **U.S. Fire Administration (USFA).** This administration provides training to firefighters and other first responders through the NFA in conjunction with the Preparedness, Training, and Exercises Directorate. The NFA offers courses pertaining to preparedness and response to terrorist events.

B. DEPARTMENT OF JUSTICE (DOJ)

Web site: **www.usdoj.gov**

Federal Bureau of Investigation. The FBI is the lead agency for crisis management and investigation of all terrorism-related matters, including incidents involving a WMD. Within FBI's role as LFA, the FBI Federal On-Scene Commander (OSC) coordinates the overall Federal response until the Attorney General transfers the LFA role to FEMA.

Web site: **www.fbi.gov**

1. **FBI Domestic Terrorism/Counterterrorism Planning Section (DTCTPS).** Within the FBI Counter Terrorism Division is a specialized section containing the Domestic Terrorism Operations Unit, the Weapons of Mass Destruction Operations Unit, the Weapons of Mass Destruction Countermeasures Unit, and the Special Event Management Unit. Each of these units has specific responsibilities in investigations of crimes or allegations of crimes committed

by individuals or groups in violation of the Federal terrorism and/or Weapons of Mass Destruction statutes. The DTCTPS serves as the point of contact (POC) to the FBI field offices and command structure as well as other Federal agencies in incidences of terrorism, the use or suspected use of WMD and/or the evaluation of threat credibility. If the FBI's Strategic Information and Operations Center (SIOC) is operational for exercises or actual incidents, the DTCTPS will provide staff personnel to facilitate the operation of SIOC.

During an incident, the FBI DTCTPS will coordinate the determination of the composition of the Domestic Emergency Support Teams (DEST) and/or the Foreign Emergency Support Teams (FEST). All incidents wherein a WMD is used will be coordinated by the DTCTPS WMD Operations Unit.

2. **FBI Laboratory Division.** Within the FBI's Laboratory Division reside numerous assets, which can deploy to provide assistance in a terrorism/WMD incident. The Hazardous Materials Response Unit (HMRU) personnel are highly trained and knowledgeable and are equipped to direct and assist in the collection of hazardous and/or toxic evidence in a contaminated environment. Similarly, the Evidence Response Team Unit (ERTU) is available to augment the local assets and have been trained in the collection of contaminated evidence. The Crisis Response Unit (CRU) is able to deploy to provide communications support to an incident. The Bomb Data Center (BDC) provides the baseline training to public safety bomb disposal technicians in the United States. BDC is the certification and accreditation authority for public safety agencies operating bomb squads and is in possession of equipment and staff that can be deployed to assist in the resolution of a crisis involving suspected or identified explosive devices. The Explosives Unit (EU) has experts who can assist in analyzing the construction of suspected or identified devices and recommend procedures to neutralize those items.

3. **FBI Critical Incident Response Group (CIRG).** CIRG has developed assets that are designed to facilitate the resolution of crisis incidents of any type. Notably, the Crisis Management Unit (CMU), which conducts training and exercises for the FBI and has developed the concept of the Joint Operations Center (JOC), is available to provide on-scene assistance to the incident and integrate the concept of the JOC and the Incident Command System (ICS) to create efficient management of the situation. CIRG coordinates a highly trained group of skilled negotiators who are adroit in techniques to de-escalate volatile situations. The Hostage Rescue Team (HRT) is a tactical asset, trained to function in contaminated or toxic hazard environments, that is available to assist in the management of the incident.

4. **National Domestic Preparedness Office (NDPO).** NDPO is to coordinate and facilitate all Federal WMD efforts to assist State and local emergency responders with planning, training, equipment, exercise, and health and medical issues necessary to respond to a WMD event. The NDPO's program areas encompass the six broad areas of domestic preparedness requiring coordination and assistance: Planning, Training, Exercises, Equipment, Information Sharing, and Public Health and Medical Services.

Office for State and Local Domestic Preparedness Support (OSLDPS). This office, within the Office of Justice Programs (OJP), has a State and Local Domestic Preparedness Technical Assistance Program that provides technical assistance in three areas: (1) general technical assistance; (2) State strategy technical assistance, and (3) equipment technical assistance. The purpose of this program is to provide direct assistance to State and local jurisdictions in enhancing their capacity and preparedness to respond to WMD terrorist incidents. The program goals are to:

- Enhance the ability of State and local jurisdictions to develop, plan, and implement a program for WMD preparedness; and

- Enhance the ability of State and local jurisdictions to sustain and maintain specialized equipment.

Technical assistance available from OSLDPS is provided without charge to requesting State or local jurisdiction. The following organizationa are eligible for the State and Local Domestic Preparedness Technical Assistance Program:

- General technical assistance: units and agencies of State and local governments.

- State strategy technical assistance: State administrative agencies, designated by the governor, under the Fiscal Year 1999 State Domestic Preparedness Equipment Program.

- Equipment technical assistance: units and agencies of State and local governments that have received OSLDPS funding to acquire specialized equipment.

Web site: www.ojp.usdoj.gov/osldps/

1. **General Technical Assistance.** OSLDPS provides general overall assistance to State and local jurisdictions for preparedness to respond to WMD terrorist incidents. This technical assistance includes:

    - Assistance in developing and enhancing WMD response plans.

    - Assistance with exercise scenario development and evaluation.

    - Provision of WMD experts to facilitate jurisdictional working groups.

    - Provision of specialized training.

2. **State Strategy Technical Assistance.** OSLDPS provides assistance to States in meeting the needs assessment and comprehensive planning requirements under OSLDPS' Fiscal Year 1999 State Domestic Preparedness Equipment Support Program. Specifically, OSLDPS:

    - Assists States in developing their three-year statewide domestic preparedness strategy.

    - Assists States in utilizing the assessment tools for completion of the required needs and threat assessments.

3. **Equipment Technical Assistance.** OSLDPS provides training by mobile training teams on the use and maintenance of specialized WMD response equipment under OSLDPS' Domestic Preparedness Equipment Support Program. This assistance will be delivered on site in eligible jurisdictions. Specifically, OSLDPS:

- Provides training on using, sustaining, and maintaining specialized equipment.

- Provides training to technicians on maintenance and calibration of test equipment.

- Provides maintenance and/or calibration of equipment.

- Assists in refurbishing used or damaged equipment.

C. DEPARTMENT OF DEFENSE (DoD)

Web site: **www.defenselink.mil**

In the event of a terrorist attack or act of nature on American soil resulting in the release of chemical, biological, radiological, nuclear material or high-yield explosive (CBRNE) devices, the local law enforcement, fire, and emergency medical personnel who are first to respond may become quickly overwhelmed by the magnitude of the attack. The Department of Defense (DoD) has many unique warfighting support capabilities, both technical and operational, that could be used in support of State and local authorities, if requested by FEMA, as the Lead Federal Agency, to support and manage the consequences of such a domestic event.

Due to the increasing volatility of the threat and the time sensitivity associated with providing effective support to FEMA in domestic CBRNE incident, the Secretary of Defense appointed an Assistant to the Secretary of Defense for Civil Support (ATSD[CS]). The ATSD(CS) serves as the principal staff assistant and civilian advisor to the Secretary of Defense and Deputy Secretary of Defense for the oversight of policy, requirements, priorities, resources, and programs related to the DoD role in managing the consequences of a domestic incident involving the naturally occurring, accidental, or deliberate release of chemical, biological, radiological, nuclear material or high-yield explosives.

When requested, the DoD will provide its unique and extensive resources in accordance with the following principles. First, DoD will ensure an unequivocal chain of responsibility, authority, and accountability for its actions to ensure the American people that the military will follow the basic constructs of lawful action when an emergency occurs. Second, in the event of a catastrophic CBRNE event, DoD will always play a supporting role to the LFA in accordance with all applicable law and plans. Third, DoD support will emphasize its natural role, skills, and structures to mass mobilize and provide logistical support. Fourth, DoD will purchase equipment and provide support in areas that are largely related to its warfighting mission. Fifth, reserve component forces are DoD's forward-deployed forces for domestic consequence management.

All official requests for DoD support to CBRNE consequence management (CM) incidents are made by the LFA to the Executive Secretary of the Department of Defense. While the LFA may submit the requests for DoD assistance through other DoD channels, immediately upon receipt, any request that comes to any DoD element shall be forwarded to the Executive Secretary. In each instance the Executive Secretary will take the necessary action so that the Deputy Secretary can determine whether the incident warrants special operational management. In such instances, upon issuance of Secretary of Defense guidance to the Chairman of the Joint Chiefs of Staff (CJCS), the Joint Staff will translate the Secretary's decisions into military orders for these CBRNE-CM events, under the policy oversight of the ATSD(CS). If the Deputy Secretary of Defense determines that DoD support for a particular CBRNE-CM incident does not require special consequence management procedures, the Secretary of the Army will exercise authority as the DoD Executive Agent through normal Director of Military Support, Military Support to Civil Authorities (MSCA) procedures, with policy oversight by the ATSD(CS).

As noted above, DoD assets are tailored primarily for the larger warfighting mission overseas. But in recognition of the unique challenges of responding to a domestic CBRNE incident, the Department established a standing Joint Task Force for Civil Support (JTF-CS) headquarters at the United States Joint Forces Command, to plan for and integrate DoD's consequence management support to the LFA for events in the continental United States. The United States Pacific Command and United States Southern Command have parallel responsibilities for providing military assistance to civil authorities for States, territories, and possessions outside the continental United States. Specific units with skills applicable to a domestic consequence management role can be found in the Rapid Response Information System (RRIS) database maintained by FEMA. Capabilities include detection, decontamination, medical, and logistics.

Additionally, DoD has established 10 Weapons of Mass Destruction Civil Support Teams (WMD-CST), each composed of 22 well-trained and equipped full-time National Guard personnel. Upon Secretary of Defense certification, one WMD-CST will be stationed in each of the 10 FEMA regions around the country, ready to provide support when directed by their respective governors. Their mission is to deploy rapidly, assist local responders in determining the precise nature of an attack, provide expert technical advice, and help pave the way for the identification and arrival of follow-on military assets. By Congressional direction, DoD is in the process of establishing and training an additional 17 WMD-CSTs to support the U.S. population. Interstate agreements provide a process for the WMD-CST and other National Guard assets to be used by neighboring states. If national security requirements dictate, these units may be transferred to Federal service.

D. DEPARTMENT OF ENERGY (DOE)

Through its Office of Emergency Response, the DOE manages radiological emergency response assets that support both crisis and consequence management response in the event of an incident involving a WMD. The DOE is prepared to respond immediately to any type of radiological accident or incident with its radiological emergency response assets.* Through its Office of Nonproliferation and National Security, the DOE coordinates activities in nonproliferation, international nuclear safety, and communicated threat assessment. DOE maintains the following capabilities that support domestic terrorism preparedness and response.

Web site: www.dp.doe.gov/emergencyresponse/

1. **Aerial Measuring System (AMS).** Radiological assistance operations may require the use of aerial monitoring to quickly determine the extent and degree of the dispersal of airborne or deposited radioactivity or the location of lost or diverted radioactive materials. The AMS is an aircraft-operated radiation detection system that uses fixed-wing aircraft and helicopters equipped with state-of-the-art technology instrumentation to track, monitor, and sample airborne radioactive plumes and/or detect and measure radioactive material deposited on the ground. The AMS capabilities reside at both Nellis Air Force Base near Las Vegas, Nevada, and Andrews Air Force Base near Washington, D.C. The fixed-wing aircraft provide a rapid assessment of the contaminated area, whereas the helicopters provide a slower, more detailed and accurate analysis of the contamination.

2. **Atmospheric Release Advisory Capability (ARAC).** Radiological assistance operations may require the use of computer models to assist in estimating early phase radiological consequences of radioactive material accidentally released into the atmosphere. The ARAC is a computer-based atmospheric dispersion and deposition modeling capability operated by Lawrence Livermore National Laboratory (LLNL). The ARAC's role in an emergency begins when a nuclear, chemical, or other hazardous material is, or has the potential of being, released into the atmosphere. The ARAC's capability consists of meteorologists and other technical staff using three-dimensional computer models and real-time weather data to project the dispersion and deposition of radioactive material in the environment. The ARAC's computer output consists of graphical contour plots showing predicted estimates for instantaneous air and ground contamination levels, air immersion and ground-level exposure rates, and integrated effective dose equivalents for individuals or critical populations. The plots can be overlaid on local maps to assist emergency response officials in deciding what protective actions are needed to effectively protect people and the environment. Protective actions could impact distribution of food and water sources and include sheltering and evacuating critical population groups. The ARAC's response time is typically 30 minutes to 2 hours after notification of an incident.

3. **Accident Response Group (ARG).** ARG is DOE's primary emergency response capability for responding to emergencies involving United States nuclear weapons. The ARG, which is managed by the DOE Albuquerque Operations Office, is composed of a cadre of approximately 300 technical and scientific experts, including senior scientific advisors, weapons engineers and technicians, experts in nuclear safety and high-explosive safety, health physi-

cists, radiation control technicians, industrial hygienists, physical scientists, packaging and transportation specialists, and other specialists from the DOE weapons complex. ARG members will deploy with highly specialized, state-of-the-art equipment for weapons recovery and monitoring operations. The ARG deploys on military or commercial aircraft using a time-phased approach. The ARG advance elements are ready to deploy within four hours of notification. ARG advance elements focus on initial assessment and provide preliminary advice to decision makers. When the follow-on elements arrive at the emergency scene, detailed health and safety evaluations and operations are performed and weapon recovery operations are initiated.

4. **Federal Radiological Monitoring and Assessment Center (FRMAC).** For major radiological emergencies impacting the United States, the DOE establishes a FRMAC. The center is the control point for all Federal assets involved in the monitoring and assessment of offsite radiological conditions. The FRMAC provides support to the affected states, coordinates Federal offsite radiological environmental monitoring and assessment activities, maintains a technical liaison with Tribal nations and State and local governments, responds to the assessment needs of the LFA, and meets the statutory responsibilities of the participating Federal agency.

5. **Nuclear Emergency Search Team (NEST).** NEST is DOE's program for dealing with the technical aspects of nuclear or radiological terrorism. A NEST consists of engineers, scientists, and other technical specialists from the DOE national laboratories and other contractors. NEST resources are configured to be quickly transported by military or commercial aircraft to worldwide locations and prepared to respond 24 hours a day using a phased and flexible approach to deploying personnel and equipment. The NEST is deployable within four hours of notification with specially trained teams and equipment to assist the FBI in handling nuclear or radiological threats. Response teams vary in size from a five person technical advisory team to a tailored deployment of dozens of searchers and scientists who can locate and then conduct or support technical operations on a suspected nuclear device. The NEST capabilities include intelligence, communications, search, assessment, access, diagnostics, render-safe operations, operations containment/damage mitigation, logistics, and health physics.

6. **Radiological Assistance Program (RAP).** Under the RAP, the DOE provides, upon request, radiological assistance to DOE program elements, other Federal agencies, State, Tribal, and local governments, private groups, and individuals. RAP provides resources (trained personnel and equipment) to evaluate, assess, advise, and assist in the mitigation of actual or perceived radiation hazards and risks to workers, the public, and the environment. RAP is implemented on a regional basis, with regional coordination between the emergency response elements of the States, Tribes, other Federal agencies, and DOE. Each RAP Region maintains a minimum of three RAP teams, which are comprised of DOE and DOE contractor personnel, to provide radiological assistance within their region of responsibility. RAP teams consist of volunteer members who perform radiological assistance duties as part of

their formal employment or as part of the terms of the contract between their employer and DOE. A fully configured team consists of seven members, to include one Team Leader, one Team Captain, four health physics survey/support personnel, and one Public Information Officer. A RAP team may deploy with two or more members depending on the potential hazards, risks, or the emergency or incident scenario. Multiple RAP teams may also be deployed to an accident if warranted by the situation.

7. **Radiation Emergency Assistance Center/Training Site (REAC/TS).** The REAC/TS is managed by DOE's Oak Ridge Institute for Science and Education in Oak Ridge, Tennessee.. The REAC/TS maintains a 24-hour response center staffed with personnel and equipment to support medical aspects of radiological emergencies. The staff consists of physicians, nurses, paramedics, and health physicists who provide medical consultation and advice and/or direct medical support at the accident scene. The REAC/TS capabilities include assessment and treatment of internal and external contamination, whole-body counting, radiation dose estimation, and medical and radiological triage.

8. **Communicated Threat Credibility Assessment.** DOE is the program manager for the Nuclear Assessment Program (NAP) at LLNL. The NAP is a DOE-funded asset specifically designed to provide technical, operational, and behavioral assessments of the credibility of communicated threats directed against the U.S. Government and its interests. The assessment process includes one-hour initial and four-hour final products which, when integrated by the FBI as part of its threat assessment process, can lead to a "go/no go" decision for response to a nuclear threat.

E. DEPARTMENT OF HEALTH AND HUMAN SERVICES (HHS)

The Department of Health and Human Services (HHS), as the lead Federal agency for Emergency Support Function (ESF) #8 (health and medical services), provides coordinated Federal assistance to supplement State and local resources in response to public health and medical care needs following a major disaster or emergency. Additionally, HHS provides support during developing or potential medical situations and has the responsibility for Federal support of food, drug, and sanitation issues. Resources are furnished when State and local resources are overwhelmed and public health and/or medical assistance is requested from the Federal government.

HHS, in its primary agency role for ESF #8, coordinates the provision of Federal health and medical assistance to fulfill the requirements identified by the affected State/local authorities having jurisdiction. Included in ESF #8 is overall public health response; triage, treatment, and transportation of victims of the disaster; and evacuation of patients out of the disaster area, as needed, into a network of Military Services, Veterans Affairs, and pre-enrolled non-Federal hospitals located in the major metropolitan areas of the United States. ESF #8 utilizes resources primarily available from (1) within HHS, (2) ESF #8 support agencies, (3) the National Disaster Medical System, and (4) specific non-Federal sources (major pharmaceutical suppliers, hospital supply vendors, international disaster response organizations, and international health organizations).

Web site: **www.hhs.gov**

1. **Office of Emergency Preparedness (OEP).** OEP manages and coordinates Federal health, medical, and health-related social service response and recovery to Federally declared disasters under the Federal Response Plan. The major functions of OEP include:

    a. Coordination and delivery of Department-wide emergency preparedness activities, including continuity of government, continuity of operations, and emergency assistance during disasters and other emergencies;

    b. Coordination of the health and medical response of the Federal government, in support of State and local governments, in the aftermath of terrorist acts involving WMD; and

    c. Direction and maintenance of the medical response component of the National Disaster Medical System, including development and operational readiness capability of Disaster Medical Assistance Teams and other special teams that can be deployed as the primary medical response teams in case of disasters.

2. **Centers for Disease Control and Prevention (CDC).** CDC is the Federal agency responsible for protecting the public health of the country through prevention and control of diseases and for response to public health emergencies. CDC works with national and international agencies to eradicate or control communicable diseases and other preventable conditions. The CDC Bioterrorism Preparedness and Response Program oversees the agency's effort to prepare State and local governments to respond to acts of bioterrorism. In addition, CDC has designated emergency response personnel throughout the agency who are responsible for responding to biological, chemical, and radiological terrorism. CDC has epidemiologists trained to investigate and control outbreaks or illnesses, as well as laboratories capable of quantifying an individual's exposure to biological or chemical agents. CDC maintains the National Pharmaceutical Stockpile to respond to terrorist incidents within the United States.

Web site: **www.cdc.gov**

3. **National Disaster Medical System (NDMS).** NDMS is a cooperative asset-sharing partnership between HHS, DoD, the Department of Veterans Affairs (VA), FEMA, State and local governments, and the private sector. The System has three components: direct medical care, patient evacuation, and the non-Federal hospital bed system. NDMS was created as a nationwide medical response system to supplement State and local medical resources during disasters and emergencies, provide backup medical support to the military and VA health care systems during an overseas conventional conflict, and to promote development of community-based disaster medical service systems. This partnership includes DoD and VA Federal Coordinating Centers, which provide patient beds, as well as 1,990 civilian hospitals. NDMS is also com-

prised of over 7,000 private-sector medical and support personnel organized into many teams across the nation. These teams and other special medical teams are deployed to provide immediate medical attention to the sick and injured during disasters, when local emergency response systems become overloaded.

- a. **Disaster Medical Assistance Team (DMAT).** A DMAT is a group of professional and paraprofessional medical personnel (supported by a cadre of logistical and administrative staff) designed to provide emergency medical care during a disaster or other event. During a WMD incident, the DMAT provides clean area medical care in the form of medical triage and patient stabilization for transport to tertiary care.

- b. **National Medical Response Team–Weapons of Mass Destruction (NMRT-WMD).** The NMRT-WMD is a specialized response force designed to provide medical care following a nuclear, biological, and/or chemical incident. This unit is capable of providing mass casualty decontamination, medical triage, and primary and secondary medical care to stabilize victims for transportation to tertiary care facilities in a hazardous material environment. There are four such teams geographically dispersed throughout the United States.

- c. **Disaster Mortuary Operational Response Team (DMORT).** The DMORT is a mobile team of mortuary care specialists who have the capability to respond to incidents involving fatalities from transportation accidents, natural disasters, and/or terrorist events. The team provides technical assistance and supports mortuary operations as needed for mass fatality incidents.

F. ENVIRONMENTAL PROTECTION AGENCY (EPA)

EPA is chartered to respond to WMD releases under the National Oil and Hazardous Substances Pollution Contingency Plan (NCP) regardless of the cause of the release. EPA is authorized by the Comprehensive Environmental Response, Compensation, and Liability Act (CERCLA); the Oil Pollution Act; and the Emergency Planning and Community-Right-to Know Act to support Federal, State, and local responders in counterterrorism. EPA will provide support to the FBI during crisis management in response to a terrorist incident. In its crisis management role, the EPA On-Scene Commander (OSC) may provide the FBI Special Agent in Charge (SAC) with technical advice and recommendations, scientific and technical assessments, and assistance (as needed) to State and local responders. The EPA OSC will support FEMA during consequence management for the incident. EPA carries out its response according to the FRP, ESF #10, Hazardous Materials. The OSC may request an Environmental Response Team that is funded by EPA if the terrorist incident exceeds available local and regional resources. EPA is the chair for the National Response Team (NRT).

The following EPA reference material and planning guidance is recommended for State, Tribal, and local planners:

- Thinking About Deliberate Releases: Steps Your Community Can Take, 1995 (EPA 550-F-95-001).
- Environmental Protection Agency's Role in Counterterrorism Activities, 1998 (EPA 550-F-98-014).

Web site: **www.epa.gov**

G. DEPARTMENT OF AGRICULTURE

It is the policy of the U.S. Department of Agriculture (USDA) to be prepared to respond swiftly in the event of national security, natural disaster, technological, and other emergencies at the national, regional, State, and county levels to provide support and comfort to the people of the United States. USDA has a major role in ensuring the safety of food for all Americans. One concern is bio-terrorism and its effect on agriculture in rural America, namely crops in the field, animals on the hoof, and food safety issues related to food in the food chain between the slaughter house and/or processing facilities and the consumer.

Web site: **www.usda.gov**

1. **The Office of Crisis Planning and Management (OCPM).** This USDA office coordinates the emergency planning, preparedness, and crisis management functions and the suitability for employment investigations of the Department. It also maintains the USDA Continuity of Operations Plan (COOP).

2. **USDA State Emergency Boards (SEBs).** The SEBs have responsibility for coordinating USDA emergency activities at the State level.

3. **The Farm Service Agency.** This USDA agency develops and administers emergency plans and controls covering food processing, storage, and wholesale distribution; distribution and use of seed; and manufacture, distribution, and use of livestock and poultry feed.

4. **The Food and Nutrition Service (FNS).** This USDA agency provides food assistance in officially designated disaster areas upon request by the designated State agency. Generally, the food assistance response from FNS includes authorization of Emergency Food Stamp Program benefits and use of USDA-donated foods for emergency mass feeding and household distribution, as necessary. FNS also maintains a current inventory of USDA-donated food held in Federal, State, and commercial warehouses and provides leadership to the FRP under ESF #11, Food.

5. **Food Safety and Inspection Service.** This USDA agency inspects meat/meat products, poultry/poultry products, and egg products in slaughtering and processing plants; assists the Food and Drug Administration in the inspection of other food products; develops plans and procedures for radiological emergency response in accordance with the Federal Radiological Emergency Response Plan (FRERP); and provides support, as required, to the FRP at the national and regional levels.

6. **Natural Resources Conservation Service.** This USDA agency provides tech-

nical assistance to individuals, communities, and governments relating to proper use of land for agricultural production; provides assistance in determining the extent of damage to agricultural land and water; and provides support to the FRP under ESF #3, Public Works and Engineering.

7. **Agricultural Research Service (ARS).** This USDA agency develops and carries out all necessary research programs related to crop or livestock diseases; provides technical support for emergency programs and activities in the areas of planning, prevention, detection, treatment, and management of consequences; provides technical support for the development of guidance information on the effects of radiation, biological, and chemical agents on agriculture; develops and maintains a current inventory of ARS controlled laboratories that can be mobilized on short notice for emergency testing of food, feed, and water safety; and provides biological, chemical, and radiological safety support for USDA.

8. **Economic Research Service.** This USDA agency, in cooperation with other departmental agencies, analyzes the impacts of the emergency on the U.S. agricultural system, as well as on rural communities, as part of the process of developing strategies to respond to the effects of an emergency.

9. **Rural Business-Cooperative Service.** This USDA agency, in cooperation with other government agencies at all levels, promotes economic development in affected rural areas by developing strategies that respond to the conditions created by an emergency.

10. **Animal and Plant Health Inspection Service.** This USDA agency protects livestock, poultry, crops, biological resources, and products thereof, from diseases, pests, and hazardous agents (biological, chemical, and radiological); assesses the damage to agriculture of any such introduction; and coordinates the utilization and disposal of livestock and poultry exposed to hazardous agents.

11. **Cooperative State Research, Education and Extension Service (CSREES).** This USDA agency coordinates use of land-grant and other cooperating State college, and university services and other relevant research institutions in carrying out all responsibilities for emergency programs. CSREES administers information and education services covering (a) farmers, other rural residents, and the food and agricultural industries on emergency needs and conditions; (b) vulnerability of crops and livestock to the effects of hazardous agents (biological, chemical, and radiological); and (c) technology for emergency agricultural production. This agency maintains a close working relationship with the news media. CSREES will provide guidance on the most efficient procedures to assure continuity and restoration of an agricultural technical information system under emergency conditions.

12. **Rural Housing Service.** This USDA agency will assist the Department of Housing and Urban Development by providing living quarters in unoccupied rural housing in an emergency situation.

13. **Rural Utilities Service.** This USDA agency will provide support to the FRP under ESF #12, Energy, at the national level.

14. **Office of Inspector General (OIG).** This USDA office is the Department's principal law enforcement component and liaison with the FBI. OIG, in concert with appropriate Federal, State, and local agencies, is prepared to investigate any terrorist attacks relating to the nation's agriculture sector, to identify subjects, interview witnesses, and secure evidence in preparation for Federal prosecution. As necessary, OIG will examine USDA programs regarding counterterrorism-related matters.

15. **Forest Service (FS).** This USDA agency will prevent and control fires in rural areas in cooperation with State, local, and Tribal governments, and appropriate Federal departments and agencies. They will determine and report requirements for equipment, personnel, fuels, chemicals, and other materials needed for carrying out assigned duties. The FS will furnish personnel and equipment for search and rescue work and other emergency measures in national forests and on other lands where a temporary lead role will reduce suffering or loss of life. The FS will provide leadership to the FRP under ESF #4, Firefighting, and support to the Emergency Support Functions, as required, at the national and regional levels. FS will allocate and assign radio frequencies for use by agencies and staff offices of USDA. FS will also operate emergency radio communications systems in support of local, regional, and national firefighting teams. Lastly, the FS law enforcement officers can serve as support to OIG in major investigations of acts of terrorism against agricultural lands and products.

H. NUCLEAR REGULATORY COMMISSION

The Nuclear Regulatory Commission (NRC) is the Lead Federal Agency (in accordance with the Federal Radiological Emergency Response Plan) for facilities or materials regulated by the NRC or by an NRC Agreement State. The NRC's counterterrorism-specific role, at these facilities or material sites, is to exercise the Federal lead for radiological safety while supporting other Federal, State and local agencies in Crisis and Consequence Management.

Web site: www.nrc.gov

1. **Radiological Safety Assessment.** The NRC will provide the facility (or for materials, the user) technical advice to ensure onsite measures are taken to mitigate offsite consequences. The NRC will serve as the primary Federal source of information regarding on-site radiological conditions and off-site radiological effects. The NRC will support the technical needs of other agencies by providing descriptions of devices or facilities containing radiological materials and assessing the safety impact of terrorist actions and of proposed tactical operations of any responders. Safety assessments will be coordinated through NRC liaison at the Domestic Emergency Support Team (DEST), Strategic Information and Operations Center (SIOC), Command Post (CP), and Joint Operations Center (JOC).

2. **Protective Action Recommendations.** The licensee and State have the pri-

mary responsibility for recommending and implementing, respectively, actions to protect the public. They will, if necessary, act, without prior consultation with Federal officials, to initiate protective actions for the public and responders. The NRC will contact State and local authorities and offer advice and assistance on the technical assessment of the radiological hazard and, if requested, provide advice on protective actions for the public. The NRC will coordinate any recommendations for protective actions through NRC liaison at the CP or JOC.

3. **Responder Radiation Protection.** The NRC will assess the potential radiological hazards to any responders and coordinate with the facility radiation protection staff to ensure that personnel responding to the scene are observing the appropriate precautions.

4. **Information Coordination.** The NRC will supply other responders and government officials with timely information concerning the radiological aspects of the event. The NRC will liaison with the Joint Information Center to coordinate information concerning the Federal response.

TAB C

HOTLINES AND ONLINE RESOURCES

Note: The Internet sites listed here are current as of April 2001. Users of this Tab should be aware that the Internet is a changing environment. New sites are added frequently. Sites also may be relocated or discontinued. Updated information on online resources will be provided through the FEMA web site, http://www.fema.gov.

A. TELEPHONE HOTLINES

Domestic Preparedness Chemical/Biological HelpLine (phone: 800-368-6498, fax: 410-612-0715, Web: http://www.nbc-prepare.org or http://dp.sbccom.army.mil, e-mail: cbhelp@sbccom.apgea.army.mil) This service provides technical assistance during business hours to eligible State and local emergency responders and their organizations.

National Response Center Hotline (800-424-8802) A service that receives reports of oil, chemical, biological, and radiological releases and actual or potential domestic terrorism; provides technical assistance to emergency responders; and connects callers with appropriate Federal resources. The hotline operates 24 hours a day, 365 days a year.

Nuclear Regulatory Commission Operations Center (301-816-5100, collect calls accepted) Accepts reports of accidents involving radiological materials.

B. INTERNET REFERENCE ADDRESSES

Army Training Support Center (http://www.atsc.army.mil) Provides a digital library with approved training and doctrine information. Files include Field Manuals, Mission Training Plans, Soldier Training Pubs, and more.

Centers for Disease Control and Prevention (CDC) (http://www.bt.cdc.gov) Information regarding infectious diseases.

Soldier and Biological Chemical Command (SBCCOM) (http://www.apgea.army.mil) Information on chemical/biological defense equipment and chemical agents.

CBIAC: Chemical and Biological Defense Information and Analysis Center (http://www.cbiac.
apgea.army.mil) Collects, reviews, analyzes, and summarizes chemical warfare/contraband detection (CW/CBD) information.

Chemical and Biological Warfare – Health and Safety (http://www.ntis.gov/health/health.html) Department of Commerce National Technical Information Service (NTIS) site has information on chemical and biological agents, Government research, detoxification and decontamination studies, developing immunizations, and drug theories.

Chemical Emergency Preparedness and Prevention Office (CEPPO) (http://www.epa.gov/
ceppo/) Information on the CEPPO office, upcoming events, publications, legislation and regulations, and links to outside resources. Also contains information on accident prevention and risk management planning.

Chemical Transportation Emergency Center (CHEMTREC) (http://www.cmahq.com). Source of technical assistance from chemical product safety specialists, emergency response coordinators, toxicologists and other hazardous materials (HazMat) specialists.

Disaster Management Central Resource (DMCR) (http://206.39.77.2/DMCR/dmrhome.html) Lackland Air Force Base (AFB) site with information on civilian support resources, triage of mass casualty situations, medicine and terrorism, terrorism injuries, and WMD medical library.

FEMA – Bio, Toxic Agents, and Epidemic Hazards Reference (www.fema.gov/emi/edu/
biblol1.html) Emergency management-related bibliography on biological, toxic agents, and epidemic hazards.

FEMA – Emergency Management – Related Bibliography (http://www.fema.gov/emi/edu/
biblo12.htm) Currently 35 links to various emergency management-related bibliographies. At least 10 of these relate to WMD.

Federal Radiological Emergency Response Plan (http://www.nrc.gov/NRC/AEOD/FRERP/ downld.html)

U.S. Army Center for Health Promotion and Preventive Medicine (CHPPM) (http://chppm-www.apgea.army.mil) Home Page providing links especially requests for CHPPM services. Links connect to Directorates of Environmental

Health Engineering, Health Promotion and Wellness, Laboratory Sciences, Occupational Health, and Toxicology.

U.S. Army Medical Research and Development (R&D) Command (http://MRMC-www.army.mil) Links include military infectious disease, chemical and biological links, scientific and technical reports, and Web site links.

U.S. Army Medical Research Institute of Chemical Defense (http://chemdef.apgea.army.mil) Provides data links to open literature for medical management of chemical casualties and assay techniques for chemical agents.

U.S. Army Medical Research Institute of Infectious Diseases (http://www.usamriid.army.mil) Provides links to Medical Command (MED-COM), Ebola site, outbreak reporting site, CDC, Defense Technical Information Center (DTIC), U.S. Army, and more.

C. CROSS-REFERENCE WEB SITES

1. Federal Departments/Agencies

a. Environmental Protection Agency (EPA)

(1) EPA's Chemical Emergency and Prevention Office (CEPPO). CEPPO provides leadership, advocacy, and assistance to prevent and prepare for chemical emergencies, respond to environmental crises, and inform the public about chemical hazards in their community. *http://www.epa.gov/ceppo/*

(2) EPA's Environmental Response Team (ERT). The ERT is a group of skilled experts in environmental emergencies who provide on-scene assistance on a "round-the-clock" basis to deal with environmental disasters. *http://www.ert.org/*

(3) EPA's Role in Counterterrorism. This Web site describes EPA's counterterrorism efforts and shares relevant counterterrorism information and resources. *http://www.epa.gov/ceppo/cntr-ter.html*

b. Department of Defense (DoD)

(1) DoD's Chemical and Biological Defense Information Analysis Center. This Web site is DoD's focal point for chemical and biological warfare information. *http://www.cbiac.apgea.army.mil*

(2) DoD's Counterproliferation: Chem Bio Defense. This is a DoD "webnetwork" on nuclear, biological, and chemical (NBC) defense. *http://www.acq.osd.mil/cp/*

(3) DoD's Hazardous Technical Information Services (HTIS). HTIS is a service of the Defense Logistics Agency, located in Richmond, Virginia. *http://www.dscr.dla.mil/htis/htis.htm*

(4) DoD's Medical (Army Surgeon General). This Web site contains extensive medical documents, training materials, audiovisual clips, a search engine, and links to other sites. *http://www.nbc-med.org*

c. Department of Justice (DOJ)

(1) Federal Bureau of Investigation (FBI)

(a) Awareness of National Security Issues and Response Program (ANSIR). The ANSIR is the "public voice" of the FBI for espionage, cyber and physical infrastructure protection. *http://www.fbi.gov/hq/nsd/ansir/ansir.htm*

(b) National Domestic Preparedness Office (NDPO). The NDPO Web site provides a location for information regarding the available Federal training and programs intended to enhance the capabilities of the public safety community in dealing with weapons of mass destruction (WMD). The NDPO mission, members, services, newsletter, and recommended links are contained on this site. *http://www.ndpo.gov*

(2) Office for State and Local Domestic Preparedness Support (OSLDPS). OSLDPS provides technical assistance to States and local jurisdictions to enhance their ability to develop, plan, and implement a program for WMD preparedness. *http://www.ojp.usdoj.gov/osldps/*

d. Federal Emergency Management Agency (FEMA)

(1) Backgrounder: Terrorism. This FEMA Web site provides basic background information on terrorism-related issues.
http://www.fema.gov/library/terror.htm

(2) Terrorism Annex to the Federal Response Plan. The site includes the full text of the Annex in PDF format that can be downloaded and reproduced. *http://www.fema.gov/r-n-r/frp/frpterr.pdf*

(3) United States Government Interagency Domestic Terrorism Concept of Operations Plan. The link provides the full text of the plan, which is designed to provide information to Federal, State, and local agencies on how the Federal government will respond to potential or actual terrorism threats. The document is in PDF format and can be downloaded and reproduced.
http://www.fema.gov/r-n-r/conplan/

(4) FEMA's Rapid Response Information System (RRIS). This Web site provides descriptions and links to eight major chemical and biological agent resources. *http://www.fema.gov/rris/reflib2.htm#chembio*

(5) National Fire Academy. The National Fire Academy homepage provides links to the course catalog and to specific courses and job aids relating to terrorism preparedness. *http://www.usfa.fema.gov/nfa/*

(6) FEMA's Emergency Response to Terrorism Self-Study Course. This Web site provides a link to a self-study course designed to provide basic awareness training to prepare first responders to respond safely and effectively to incidents of terrorism. *http://www.usfa.fema.gov/*

nfa/tr_ertss1.htm

  e. **Department of Health and Human Services**

(1) Office of Emergency Preparedness / National Disaster Medical System – The website provides information on current and previous disaster responses, counter terrorism programs and links to other Federal sites. *http://www.oep-ndms.dhhs.gov*

(2) Centers for Disease Control and Prevention, Bioterrorism Preparedness and Response Program – The website provides information on bioterrorism preparedness issues, response planning and recent publications related to bioterrorism. *http://www.bt.cdc.gov*

The Centers for Disease Control and Prevention (CDC) also provide helpful (though not comprehensive) lists of chemical and biological agents that might be used by terrorists. These lists are included in "Biological and Chemical Terrorism: Strategic Plan for Preparedness and Response," in CDC's *Morbidity and Mortality Weekly Report*, April 21, 2000 (Vol. 49, No. RR-4), available at http://www.cdc.gov/mmwr/mmwr_rr.html.

(3) Metropolitan Medical Response System (MMRS) – Although the MMRS program is locally controlled, this website provides information which will assist any local, State or Federal planner or responder working with domestic preparedness issues. *http://www.mmrs.hhs.gov*

2. **Other Resources**

  a. Critical Infrastructure Assurance Office. This Web site provides information on the Administration's current initiatives in critical infrastructure protection. *http://www.ciao.gov*

  b. DOE's Radiation-Related Web sites. This Web site is maintained by DOE's Office of Civilian Radiation Waste Management. *http://www.rw.doe.gov/*

  c. National Response Team (NRT). The NRT Web site contains information about standing NRT committees, the Regional Response Teams (RRTs), upcoming events, and NRT publications. *http://www.nrt.org/*

TAB D

INCIDENT INDICATIONS AND FIRST RESPONDER CONCERNS

NOTE: Extensive additional information on weapons of mass destruction (WMD) hazards and response, including information addressing first responder concerns, is available from various commercial publishers.

A. BIOLOGICAL

1. **Indications.** Indicators that a WMD incident involving biological agents has taken place may take days or weeks to manifest themselves, depending on the biological toxin or pathogen involved. The Centers for Disease Control and Prevention (CDC) recently developed the following list of epidemiologic clues that may signal a bioterrorist event:

a. Large number of ill persons with a similar disease or syndrome.

b. Large numbers of unexplained disease, syndrome, or deaths.

c. Unusual illness in a population.

d. Higher morbidity and mortality than expected with a common disease or syndrome.

e. Failure of a common disease to respond to usual therapy.

f. Single case of disease caused by an uncommon agent.

g. Multiple unusual or unexplained disease entities coexisting in the same patient without other explanation.

h. Disease with an unusual geographic or seasonal distribution.

i. Multiple atypical presentations of disease agents.

j. Similar genetic type among agents isolated from temporally or spatially distinct sources.

k. Unusual, atypical, genetically engineered, or antiquated strain of agent.

l. Endemic disease with unexplained increase in incidence.

m. Simultaneous clusters of similar illness in noncontiguous areas, domestic or foreign.

n. Atypical aerosol, food, or water transmission.

o. Ill people presenting near the same time.

p. Deaths or illness among animals that precedes or accompanies illness or death in humans.

q. No illness in people not exposed to common ventilation systems, but illness among those people in proximity to the systems.

2. **First Responder Concerns**

    a. The most practical method of initiating widespread infection using biological agents is through aerosolization, where fine particles are sprayed over or upwind of a target where the particles may be inhaled. An aerosol may be effective for some time after delivery, since it will be deposited on clothing, equipment, and soil. When the clothing is used later, or dust is stirred up, responding personnel may be subject to "secondary" contamination.

    b. Biological agents may be able to use portals of entry into the body other than the respiratory tract. Individuals may be infected by ingestion of contaminated food and water, or even by direct contact with the skin or mucous membranes through abraded or broken skin. Use protective clothing or commercially available Level C clothing. Protect the respiratory tract through the use of a mask with biological high-efficiency particulate air (HEPA) filters.

Exposure to biological agents, as noted above, may not be immediately apparent. Casualties may occur minutes, hours, days, or weeks after an exposure has occurred. The time required before signs and symptoms are observed is dependent on the agent used. While symptoms will be evident, often the first confirmation will come from blood tests or by other diagnostic means used by medical personnel.

B. CHEMICAL

1. **Indications.** The following may indicate a potential chemical WMD has been released. There may be one or more of these indicators present.

    a. An unusually large or noticeable number of sick or dead wildlife. These may range from pigeons in parks to rodents near trash containers.

    b. Lack of insect life. Shorelines, puddles, and any standing water should be checked for the presence of dead insects.

    c. Considerable number of persons experiencing water-like blisters, weals (like bee-stings), and/or rashes.

    d. Numbers of individuals exhibiting serious heath problems, ranging from nausea, excessive secretions (saliva, diarrhea, vomiting), disorientation, and difficulty breathing to convulsions and death.

    e. Discernable pattern to the casualties. This may be "aligned" with the wind direction or related to where the weapon was released (indoors/outdoors).

    f. Presence of unusual liquid droplets, e.g., surfaces exhibit oily droplets or film or water surfaces have an oily film (with no recent rain).

    g. Unscheduled spraying or unusual application of spray.

    h. Abandoned spray devices, such as chemical sprayers used by landscaping crews.

    i. Presence of unexplained or unusual odors (where that particular scent or smell is not normally noted).

    j. Presence of low-lying clouds or fog-like condition not compatible with the weather.

    k. Presence of unusual metal debris—unexplained bomb/munitions material, particularly if it contains a liquid.

    l. Explosions that disperse or dispense liquids, mists, vapors, or gas.

    m. Explosions that seem to destroy only a package or bomb device.

    n. Civilian panic in potential high-profile target areas (e.g., government buildings, mass transit systems, sports arenas, etc.).

    o. Mass casualties without obvious trauma.

2. **First Responder Concerns.** The first concern must be to recognize a chemical event and protect the first responders. Unless first responders recognize the

danger, they will very possibly become casualties in a chemical environment. It may not be possible to determine from the symptoms experienced by affected personnel which chemical agent has been used. Chemical agents may be combined and therefore recognition of agents involved becomes more difficult.

C. NUCLEAR/RADIOLOGICAL

1. **Indications.** Radiation is an invisible hazard. There are no initial characteristics or properties of radiation itself that are noticeable. Unless the nuclear/radiological material is marked to identify it as such, it may be some time before the hazard has been identified as radiological.

2. **First Responder Concerns.** While there is no single piece of equipment that is capable of detecting all forms of radiation, there are several different detectors for each type of radiation. Availability of this equipment, in addition to protective clothing and respiratory equipment, is of great concern to first responders.

TAB E

POTENTIAL AREAS OF VULNERABILITY

Areas at risk may be determined by several points: population, accessibility, criticality (to everyday life), economic impact, and symbolic value. The identification of such vulnerable areas should be coordinated with the Federal Bureau of Investigation (FBI).

Traffic: Determine which roads/tunnels/bridges carry large volumes of traffic.

Identify points of congestion that could impede response or place citizens in a vulnerable area.

Note time of day and day of week this activity occurs.

Trucking and Transport Activity: Note location of hazardous materials (HazMat) cargo loading/unloading facilities.

Note vulnerable areas such as weigh stations and rest areas this cargo may transit.

Waterways: Map pipelines and process/treatment facilities (in addition to dams already mentioned).

Note berths and ports for cruise ships, roll-on/roll-off cargo vessels, and container ships.

Note any international (foreign) flagged vessels (and cargo they carry) that conduct business in the area.

NOTE: The Harbor and Port Authorities, normally involved in emergency planning, should be able to facilitate obtaining information on the type of vessels and the containers they carry.

Airports: Note information on carriers, flight paths, and airport layout.

Annotate location of air traffic control (ATC) tower, runways, passenger terminal, and parking areas.

Trains/Subways: Note location of rails and lines, interchanges, terminals, tunnels, and cargo/passenger terminals.

Note any HazMat material that may be transported via rail.

Government Facilities: Note location of Federal/State/local government offices.

Include locations of post office, law enforcement stations, fire/rescue, town/city hall, and local mayor/governor's residences.

Note judicial offices and courts as well.

Recreation Facilities: Map sports arenas, theaters, malls, and special interest group facilities.

Other Facilities: Map location of financial institutions and the business district.

Make any notes on the schedule business/financial district may follow.

Determine if shopping centers are congested at certain periods.

Military Installations: Note location and type of military installations.

HazMat Facilities, Utilities, and Nuclear Facilities: Map location of these facilities.

NOTE: Security and emergency personnel representing all of the above facilities should work closely with local and State personnel for planning and response.

TAB F

DEFINITIONS

Aerosol – Fine liquid or solid particles suspended in a gas, for example, fog or smoke.

Biological Agents – Living organisms or the materials derived from them that cause disease in or harm to humans, animals, or plants or cause deterioration of material. Biological agents may be used as liquid droplets, aerosols, or dry powders.

Chemical Agent – A chemical substance that is intended to kill, seriously injure, or incapacitate people through physiological effects. Generally separated by severity of effect: lethal, blister, and incapacitating.

Consequence Management – Measures to protect public health and safety, restore essential government services, and provide emergency relief to governments, businesses, and individuals affected by the consequences of terrorism. State and local governments exercise primary authority to respond to the consequences of terrorism. (Source: FRP Terrorism Incident Annex, page TI-2, April 1999). The Federal Emergency Management Agency (FEMA) has been designated the Lead Federal Agency (LFA) for consequence management to ensure that the Federal Response Plan is adequate to respond to terrorism. Additionally, FEMA supports the Federal Bureau of Investigation (FBI) in crisis management.

Crisis Management – This is the law enforcement aspect of an incident that involves measures to identify, acquire, and plan the resources needed to anticipate, prevent, and/or resolve a threat of terrorism. The FBI is the LFA for crisis management for such an incident. (Source: FBI) During crisis management, the FBI coordinates closely with local law enforcement authorities to provide successful law enforcement resolution to the incident. The FBI also coordinates with other Federal authorities, including FEMA. (Source: FRP Terrorism Incident Annex, April 1999)

Decontamination – The process of making people, objects, or areas safe by absorbing, destroying, neutralizing, making harmless, or removing the HazMat.

Federal Response Plan (FRP) – The FRP establishes a process and structure for the systematic, coordinated, and effective delivery of Federal assistance to address the consequences of any major disaster or emergency declared under the Robert T. Stafford Disaster Relief and Emergency Assistance Act, as amended (42 U.S. Code [USC], et seq.). The FRP Terrorism Incident Annex defines the organizational structures used to coordinate crisis management with consequence management. (Source: FRP Terrorism Incident Annex, April 1999)

1. **Lead Agency** – The Federal department or agency assigned lead responsibility under U.S. law to manage and coordinate the Federal response in a specific functional area. The FBI is the lead agency for crisis management and FEMA is the lead agency for consequence management. Lead agencies support the overall Lead Federal Agency (LFA) during all phases of the response.

2. **Lead Federal Agency (LFA)** – The agency designated by the President to lead and coordinate the overall Federal response is referred to as the LFA and is determined by the type of emergency. In general, an LFA establishes operational structures and procedures to assemble and work with agencies providing direct support to the LFA in order to provide an initial assessment of the situation, develop an action plan, monitor and update operational priorities, and ensure each agency exercises its concurrent and distinct authorities under U.S. law and supports the LFA in carrying out the President's relevant policy. Specific responsibilities of an LFA vary according to the agency's unique statutory authorities.

Mitigation – Those actions (including threat and vulnerability assessments) taken to reduce the exposure to and detrimental effects of a WMD incident.

Nonpersistent Agent – An agent that, upon release, loses its ability to cause casualties after 10 to 15 minutes. It has a high evaporation rate, is lighter than air, and will disperse rapidly. It is considered to be a short-term hazard; however, in small, unventilated areas, the agent will be more persistent.

Persistent Agent – An agent that, upon release, retains its casualty-producing effects for an extended period of time, usually anywhere from 30 minutes to several days. A persistent agent usually has a low evaporation rate and its vapor is heavier than air; therefore, its vapor cloud tends to hug the ground. It is considered to be a long-term hazard. Although inhalation hazards are still a concern, extreme caution should be taken to avoid skin contact as well.

Plume – Airborne material spreading from a particular source; the dispersal of particles, gases, vapors, and aerosols into the atmosphere.

Preparedness – Establishing the plans, training, exercises, and resources necessary to achieve readiness for all hazards, including WMD incidents.

Radiation – High-energy particles or gamma rays that are emitted by an atom as the substance undergoes radioactive decay. Particles can be either charged alpha or beta particles or neutral neutron or gamma rays.

Recovery – Recovery, in this document, includes all types of emergency actions dedicated to the continued protection of the public or promoting the resumption of normal activities in the affected area.

Response – Executing the plan and resources identified to perform those duties and services to preserve and protect life and property as well as provide services to the surviving population.

Terrorism – The unlawful use of force or violence against persons or property to intimidate or coerce a government, the civilian population, or any segment thereof, in furtherance of political or social objectives. Domestic terrorism involves groups or individuals who are based and operate entirely within the United States and U.S. territories without foreign direction and whose acts are directed at elements of the U.S. government or population.

Toxicity – A measure of the harmful effects produced by a given amount of a toxin on a living organism.

Weapons-Grade Material – Nuclear material considered most suitable for a nuclear weapon. It usually connotes uranium enriched to above 90 percent uranium-235 or plutonium with greater than about 90 percent plutonium-239.

Weapons of Mass Destruction – Any explosive, incendiary, or poison gas, bomb, grenade, rocket having a propellant charge of more than 4 ounces, or a missile having an explosive incendiary charge of more than 0.25 ounce, or mine or device similar to the above; poison gas; weapon involving a disease organism; or weapon that is designed to release radiation or radioactivity at a level dangerous to human life. (Source: 18 USC 2332a as referenced in 18 USC 921)

TAB G

ACRONYMS

AFB	Air Force Base
AMS	Aerial Measuring System
ANSIR	Awareness of National Security Issues and Response Program
ARAC	Atmospheric Release Advisory Capability

ARG	Accident Response Group
ARS	Agriculture/Research Service
ATC	Air Traffic Control
ATSD(CS)	Assistant to the Secretary of Defense for Civil Support
BDC	Bomb Data Center
CBIAC	Chemical and Biological Defense Information and Analysis Center
CBRNE	Chemical, Biological, Radiological, Nuclear Material, or High-Yield Explosive
CDC	Centers for Disease Control and Prevention
CDRG	Catastrophic Disaster Response Group
CEPPO	Chemical Emergency Preparedness and Prevention Office
CERCLA	Comprehensive Environmental Response, Compensation, and Liability Act
CHEMTREC	Chemical Transportation Emergency Center
CHPPM	Center for Health Promotion and Preventive Medicine
CIRG	Crisis Incident Response Group
CJCS	Chairman of the Joint Chiefs of Staff
CM	Consequence Management
CMU	Crisis Management Unit (CIRG)
CRU	Crisis Response Unit
CSREES	Cooperative State Research, Education and Extension Service
CST	Civil Support Teams
CW/CBD	Chemical Warfare/Contraband Detection
DEST	Domestic Emergency Support Team
DFO	Disaster Field Office
DMAT	Disaster Medical Assistance Team
DMCR	Disaster Management Central Resource
DMORT	Disaster Mortuary Operational Response Team
DoD	Department of Defense
DOE	Department of Energy
DOJ	Department of Justice
DPP	Domestic Preparedness Program

DTCTPS	Domestic Terrorism/Counter Terrorism Planning Section (FBI HQ)
DTIC	Defense Technical Information Center
EM	Emergency Management
EMI	Emergency Management Institute
EMS	Emergency Medical Services
EOC	Emergency Operations Center
EOP	Emergency Operations Plan
EPA	Environmental Protection Agency
EPCRA	Emergency Planning and Community Right-to Know Act
ERT	Emergency Response Team (FBI)
ERT-A	Emergency Response Team – Advance Element
ERTU	Evidence Response Team Unit
ESF	Emergency Support Function
EST	Emergency Support Team
EU	Explosives Unit
FBI	Federal Bureau of Investigation
FEMA	Federal Emergency Management Agency
FEST	Foreign Emergency Support Team
FNS	Food and Nutrition Service
FRERP	Federal Radiological Emergency Response Plan
FRMAC	Federal Radiological Monitoring and Assessment Center
FRP	Federal Response Plan
FS	Forest Service
HazMat	Hazardous Materials
HHS	Department of Health and Human Services
HMRU	Hazardous Materials Response Unit
HQ	Headquarters
HRT	Hostage Rescue Team (CIRG)
HTIS	Hazardous Technical Information Services (DoD)
IC	Incident Commander
ICS	Incident Command System

IND	Improvised Nuclear Device
JIC	Joint Information Center
JOC	Joint Operations Center
JTF-CS	Joint Task Force for Civil Support
LEPC	Local Emergency Planning Committee
LFA	Lead Federal Agency
LLNL	Lawrence Livermore National Laboratory
MEDCOM	Medical Command
MMRS	Metropolitan Medical Response System
MOA	Memorandum of Agreement
MSCA	Military Support to Civil Authorities
NAP	Nuclear Assessment Program
NBC	Nuclear, Biological, and Chemical
NCP	National Oil and Hazardous Substances Pollution Contingency Plan
NDMS	National Disaster Medical System
NDPO	National Domestic Preparedness Office
NEST	Nuclear Emergency Search Team
NETC	National Emergency Training Center
NFA	National Fire Academy
NMRT	National Medical Response Team
NRC	Nuclear Regulatory Commission
NRT	National Response Team
NSC	National Security Council
NTIS	National Technical Information Service
OEP	Office of Emergency Preparedness
OFCM	Office of the Federal Coordinator for Meteorology
OIG	Office of the Inspector General (USDA)
OSC	On-Scene Commander
OSLDPS	Office for State and Local Domestic Preparedness Support
PDD	Presidential Decision Directive
PHS	Public Health Service

POC	Point of Contact
PT	Preparedness, Training, and Exercises Directorate (FEMA)
R&D	Research and Development
RAP	Radiological Assistance Program
RCRA	Research Conservation and Recovery Act
RDD	Radiological Dispersal Device
REAC/TS	Radiation Emergency Assistance Center/Training Site
ROC	Regional Operations Center
RRIS	Rapid Response Information System (FEMA)
RRT	Regional Response Team
SAC	Special Agent in Charge (FBI)
SARA	Superfund Amendments and Reauthorization Act
SBCCOM	Soldier and Biological Chemical Command (U.S. Army)
SCBA	Self-Contained Breathing Apparatus
SEB	State Emergency Board
SERC	State Emergency Response Commission
SIOC	Strategic Information and Operations Center (FBI HQ)
SLG	State and Local Guide
TERC	Tribal Emergency Response Commission
TIA	Terrorist Incident Appendix
TRIS	Toxic Release Inventory System
UC	Unified Command
UCS	Unified Command System
USC	U.S. Code
USDA	U.S. Department of Agriculture

USFA	U.S. Fire Administration
VA	Department of Veterans Affairs
WMD	Weapons of Mass Destruction
WMD-CST	WMD Civil Support Team

Endnotes

[1] http://www.fema.gov/nims/

[2] http://www.fema.gov/doc/rrr/allhzpln.doc

[3] Definitions of terms and acronyms used in this document are given in Tabs F and G, respectively.

[4] Table 5 provides an overview of events likely to occur in a WMD incident. It is designed to help planners better understand the interface that State and local response will likely have with Federal response organizations. The table includes both crisis management and consequence management activities that would be operating in parallel and is intended to illustrate the complex constellation of responses that would be involved in a WMD incident.

* For facilities or materials regulated by the Nuclear Regulatory Commission (NRC), or by an NRC Agreement State, the technical response is led by NRC as the LFA (in accordance with the Federal Radiological Emergency Response Plan) and supported by DOE as needed.

EIGHT

MASS CASUALTY/FATALITY PLANNING

Courtesy of Corbis Images.

8

Overview

Chapter Eight will continue discussions started earlier in this text about the specific needs of mass casualty and mass fatality planning. Mass fatality events require very specific needs that will be different from mass casualty. While a mass casualty event will require extensive first aid, emergency, and trauma medicine and patient intake capacity, a mass fatality event will trigger the need for the safe, thorough, and solemn task of recovering, identifying, hygienically storing, accounting for and to the greatest degree possible, the safe returning of human remains to the proper next-of-kin. By both definitions a "mass fatality" or "mass casualty" event is one that will typically overwhelm and exceed the medical and mortuary capabilities of a single community.

MULTIPLE CASUALTY INCIDENTS[1]

A multiple casualty incident (MCI) is an event in which the resources available are insufficient to manage the number of casualties or the nature of the emergency. It is not uncommon for EMS to have more than one patient at a trauma scene. However, most day-to-day operational procedures are designed for the single trauma or medical patient. Paramount in all EMS activities are safety, organization and communication. When confronted with multiple patients these needs are even greater. This module will discuss the EMS response to multiple casualty incidents with an emphasis on medical command, triage, initial assessment, standards of care and debriefing. Finally, the Pine Lake disaster will be discussed.

INCIDENT COMMAND SYSTEM

The Incident Command System (ICS) was developed in Southern California in the early 1970's (Campbell, 2000). The components of an ICS include Command, Fire Suppression, Rescue/Extrication, Law Enforcement, and Medical Services.

The flexibility of an ICS enables it to be adapted to all types of emergencies including fire, rescue, law enforcement, and MCI's. An ICS can be expanded or compressed depending on the current condition of the incident.

The purpose of the ICS is to prevent independent actions and chaos at the scene of the incident. If an ICS is not established immediately, other rescuers may take independent actions, which will often be in conflict. *"Without organization and accountability, chaos will occur and too many people will attempt to command the incident. If you do not control the situation, the situation will control you"* (Campbell, 2000 pg. 342).

MEDICAL INCIDENT COMMAND[2]

One component of the ICS is the Medical sector. For the purpose of this module, only Medical Services will be discussed. The Medical sector includes: Medical Command, Triage, Treatment, Transport and Staging.

Each component does not have to have one person exclusively assigned. However, it is necessary to ensure that the function of each position is executed.

All participants of an ICS need to know their responsibilities. The following paragraphs explain the roles of the officers of an ICS.

Medical Command

Most experts agree that the responsibility of command should belong to ONE individual who has the ability to coordinate a variety of emergency activities. This is the cornerstone of the ICS structure.

The first on-scene unit assumes the role of command and directs all initial efforts. The person assuming the role of command must be familiar with the ICS structure and the operating procedures of other responding rescue vehicles. The command officer does not have to be the individual with the most medical training but must be able to manage the emergency scene.

The command officer must be clearly identified immediately, and all others at the scene must be aware that only one individual is in command. As more qualified personnel arrive the role of command may be transferred. Once established, medical command should do the following:

- Assume an effective command mode and position.
- Transmit a brief radio report to the communication center.
- Ensure that proper rescue/extrication services are activated.
- Ensure law enforcement involvement as required.
- Ensure that helicopter landing zone operations are coordinated if required.
- Determine the amount and type of additional medical resources and supplies.

- Ensure that area hospitals and Medical Direction are aware of the situation.
- Designate assistant officers and their locations.
- Maintain an appropriate scan of the scene and control.
- Work as a conduit of communications between subordinates and the Incident Commander.

EMS Staging Officer

Staging sectors are required for large incidents to prevent vehicle congestion and response delays. All emergency vehicles (fire, police, EMS) should report to this sector for direction. The staging officer also controls other agencies such as disaster relief and the media. The roles of the EMS staging officer include the following:

- Maintain a log of available units and medical supplies.
- Coordinate location of incoming resources (i.e. ambulances and helicopters).
- Coordinate incoming personnel who wish to aid at the scene. Provide updates to Medical Command as required.

Triage Officer

The third officer of the medical sector is the triage officer. The duties of the triage officer are:

- Ensure proper utilization of the Initial Assessment triage system or other local protocol for patient assessment. Some services permit opening the airway and controlling obvious bleeding.
- Ensure that the triage tags or other visual identification techniques are properly completed and secured to the patient.
- Make requests for additional resources through Medical Command.
- Provide updates to Medical Command as necessary.

Treatment Officer

The roles of the treatment officer include:

- Establishing suitable treatment areas.
- Communicating resource needs to Medical Control.
- Assigning, supervising, and coordinating treatment of patients.
- Providing updates to Medical Command as required.

EMS personnel assigned to the treatment sector are responsible for advanced care and initial stabilization until patients can be transported to a medical facility.

Transport Officer

The final component of the medical sector is the transport officer. The duties of the transport officer are as follows:

- Ensure the organized transport of patients' off-scene.
- Ensure an appropriate distribution to all hospitals to prevent hospital overloading.
- Complete a transportation log.
- Contact receiving hospitals to advise them of the number of patients and condition (may be delegated to a communication officer).
- Provide updates to Medical Command as required.

TRIAGE

Triage is a system of sorting patients to determine the order in which they will receive treatment and transport to a medical facility. In an MCI, the triage goal is to meet the needs of the most individuals possible by delaying treatment of selected patients. As a triage officer, you are to spend less than one minute doing an initial assessment to determine the priority of a patient. *"It can not be overemphasized that the triage officer does NOT render any treatment to a patient"* (Campbell, 2000, pg. 344). The treatment of patients is to be performed by the treatment officer. If the triage officer allows himself to provide treatment to victims the function of the triage must be reassigned. The Basic Trauma Life Support decision tree assists in determining medical priority. Once medical priority is determined the triage officer should affix a completed triage tag or other visual identification technique to the victim and then move to the next victim.

Triage Categories

1. *Priority 1: Red Tag* - Critical condition, unstable but salvageable with timely and appropriate intervention. Patients normally categorized as CRITICAL include those with airway problems or respiratory distress (Tension pneumothorax, upper airway obstruction, flail chest, open chest wound), possible cardiac injury (tamponade, severe contusion), uncontrolled hemorrhage (including internal), and altered mental or neurological status (concussion, skull fracture, spinal cord injury).

2. *Priority 2: Yellow Tag* - Serious condition/potentially unstable. These patients require timely transport, but only after critical patients are attended to. These patients may tolerate a one hour delay in transport. Patients in this category include those with major extremity or soft tissue injury, burns without an airway compromise, burns, electrical injuries and blunt abdominal or thoracic trauma.

3. ***Priority 3: Green Tag***: Stable condition/minor injuries. These patients are often referred to as the "walking wounded" and are transported after red and yellow-tagged patients. Patients with simple fractures, lacerations, small burns and sprains fall into this category. An ambulance may not be required to transport these patients. For example, they may be transported in a bus.

4. ***Priority 4: Black Tag:*** Dead or alive but nonsalvageable. These patients require excessive manpower and resources to survive. Most patients in cardiac arrest are considered low priority in MCI situations.

INITIAL ASSESSMENT

During an MCI there is a tendency to over triage and this must be avoided. Over triage has a determintal impact on available EMS resources. The triage assessment needs to be accurate.

The following three basic human systems need to be quickly evaluated to determine the patient's medical priority:

- Respiratory system
- Circulatory system
- Neurological system

BTLS International recommends using the BTLS Initial Assessment during the triage phase and the Rapid Trauma Assessment or the Focused Assessment in the treatment phase will enable EMS providers to complete accurate assessments. The use of these assessment tools will provide the greatest amount of good to the greatest number of patients. Some EMS systems use other assessment protocols but the goal is the same: to rapidly assess , treat and transport patients.

The components of the initial assessment are general impression, level of consciousness, airway, breathing and circulation.

General Impression (Patient Overview)

- What is the victim's approximate age?
- What position are they in?
- What is their activity (aware of surroundings, anxious, in distress)?
- Are they perfusing (skin color)?
- Are there any major injuries or bleeding?

Level Of Consciousness (AVPU)

Is classified as:

- Alert
- Responds to verbal stimuli
- Responds to painful stimuli
- Unresponsive

Airway

- Is it open and self-maintained?
- Is it compromised?

Breathing

- Is the victim breathing?
- What is the rate and quality?

Circulation

- Is there a pulse?
- What is the rate and quality?

After the Initial Assessment a survivability factor will be determined and the patient will be prioritized accordingly. An example of applying the survivability factor would be the situation when you are presented with a pediatric patient and a geriatric patient with similar injuries. You have enough resources to care for only one patient. Which patient do you choose and why? This decision is based on an objective evaluation rather than on emotions.

Standard of Care

Reviewing the care that patients receive during an MCI is important because it reinforces the principles of an MCI. The adverse circumstances which EMS were operating under must be taken into consideration. During normal day-to-day operations, patients are treated according to standard protocols, thus many patients are over treated in anticipation of deterioration. However, during an MCI or disaster, this inefficient use of manpower and resources may be catastrophic. The primary principle in triage and treatment of victims of an MCI is to do *the greatest good for the greatest number of patients with the least depletion of available resources.*

Critique and Debriefing of the Incident

Whether an MCI response was real or a practice, it is imperative that all involved meet and talk about the incident afterward. The primary focus of the critique is on what worked, what did not work, and what could be better. All personnel should be honest and learn from the experience. A MCI/Disaster

Plan is a dynamic document and should be modified when a problem is identified. Discussion leaders achieve the best results by encouraging both openness and constructive attitudes. *The goal of the critique session is to learn, not to place blame.*

CRITICAL INCIDENT STRESS DEBRIEFING

To a casual observer, the victims of an MCI are the people who were injured. However, the rescuers themselves can often become victims as well. The tragedy, the suffering, the extensive injuries and the unfairness of the situation may be replayed on the minds of the rescuers long after the disaster is over. The resolution of emotional trauma may be more complex than the healing of physical injuries.

Critical incidents are extraordinary events that interfere, or have the potential to interfere with an individual's psychological ability to cope with stress. The concept of critical incident stress is often associated with large-scale disasters or MCI's. However, most critical events involve only one patient. A critical incident is defined as an event that exceeds the rescuer's ability to cope psychologically.

Certain events are classified as critical incidents automatically. These include:

- Death or serious injury of an emergency co-worker in the line of duty
- MCI resulting in serious injury or death
- Suicide of an emergency worker
- Death of a civilian as a result of emergency service or law enforcement operations.

The reactions of rescuers may range from simple anxiety, short-term depression, significant depression or even suicide. Rescuers may question their own actions and feel responsible for injuries or death that were beyond their control.

A formal system must be established immediately following the disaster to identify those responders with stress-related problems. This system must provide access to professional help. This system is referred to as a Critical Incident Stress Management System (CISM). CISM is a structured group meeting that allows emergency and rescue personnel the opportunity to discuss their feelings and other reactions after the incident. This is not psychotherapy or psychological treatment. CISM meetings are designed to reduce the impact of a critical event and to accelerate the normal recovery of normal people. This is not an operational debriefing.

It is normal to suffer painful reactions following an abnormal event. An abnormal reaction occurs when such feelings are not shared. Every EMS service should offer CISM or similar programs to personnel who encounter a critical incident. Many communities have formal debriefing programs with volunteers trained in CISM. In the event of an MCI or other critical incident these individuals are often mobilized quickly.

Case Study: PINE LAKE DISASTER

July 14, 2000 will never be forgotten in Alberta. At 1900 hours an F3 tornado stuck the Green Acers Campground. It is estimated that over 500 camper units where at the campground. The tornado's path of destruction was 1.5 miles wide and 17 miles long. Unfortunately 12 people died as a result of the tornado.

Discussion with Mr. Sterling Martin EMT-P and Acting Fire Chief Cliff Fuller provided the following information. The emergency call was received as a *"micro-burst at unit 56 with one seriously injured"*. The RCMP was first to arrive followed shortly behind by one unit from Guardian Ambulance. This first ambulance did not actually enter the campground. They transported two critically injured children from the entrance of the park. One call for help was made over a cell phone, and then the cell tower was overwhelmed with calls.

The County of Red Deer Fire Department was next to arrive. Acting Fire Chief Cliff Fuller assumed the role of Incident Command. His role included delegation of a search and rescue as well as contacting the numerous agencies that were involved. Those agencies included Guardian Ambulance, the Decontamination Unit from Edmonton, Decontamination unit from Nova Chemicals, STARS, HUSSARS (Heavy Equipment), the dive team from Calgary Fire, Red Deer Search and Rescue, RCMP, and the Military, as well as, many other agencies too numerous to mention.

Sterling Martin assumed the role of Medical Command. Within 20 minutes over 100 injured people came forward looking for help. "It was like a scene out of a war zone." He estimates that within 3 hours and 40 minutes all patients had been transported. Over 130 patients were rescued, triaged and transported to hospitals. The majority of the patients were transported to Red Deer. Hospitals in Calgary were placed on alert and prepared by discharging those patients who could safely be managed at home.

Unfortunately the tornado destroyed the computer which had the records of who was camping in Pine Lake at the time of the disaster. Therefore, a complete search of the path of destruction was required. All 17 miles were searched. This search lasted 10 days.

On day two, Cliff Fuller assumed the role of site manager. His duties included over-seeing of all other rescue efforts. He assumed this role for the next nine days.

Challenges

Mr. Martin discussed several challenges of the Pine Lake disaster. The first challenge was a lack of communication. The responding ambulances were able to communicate to with each other in the park. However, due the overwhelming number of cell phones in use at the campground after the tornado struck, they were only able to place one call for additional help. This is an important lesson for those services that use cell phones as a primary method for communication.

Secondly, he said that they had made the treatment area "too small". This area became congested and this made treatment a challenge. It is recommended that in large scale disasters that the treatment area be located far away from the actual disaster to prevent recreation of the disaster in this area.

Mr. Martin also mentioned the importance of having a formalized method of identification for the rescuers. Fortunately the rescuers at Pine Lake knew each other but he mentioned how important it is to know who is performing each role.

As well, they discovered that no one had kept a record of the patient names and where these patients were transported.

Finally, he stated that the role of those services called to respond after the initial assessment is to provide assistance but their role is not assume control. Important for all EMS services is that the grassroots EMR's, EMT's, and EMT-P's get the work done and large EMS services should avoid the tendency to want to take control from the small services, unless they are asked to provide this service.

SUMMARY

The priorities of any incident, regardless of size, should be safety, organization, and then patient care. In order to provide the most effective and efficient patient care, EMS personnel must approach each situation in a safe and organized fashion.

To have an effective disaster plan requires rescue personnel to use portions of this plan in day-to-day operations, including the small routine emergency. This rehearsal of the components of an MCI plan will develop proficiency and allow for a smooth transition into a larger more complex emergency when required. If the disaster plan is only activated on large-scale emergencies there will be a lack of familiarity with its use. Furthermore, routine activation develops confidence for all levels of command and the other agencies involved. To avoid the "paper plan syndrome", regular implementation and review of the disaster plan are essential to having a successful operation. An example of this is found in Medicine Hat where colored triage cards are used on all ambulance calls.

The key component of a disaster plan is the Incident Command System. The ICS includes fire suppression, rescue, law enforcement, and medical services. Medical Incident Command is then further broken down to include staging, triage, treatment, and transportation. Successful management of an MCI requires that all participants know their responsibilities.

Fortunately large scale MCI's are unique events that do not occur frequently. These events require responders to change roles from their normal routine. EMS providers must think of the multiple casualties, and be prepared to overlook nonsalvageable victims for less injured victims. Initial scene assessment and communication are higher priorities than patient care.

Following any critical incident a Critical Incident Stress Debriefing must occur to preserve the mental well being of those rescuers involved. Many EMS services have formalized CISM. This is separate from the operational debriefing.

Important lessons always come from a tragedy. The use of cell phones, as a primary method of communication, do not work during times of large scale disasters. Ensure that you make the treatment area large enough to avoid congestion. Have a method to identify rescuers. The importance of disaster preparedness for all EMS services is paramount. Often the arrival of other EMS services occurs well after most of the triage, treatment and transportation has begun.

NTSB and DMORT: Key Partners in Transportation Disaster Response[2]

The critical nature of mass fatality victim identification and family assistance calls for experienced, trained personnel who bring compassion and focus to their work. The importance of this work is underscored by the expectations of family members, the public, elected officials, and the media, who each demand an effective, timely response to mass fatality events.

In the past several years, the National Transportation Safety Board Office of Transportation Disaster Assistance (TDA) and the DMORT teams have built a close relationship based on this shared vision. Of the 21 DMORT deployments since 1993, twelve have been in response to transportation accidents. Although each has posed unique challenges, all were concluded with a focus on professionalism. Through these responses, NTSB staff and DMORT team members contribute to effective disaster responses. The NTSB views DMORT as our designated official partner in victim identification, morgue operations, and related family assistance support. The partnership is strong and will continue to strengthen in years ahead.

DHS/FEMA and NTSB staff met recently to reaffirm our partnership and to discuss response and training issues. We are updating the memorandum of understanding between DHS and NTSB to reflect changes driven by the move to DHS and by federal directives for disaster response. NTSB is planning a meeting with DMORT team leaders in the near future to examine our relationship and update our roles and responsibilities.

Currently, the NTSB has very specific requirements for the support a DMORT team provides in the event of a transportation accident. These are:

- Assist the medical examiner in victim identification and mortuary services.
- Provide a morgue facility, the DPMU, and the necessary equipment and supplies.
- Collect and monitor the status of incoming dental and medical records and radiographs.
- Manage antemortem and postmortem data.
- Provide a Family Assistance Center (FAC) team to interview family members for antemortem identification information and disposition of remains information.

- Assist in the collection of DNA reference samples from family members.
- Ensure DMORT, DPMU, and FAC teams are staffed with experienced, trained personnel and that DMORT have written standard operating procedure for morgue operations, victim identification, and associated medicolegal interpretations and operations.

DMORT: TRIAGE PROTOCOL

Open bags delivered from scene.

Team sorts through materials to separate tissue from other material.

Airplane parts and other hardware are routed to [AGENCY]

Any items of evidentiary value (e.g., possible fragments of weapons) are route to [AGENCY]

Isolated personal effects are routed to [AGENCY]

Review human remains for further analysis.

- Tissue which will not yield any information in one or more of the following areas
- Pathology (identification, pathology, injury, etc.)
- Anthropology (age, sex, stature, ancestry, etc.)
- Dental
- Fingerprint analysis
- DNA/identification — Consult DNA Section for guidelinesis placed in a red barrel to be included in the common tissue disposition.
- Tissue with the potential for further identification is placed in a bag and the Victim Processing Record is checked in the left margin to indicate the stations where the specimen should be routed. All specimens go to Photography, Radiography, and DNA.
- The triage scribe signs and dates the Victim Processing Record. The specimen is routed to Admitting.

Human remains associated with personal effects are treated as follows:

- The PE is removed from the human remains as long as removal will not damage or compromise the remains. Notation about the clothing is entered into the comments section of the Victim Processing Record and the clothing is turned over to [AGENCY].
- If the human remains are suitable for further analysis, they are processed through the regular channels. [AGENCY] is informed that the human remains will be assigned a morgue number in Admitting, and they should accompany the remains to Admitting to obtain that number.
- If the remains are not suitable for further analysis, they are placed in the common tissue red barrel.

- For remains that the PE cannot be removed without possible damage, notify the [AGENCY], and leave effects associated with tissue. Mark "[AGENCY]" in red marker on the Victim Processing Record. Send specimen through procedures described above. [AGENCY] receives the specimen after all other relevant stations have signed off. These specimens may be expedited through the systems at the request of the [AGENCY].

ADMITTING PROTOCOL

Receive human remains in plastic bag from Triage.

Record next sequential morgue number available and provenience (or other identifying information from the scene) on the flip chart.

Label bag of tissue with assigned morgue number. Do not put provenience or scene information on the bag—only the morgue number.

Create a folder with the assigned morgue number. Place appropriate paperwork, as indicated on the Victim Processing Record, inside the folder. Number all paperwork appropriately.

Create a Tyvek tag with the assigned morgue number. Place the Tyvek tag in the folder, NOT in the bag.

Staple Victim Processing Record to front of file. Initial this form.

If remains are determined, at any station, to be unrelated, they will be separated and returned to Triage for assessment.

- One specimen will be designated by the original morgue number.
- The second specimen will be admitted into the identification process according to the above procedures and receive new paperwork.
- All paperwork for each specimen should have reference to the morgue number of the other specimen.

RADIOLOGY PROTOCOL

Turn on processor at beginning of the day. Hit the run button. Processor will be ready in approximately 15 minutes. Ready light will come on when processor is ready.

Gloves will be worn at all times

Place cassette inside plastic cover. Place remains on cassette.

- Depending on the size of the cassette, several items may be radiographed on the same film.
- Label the remains with the corresponding morgue number with the lead numbers provided. Lead numbers should be placed as close to the specimen as possible. Do not place multiple specimens together if the morgue numbers run from 0 to 1 (example, 60 with 61).

- Attempt to place the remains in anatomical position when possible. The Anthropology or Pathology team will assist as needed.

Process the film in the darkroom. Return the film to the imaging area. It is useful to have one person outside the morgue to transport and process the film from the morgue door. This eliminates signing in and out, and having to put on and take off PPE.

Review the film for adequate exposure and proper labeling.

- Film should have proper morgue number label along with MFI name on the film.
- If remains need to be repositioned to reflect anatomical position of the body part, take an additional radiograph.
- [AGENCY] is notified of any unusual findings (e.g., possible weapons or pieces of weapons).

The scribe

- completes the Radiology Victim Processing Record
- initials logbook
- places films in corresponding x-ray folder.
- If multiple remains are included on one film, note on outside of x-ray folder. Place up to twenty specimen radiographs in one envelope.
- signs the Victim Processing Record on the front of the folder
- If remains are not received in numerical order, note missing remains for future reference. If specimen is not received by the end of the day, contact Morgue Manager.

End of day clean-up

- Use disinfectant spray or wipes on all equipment, cassettes and table.
- Turn off x-ray equipment and processor. Lift the lid of the processor for ventilation

When radiograph is requested for review by another area, a representative of that area will sign it out with the date and time and sign back.

PHOTOGRAPHY PROTOCOL

Specimen is received and placed on white background (photo copy stand).

Place right-angle metric rule next to specimen. Add extension ruler if required.

Place Tyvek tag (from folder) with morgue number next to specimen.

Shoot photo.

Record in Photodocument Log:

- Date
- Film roll number
- Photo number
- Morgue number. Add a letter suffix for subsequent photos of the same specimen, e.g., 32, 32A, 32B.
- Camera settings
- Notes, as f needed
- Dental
- Perforations in tissue
- Correlation of morgue numbers
- "PE" if personal effects are in the photo

Sign and date the Victim Processing Record.

Label:

- Exposed (used) film with the photo roll number and the MFI name
- Exterior of film canister lid with film roll number
- Inside of film canister lid lid with MFI name
- Ziplock baggie with MFI name and photo roll number

Place photo roll and Photodocument Log in baggie and seal.

Document Photo roll numbers and specimen shots in the Photo Log.

If a specimen comes back to be re-photographed, look up the morgue number in the Photodocument Log to determine the last number/suffix used so that the new photograph can receive the correct sequential number. For example, if specimen 32 had three photographs taken when it originally came through, those photos are numbered 32, 32A, and 32B. If 32 comes back for more photos, they should be labeled 32 C, 32 D, etc.

Trooper takes possession of film and the original Photodocument Log. He will hand-deliver this directly to [AGENCY]. The Trooper will keep a copy the Photodocument Log for himself. A copy should also be retained in the Photography Section.

One set of photographs was made to place in the original postmortem files. A second set is placed in the postmortem file copies maintained by DMORT.

[AGENCY] maintains all original film negatives.

ANTHROPOLOGY PROTOCOL

The Anthropological Analysis is completed by a team consisting of 2 anthropologists and 1 scribe.

The human remains are received and placed examination table. Morgue number is verified on file and specimen bag.

Anthropologists:

- Assess biological parameters
- Review Pathology and Dental forms for consistency (bone, side, biological parameters, etc.) with Anthropology assessment. If there is a discrepancy, the team consult the other teams(s) and reach a consensus on the assessment.

Scribe:

- Locates X-ray and places it on light box for review.
- Transcribes information dictated by Anthropologists to Anthropology Examination form.
- Completes Anthro Log for each specimen

Anthropologist signs and dates Anthropology Examination Form and the Victim Process Record.

Anthropology Specimen Cleaning Protocol

During the processing of specimens by anthropology, it may be necessary to remove the tissue from bone features used for analysis of age, sex or pathology in order to observe subtle features. All attempts are made to remove the adherent tissue using scalpels, scissors and/or periosteal elevators. If additional tissue removal is necessary, the following procedures are observed:

- Runner takes specimen to DNA for immediate sectioning. If DNA requires a section of the bone, indicate which part is still needed for anthropological analysis and return this portion to Anthropology after sectioning.
- Process bone as needed:
- Place the bone in microwave-safe container and fill with water so bone is barely submerged. Microwave for 5 minute intervals and continue to clean the bone manually (up to 30 minutes total).
- If tissue is still present after heating and cleaning, soak in bleach solution (50% bleach, 50% water) for 1 hour. Increase bleach concentration for second soaking if necessary.
- Indicate cleaning procedures used (microwave, bleach, etc.) in comments of Anthropology Examination Form.

PATHOLOGY PROTOCOL

The Pathology Analysis is completed by a team consisting of a pathologist and a scribe.

The human remains are received and placed examination table. Morgue number is verified on file and specimen bag.

Pathologist:

- Assesses appropriate dimensions and features of each specimen
- Notifies [AGENCY] of any unusual findings, e.g., possible wounds
- If specimen cannot be analyzed, the forms must still be completed. A notation of "no analysis" or "no pathology" should be made.

Scribe:

- Locates X-ray and places it on light box for review.
- Transcribes information dictated by Pathologist to Pathology Examination Form.
- Completes Pathology Log for each specimen

Pathologist signs and dates Pathology Examination Form and the Victim Processing Record.

DENTAL ID PROTOCOL (ANTEMORTEM)

Obtain list of possible victims.

Contact last known treating dentist.

Record antemortem dental records (PADIT Manual Page III 8-9).

Deliver information to antemortem file (See PADIT MANUAL III 12-14).

DENTAL ID PROTOCOL (POSTMORTEM)

Receive dental remains from previous station.

Clean remains.

Examine and chart remains according to PADIT Manual Page III 3-5.

X-ray dental remains (conventional/digital), PADIT Manual pages III 10-11.

Digital photograph of remains (if authorized by DMORT Commander).

Complete and copy all postmortem records.

Deliver and log postmortem record to dental comparison section (see PADIT Manual Page III 12-14).

DNA PROTOCOL

Set up computer from AFDIL with AFDIL incident number and initials of AFDIL personnel present.

Set up station for DNA recovery.

- Scalpels
- Stryker saws
- Diluted (10%) bleach solution

- Disposable covers (12 X 12 Bench Kote)
- 4 X 4's to wipe instruments
- Collection tubes
- Evidence bags

Specimens should come to DNA station LAST. If the Victim Identification form indicates that a station has been skipped, a runner should be directed to return the specimen and file to that station. Exceptions can be made for special treatment of the specimens by request from the [AGENCY].

The DNA recovery team examines the specimen to determine whether a sample will be taken, as per AFDIL guidelines:

- 5-10 grams of deep skeletal muscle (avoid tissues that may have been crushed together by incident impact or blast forces)
- 1-2 cm x 4-6 cm x 0.5-1 cm of cortical bone (avoid anthropological landmarks and articular margins, as well as fresh-broken margins, when possible; cut windows in long bones and crania)
- Upper or lower canine or other intact tooth without restorations
- Other portion of soft or hard tissue that fits into a 50 mL conical tube

The morgue number of the specimen is noted on the DNA Log, along with a YES or NO indication for sampling. Start log with date, page number, and MFI name.

If a sample is taken, the specimen is placed into a specimen tube that has been pre-labeled, by hand, with the AFDIL number AND the morgue number. The numbers should appear on the tube itself AND on the lid.

The specimen tube is given to the computer operator. The computer operator:

- Enters the morgue number of the specimen, the type of material, and the exact nature of the specimen.
- Generates two labels
- The first label is placed on the tube on the opposite side of the hand-written numbers, as close to the lid as possible.
- The second label is placed on the plastic evidence bag
- Inserts the labeled tube into the labeled bag.

The bag is heat-sealed and placed into a cooler or a -20° freezer until it is released to AFDIL. Once a specimen is frozen, it should remain frozen.

The specimens should be kept cold while awaiting sampling. If there is an extended break, or if the sampling takes longer than usual, the specimens should be returned to the truck temporarily.

Completed samples are released to AFDIL by the coroner/medical examiner.

The Victim Processing Record is initialed, and a YES or NO is written to indicate sampling.

RUNNER PROTOCOL

GENERAL

- Help locate files and specimens as needed.
- Help keep specimens moving from one Section to the next.
- Make sure that files and/or specimens removed from any Section are logged out and back in appropriately.
- Specimens of interest to the [AGENCY] are given priority.

PHOTOGRAPHY RUNNER

- Pick up files and specimens.
- Verify that specimen and file numbers match.
- Verify that Photography has signed off on the Victim Processing Record.
- Deliver to next appropriate station. Check the Victim Processing Record to determine station. This is usually Pathology but may be Dental, Fingerprinting, or Anthropology

PATHOLOGY RUNNER

- Keep Pathology specimens in order in container labeled "PATHOLOGY - TO BE DONE"
- Keep files that correspond to specimens in numerical order
- Once analysis is completed, deliver the specimen and file to next station

ANTHROPOLOGY RUNNER

- Keep Anthropology specimens in order in container labeled "ANTHROPOLOGY TO BE DONE"
- Set out several specimens (as space permits) in sequence with their associated x-rays and files
- Keep files that correspond to specimens in numerical order
- When analysis is completed
- Return x-ray back to x-ray file
- Deliver specimens and file to next station

DNA RUNNER

- Place specimens in container labeled "DNA - TO BE DONE"
- Verify that specimens have been examined by all Sections before bringing to DNA for review.

SPECIAL WALK-THROUGH PROTOCOL

[AGENCY] determines when a specimen becomes evidence. Whenever a Section processing remains feels there is a significant clue has been found, [AGENCY] is notified immediately to make the final determination. Once [AGENCY] advises DMORT that a particular specimen is to be treated as evidence, a Runner is assigned to walk the specimen(s) through any remaining stations. The Runner observes the following procedures:

- Stays within view of the specimen(s) while they are processed or transfers temporary custody of them if required to leave the specimens(s) for any reason.

- Ensures that the [AGENCY] and appropriate DMORT staff signs the Specimen Removal Form AFTER all Sections have completed their analysis

- Copies the specimen file, including the Specimen Removal Form, for DMORT records

- Hand delivers the copied file to the Records Room Supervisor for data entering and permanent filing.

MORGUE SPECIMEN IN-PROCESSING AND SANITATION PROTOCOL

SPECIMEN IN-PROCESSING

Remains are delivered in biohazard bags from the disaster site. These are weighed and placed in refrigerator truck #1.

As the Triage Station is ready, bags of remains are brought in at a gradual rate so that materials do not sit out in ambient temperature any longer than necessary.

After the specimens are processed through all Sections, they are stored in sequential morgue number in refrigerator truck #2.

For Quality Assurance, the Common Tissue remains are x-rayed and reviewed by the pathology station. Specimens identified as important may be pulled from the bags and returned to Triage for re-assessment. Once Quality Assurance has been completed, the Common Tissue remains are bio-sealed in larger bags. These bags are numbered and dated, and a log is maintained. Once labeled, the bags are placed in the refrigerator truck #2.

SANITATION PROTOCOL

Temporary sinks are periodically checked for spillage and overflow of drainage.

At the cessation of each day's morgue operations, the following sanitation measures are taken:

- All tissue specimens, whether processed or unprocessed, are returned to refrigerator trucks. These are sorted into containers labeled for the Section that they were awaiting examination so that they can be returned to those Sections the next day.

- All biohazard materials are collected and sealed for pickup. New biohazard containers are prepared and placed within the morgue.
- All sinks, processing surfaces, and processing areas are disinfected with bleach or other disinfectants.
- All body fluid and processing waste buckets empty and properly disposed of. The buckets are then treated with bleach or other disinfectant.
- Floor areas are cleared and mopped with disinfectant
- Refrigerator trucks are checked and locked.
- All morgue entrances are secured.

TEMPORARY SPECIMEN REMOVAL PROTOCOL

Use the Temporary Removal Form when [AGENGY] needs to examine a specimen temporarily.

[AGENCY] should be written in red marker in the Comments section of the Victim Processing Record and signed by the [AGENYC]

Keep the form in your section until the [AGENCY] returns the specimen and signs it back in.

TEMPORARY REMOVAL FORM

Specimen # _____

Date:_____

Time:_____

Checked out from Station _____

By _____

Returned date:_____

Returned time:_____

By:_____

PERMANENT SPECIMEN REMOVAL PROTOCOL

Use the Permanent Removal Form when [AGENCY] needs to permanently take custody of a specimen for analysis.

[AGENCY] should be written in red marker in the Comments section of the Victim Processing Record and signed by the [AGENCY]

A copy of the paperwork will be made to travel with the specimen.

PERMANENT REMOVAL FORM

Specimen # _____

Date:_____

Time:_____

Checked out from Station _____

By ([AGENCY] AGENT)_____

Morgue Representative Signature _____

LIBRARY PROTOCOL

GENERAL

All antemortem data, except dental, are entered at the Family Assistance Center (FAC). Information being generated at FAC will be merged into the computer at IRC (Information Resource Center).

All postmortem folders are data entered and filed in IRC. All charts must be signed in and out.

All antemortem records (other than dental and medical) are to be filed in IRC and must be signed in and out.

ANTEMORTEM RECORDS - DENTAL AND MEDICAL

All dental and medical records are logged in at the IRC. Faxed records are received by [AGENCY] and recorded as received by the IRC. The records are held at IRC until retrieved by the Antemortem Section Leader.

Antemortem dental records are entered in on the Antemortem Log and placed in the Unprocessed File Folder in the Antemortem Records File.

Antemortem dental records are charted according to the established Dental Protocol. Completion of charting is entered on the Antemortem Log. The records are then are placed in the box labeled "To Be Entered in WinID."

After data entry, the records are filed numerically in the Antemortem File. Completion of the data entry is entered on the Antemortem Log.

POSTMORTEM RECORDS - DENTAL

All postmortem records are hand-carried from the Dental Section to the Dental ID office where they are entered onto the Postmortem Log and placed in the box labeled "To Be Entered in WinID."

After data entry, the records are filed numerically in the postmortem file. Completion of the data entry is entered on the Postmortem Log.

All records must be signed in and out of central filing.

FILE QA/QC PROTOCOL

Unnecessary and/or blank forms are removed from the files.

Cross-check that specimens sampled for DNA analysis as indicated on the Victim Processing Record are listed on the master listed compiled by the DNA Section.

The files are reviewed to assure that there are no discrepancies, inconsistencies, or omissions. If a problem is found the following procedures are observed:

- If a Section has not signed the Victim Processing Record, the file is sent to the appropriate section for signature.

- If a Section has not processed the specimen, a blue sheet for the appropriate Section is labeled with morgue number. The file with the blue sheet and the specimen are sent to the appropriate section(s) for analysis.

- If discrepancies are found between between the Sections analyses, a blue sheet is filled out indicating the nature of the discrepancies. The file is routed to the appropriate section(s) for re-analysis and problem resolution.

- If an inconsistency is noted between the Scientific Sections (i.e., Pathology, Anthropology) and Radiology on the identification of the specimen, the Radiology form is annotated with the identification provided by the scientific section. This annotation is initialed and dated by the File QA team.

- If an inconsistency is noted within a Section report, a blue sheet is filled out noting the inconsistency and the file and/or specimen is routed to the appropriate Section for resolution.

- If a specimen has already been positively identified prior to the QA assessment and not all Sections have completed their analyses, the specimen is considered fully processed. The specimen is not re-routed for further analysis. The remaining sections are crossed off of the Victim Processing Record and initialed by the QA team.

A list is maintained of all files that are re-routed. As the files are returned to the QA team, they are crossed off the list if they pass the remaining QA standards.

Once files have been assessed for all of the above criteria and passed all quality checks, a blue "Q" is written on the lower right corner of the Victim Processing Record and marked off a master list indicating that the file has passed QA standards. The files are then sent to IRC for copies to be made.

Once copies of the files have been made, any additions or changes must be made on orange (not blue) paper. These orange sheets will be copied and filed in each copy of the files.

Endnotes

[1] http://www.collegeofparamedics.org/con_ed/2001/mci.htm

[2] http://www.dmort.org/news/August2004.htm

Adapted from: Campbell (2000). Basic Trauma Life Support (4th Edition), pg. 343.

NINE

FEDERAL RESPONSE PLANNING (FRP)

9

Overview

As stated earlier in this text, the political culture of the United States is one that deliberately separates and respects the authority of municipalities, states, and the Federal government. While these divisions are extremely important to maintain and respect, and while this system has worked extremely well for over two centuries, it does pose certain challenges that need to be addressed through very specific legal plans. For the Federal government to intervene in emergency response at a state or local levels, certain causes and conditions must be identified and satisfied. In this chapter we will study the National Response Plan from the U.S. Department of Homeland Security as a case study in how Federal resources and personnel may be legally activated to assist in a community-wide emergency anywhere within the United States.

National Response Plan[1]

The National Response Plan establishes a comprehensive all-hazards approach to enhance the ability of the United States to manage domestic incidents. The plan incorporates best practices and procedures from incident management disciplines—homeland security, emergency management, law enforcement, firefighting, public works, public health, responder and recovery worker health and safety, emergency medical services, and the private sector—and integrates them into a unified structure. It forms the basis of how the federal government coordinates with state, local, and tribal governments and the private sector during incidents. It establishes protocols to help

- Save lives and protect the health and safety of the public, responders, and recovery workers;
- Ensure security of the homeland;
- Prevent an imminent incident, including acts of terrorism, from occurring;
- Protect and restore critical infrastructure and key resources;

- Conduct law enforcement investigations to resolve the incident, apprehend the perpetrators, and collect and preserve evidence for prosecution and/or attribution;

- Protect property and mitigate damages and impacts to individuals, communities, and the environment; and

- Facilitate recovery of individuals, families, businesses, governments, and the environment.

Concept of Operations[2]

A. Mission

The overall Lead Federal Agency, in conjunction with the lead agencies for crisis and consequence management response, and State and local authorities where appropriate, will notify, activate, deploy and employ Federal resources in response to a threat or act of terrorism. Operations will be conducted in accordance with statutory authorities and applicable plans and procedures, as modified by the policy guidelines established in PDD-39 and PDD-62. The overall LFA will continue operations until the crisis is resolved. Operations under the CONPLAN will then stand down, while operations under other Federal plans may continue to assist State and local governments with recovery.

B. Command and Control

Command and control of a terrorist threat or incident is a critical function that demands aunified framework for the preparation and execution of plans and orders. Emergency response organizations at all levels of government may manage command and control activities somewhat differently depending on the organization's history, the complexity of the crisis, and their capabilities and resources. Management of Federal, State and local response actions must, therefore, reflect an inherent flexibility in order to effectively address the entire spectrum of capabilities and resources across the United States. The resulting challenge is to integrate the different types of management systems and approaches utilized by all levels of government into a comprehensive and unified response to meet the unique needs and requirements of each incident.

1. Consequence Management

State and local consequence management organizations are generally structured to respond to an incident scene using a modular, functionally-oriented ICS that can be tailored to the kind, size and management needs of the incident. ICS is employed to organize and unify multiple disciplines with multi-jurisdictional responsibilities on-scene under one functional organization. State and local emergency operations plans generally establish direction and control procedures for their agencies' response to disaster situations. The organization's staff is built from a "top-down" approach with responsibility and authority placed initially with an Incident Commander who determines which local resources will be deployed. In many States, State law or local jurisdiction ordinances will identify by organizational position the person(s) that will be responsible for serving as the incident commander. In most cases, the incident commander will come from the State or local organization that has primary responsibility for managing the emergency situation.

When the magnitude of a crisis exceeds the capabilities and resources of the local incident commander or multiple jurisdictions become involved in order to resolve the crisis situation, the ICS command function can readily evolve into a Unified Command.... Under Unified Command, a multi-agency command post is established incorporating officials from agencies with jurisdictional responsibility at the incident scene. Multiple agency resources and personnel will then be integrated into the ICS as the single overall response management structure at the incident scene.

Multi-agency coordination to provide resources to support on-scene operations in complex or multiple incidents is the responsibility of emergency management. In the emergency management system, requests for resources are filled at the lowest possible level of government. Requests that exceed available capabilities are progressively forwarded until filled, from a local Emergency Operations Center (EOC), to a State EOC, to Federal operations centers at the regional or national level.

State assistance may be provided to local governments in responding to a terrorist threat or recovering from the consequences of a terrorist incident as in any natural or man-made disaster. The governor, by State law, is the chief executive officer of the State or commonwealth and has full authority to discharge the duties of his office and exercise all powers associated with the operational control of the State's emergency services during a declared emergency. State agencies are responsible for ensuring that essential services and resources are available to the local authorities and Incident Commander when requested. When State assistance is provided, the local government retains overall responsibility for command and control of the emergency operations, except in cases where State or Federal statutes transfer authority to a specific State or Federal agency. State and local governments have primary responsibility for consequence management. FEMA, using the FRP, directs and coordinates all Federal response efforts to manage the consequences in domestic incidents, for which the President has declared, or expressed an intent to declare, an emergency.

2. Crisis Management

As the lead agency for crisis management, the FBI manages a crisis situation from an FBI command post or JOC, bringing the necessary assets to respond and resolve the threat or incident. These activities primarily coordinate the law enforcement actions responding to the cause of the incident with State and local agencies.

During a crisis situation, the FBI Special Agent In Charge (SAC) of the local Field Division will establish a command post to manage the threat based upon a graduated and flexible response. This command post structure generally consists of three functional groups, Command, Operations, and Support, and is designed to accommodate participation of other agencies, as appropriate ....

When the threat or incident exceeds the capabilities and resources of the local FBI Field Division, the SAC can request additional resources from the FBI's Critical Incident Response Group, located at Quantico, VA, to augment existing crisis management capabilities. In a terrorist threat or incident that may involve a WMD, the traditional FBI command post is expanded into a JOC incorporating a fourth functional entity, the Consequence Management Group.

Requests for DOD assistance for crisis management during the incident come from the Attorney General to the Secretary of Defense through the DOD Executive Secretary. Once the Secretary has approved the request, the order will be transmitted either directly to the unit involved or through the Chairman of the Joint Chiefs of Staff.

C. Unification of Federal, State and Local Response

1. Introduction

Throughout the management of the terrorist incident, crisis and consequence management components will operate concurrently.... The concept of operations for a Federal response to a terrorist threat or incident provides for the designation of an LFA to ensure multi-agency coordination and a tailored, time-phased deployment of specialized Federal assets. It is critical that all participating Federal, State, and local agencies interact in a seamless manner.

2. National Level Coordination

The complexity and potential catastrophic consequences of a terrorist event will require application of a multi-agency coordination system at the Federal agency headquarters level. Many critical on-scene decisions may need to be made in consultation with higher authorities. In addition, the transfer of information between the headquarters and field levels is critical to the successful resolution of the crisis incident.

Upon determination of a credible threat, FBI Headquarters (FBIHQ) will activate its Strategic Information and Operations Center (SIOC) to coordinate and manage the national level support to a terrorism incident. At this level, the SIOC will generally mirror the JOC structure operating in the field. The SIOC is staffed by liaison officers from other Federal agencies that are required to provide direct support to the FBI, in accordance with PDD-39. The SIOC performs the critical functions of coordinating the Federal response and facilitating Federal agency headquarters connectivity. Affected Federal agencies will operate headquarters-level emergency operations centers, as necessary.

Upon notification by the FBI of a credible terrorist threat, FEMA may activate its Catastrophic Disaster Response Group. In addition, FEMA will activate the Regional Operations Center and Emergency Support Team, as required.

3. Field Level Coordination

During a terrorist incident, the organizational structure to implement the Federal response at the field level is the JOC. The JOC is established by the FBI under the operational control of the Federal OSC, and acts as the focal point for the strategic management and direction of on-site activities, identification of State and local requirements and priorities, and coordination of the Federal response. The local FBI field office will activate a Crisis Management Team to establish the JOC, which will be in the affected area, possibly collocated with an existing emergency operations facility. Additionally, the JOC will be augmented by outside agencies, including representatives from the DEST (if deployed), who provide interagency technical expertise as well as inter-agency continuity during the transition from an FBI command post structure to the JOC structure.

Similar to the Area Command concept within the ICS, the JOC is established to ensure inter-incident coordination and to organize multiple agencies and jurisdictions within an overall command and coordination structure. The JOC includes the following functional groups: Command, Operations, Admin/Logistics, and Consequence Management.... Representation within the JOC includes officials from local, State and Federal agencies with specific roles in crisis and consequence management.

The Command Group of the JOC is responsible for providing recommendations and advice to the Federal OSC regarding the development and implementation of strategic decisions to resolve the crisis situation and for approving the deployment and employment of resources. In this scope, the members of the Command Group play an important role in ensuring the coordination of Federal crisis and consequence management functions. The Command Group is composed of the FBI Federal OSC and senior officials with decision making authority from local, State, and Federal agencies, as appropriate, based upon the circumstances of the threat or incident. Strategies, tactics and priorities are jointly determined within this group. While the FBI retains authority to make Federal crisis management decisions at all times, operational decisions are made cooperatively to the greatest extent possible. The FBI Federal OSC and the senior FEMA official at the JOC will provide, or obtain from higher authority, an immediate resolution of conflicts in priorities for allocation of critical Federal resources between the crisis and consequence management responses.

A FEMA representative coordinates the actions of the JOC Consequence Management Group, and expedites activation of a Federal consequence management response should it become necessary. FBI and FEMA representatives will screen threat/incident intelligence for the Consequence Management Group. The JOC Consequence Management Group monitors the crisis management response in order to advise on decisions that may have implications for consequence management, and to provide continuity should a Federal consequence management response become necessary.

Should the threat of a terrorist incident become imminent, the JOC Consequence Management Group may forward recommendations to the ROC Director to initiate limited pre deployment of assets under the Stafford Act. Authority to make decisions regarding FRP operations rests with the ROC Director until an FCO is appointed. The senior FEMA official in the JOC ensures appropriate coordination between FRP operations and the JOC Command Group.

4. On-Scene Coordination

Once a WMD incident has occurred (with or without a pre-release crisis period), local government emergency response organizations will respond to the incident scene and appropriate notifications to local, State, and Federal authorities will be made. Control of this incident scene will be established by local response authorities (likely a senior fire or law enforcement official). Command and control of the incident scene is vested with the Incident Commander/Unified Command. Operational control of assets at the scene is retained by the designated officials representing the agency (local, State, or Federal) providing the assets. These officials manage tactical operations at the scene in coordination with the UC as directed by their agency counterparts at field-level operational centers, if used. As mutual aid partners, State and Federal responders arrive to augment the local responders. The incident command structure that was initially established will likely transition into a Unified Command (UC). This UC structure will facilitate both crisis and consequence management activities. The UC structure used at the scene will expand as support units and agency representatives arrive to support crisis and consequence management operations. On-scene consequence management activities will be supported by the local and State EOC, which will be augmented by the ROC or Disaster Field Office, and the Emergency Support Team, as appropriate.

When Federal resources arrive at the scene, they will operate as a Forward Coordinating Team (FCT). The senior FBI representative will join the Unified Command group while the senior FEMA representative will coordinate activity of Federal consequence management liaisons to the Unified Command. On-scene Federal crisis management resources will be organized into a separate FBI Crisis Management Branch within the Operations Section, and an FBI representative will serve as Deputy to the Operations Section Chief. Federal consequence management resources will assist the appropriate ICS function, as directed....

Throughout the incident, the actions and activities of the Unified Command at the incident scene and the Command Group of the JOC will be continuously and completely coordinated.

Endnotes

[1] www.dhs.gov

[2] http://www.fema.gov/rrr/conplan/conpln4p.shtm

TEN

ON-SCENE COMPLICATION PLANNING

10

Overview

"On-Scene Complication" is almost a redundant statement—every scene is, by its very nature, a complication and in terms of planning, it is safe to assume that every crisis scene will produce complications and factors not adhering to initial expectations.

Expecting the unexpected is, basically, the on-going effort of every emergency manager. No matter how solid pre-event planning is, it will invariably be complicated in the real world by uncontrollable factors: equipment malfunctioning, multiple events occurring simultaneously, unexpected weather, and so forth. Incident commanders must be flexible and quick-thinking in their ability to adjust and adapt to scene complications. Emergency managers should have concrete ideas and plans for what additional resources or alternative options could be called upon in the event that a situation suddenly calls for a departure from all previously devised response protocols.

What Can Go Wrong—Will!

The famous axiom of "Murphy's Law" is pretty much a given when discussing a catastrophic event. No matter how thorough an emergency planning team may think they have been, chances are the forces of chaos and catastrophe can devise a way to defeat them. Effective catastrophic event planning has to employ a certain amount of imagination and willingness to participate in "zero-sum game" exercises—collaborative brainstorming to envision the absolute worst thing that nature or an enemy could throw at you and how you would protect yourself from it. This doesn't mean that considerable time and money has to be invested in every fantastic possibility, but it does imply that catastrophic event planning should consider the thing that hasn't happened yet—looking at the trends and extrapolating what it may mean in the future.

Case in point: By the dawn of the new millennium, western security planning was largely focused on worse-case scenarios we knew from the past. Radicalist hijackings in the sixties, seventies and eighties led to metal-detectors and security screeners in airports, looking for passengers carrying guns, and to the creation of hijack policies for airlines that largely dictated negotiation and cooperation with hijacker demands.

Airline bombings led to a new focus on bomb detection and the positive matching of all passengers with their checked luggage.

Because a B-25 bomber lost in bad weather had accidentally crashed into the Empire State Building in 1945, the Twin Towers of Manhattan's World Trade Center had been designed to withstand the impact of a lost aircraft accidentally striking it as well. Based on what had gone wrong in the past, process improvements had been implemented.

Unfortunately, the threat assessment of previous planners had not been updated, nor had it been extended forward to anticipate the new and emerging threats in our world.

What if hijackers used methods other than guns to seize an airplane (e.g., legal chemical mace, small knives, and box cutters)?

What if the hijackers had trained as pilots so that they could fly the aircraft themselves after killing the legitimate flight crew?

What if the hijackers were not interested in negotiation but were on a suicide mission?

What if when the airplane strikes the World Trade Center it is a not a lost 707 — the biggest airline of 1970 when the towers were built — flying slowly in the fog and low-on-fuel as it readies to land, but rather a much larger 767 traveling at 400 mph and loaded with jet fuel?

One-by-one, on the morning of September 11, 2001 our security and emergency response plans were rendered ineffective by the guile and devious creativity of the hijackers. Because the hijackers were willing to consider modern options that emergency planners had not, efforts to respond and rescue were maddeningly frustrated and fruitless.

The Psychology of Command[1]

The psychology of command is beginning to emerge as a distinct research topic for psychologists interested in selection, training, competence assessment, decision making, stress management, leadership and team working. The following overview of recent research into decision making, stress and leadership is based on Flin (1996) which gives a more detailed examination of these issues.

A1.1 Decision Making

The decision making skill of the Incident Commander is one of the essential components of effective command and control in emergency response. Despite the importance of high speed decision making in the fire service and a number of other occupations, it has only been very recently that research psychologists have begun to investigate leaders' decision making in demanding, time-pressured situations.

The traditional decision-making literature from management, statistics and economics is very extensive but it offers little of relevance to the Incident Commander, as it tends to be derived from studies of specified problems (often artificial in nature), inexperienced decision makers and low stake payoffs. Moreover, it is rarely concerned with ambiguous, dynamic situations, life threatening odds, or high time pressure, all important features of a fire or rescue environment.

If we turn to the standard psychological literature on decision-making it tells us almost nothing of emergency decision making, as so much of it is based on undergraduates performing trivial tasks in laboratories. Similarly, the management research is concerned with individuals making strategic decisions when they have several hours or days to think about the options, carefully evaluating each one in turn against their business objectives using decision analysis methods. These provide a range of explanatory frameworks which may have value for managers' decision making where they are encouraged to emulate an analytical style of decision making. At its simplest form this usually incorporates the following stages.

- Identify the problem
- Generate a set of options for solving the problem/ choice alternatives
- Evaluate these options concurrently using one of a number of strategies, such as weighting and comparing the relevant features of the options.
- Choose and implement the preferred option.

In theory, this type of approach should allow you to make the 'best' decision, provided that you have the mental energy, unlimited time and all the relevant information to carry out the decision analysis. This is typically the method of decision making in which managers are trained. But we know from our everyday experience that when we are in a familiar situation, we take many decisions almost automatically on the basis of our experience. We do not consciously generate and evaluate options, we simply know the right thing to do. This may be called intuition or gut feel' but in fact to achieve these judgements some very sophisticated mental activity is taking place. So we can compare these two basic types of decision making, the slower but more analytic comparison and the faster, intuitive judgement. Which style do commanders use when deciding what to do at the scene of an incident?

A1.2 Naturalistic Decision Making (NDM)

In the last ten years there has been a growing interest by applied psychologists into naturalistic decision making (NDM) which takes place in complex real world settings (Klein et al, 1993; Zsambok & Klein, 1997; Flin et al, 1997). These researchers typically study experts' decision making in dynamic environments such as flight decks, military operations, firegrounds, hospital trauma centres/ intensive care units and high hazard industries, for example nuclear plant control rooms. This NDM research has enormous significance for the understanding of how commanders and their teams make decisions at the scene of an incident as it offers descriptions of what expert commanders actually do when taking operational decisions in emergencies.

Ten factors characterize decision making in naturalistic settings:

- Ill defined goals and ill structured tasks.
- Uncertainty, ambiguity and missing data.
- Shifting and competing goals.
- Dynamic and continually changing conditions.
- Action feedback loops (real-time reactions to changed conditions).
- Time stress.
- High Stakes.
- Multiple players (team factors).
- Organizational goals and norms.
- Experienced decision makers.

In typical NDM environments information comes from many sources, is often incomplete, can be ambiguous, and is prone to rapid change. In an emergency, the Incident Commander and her or his team are working in a high stress, high risk, time pressured setting and the lives of those affected by the emergency, (including their own rescue personnel) may be dependent on their decisions.

How then do they decide the correct courses of action? In the view of the NDM researchers, traditional, normative models of decision making which focus on the process of option generation and simultaneous evaluation to choose a course of action do not frequently apply in NDM settings. There are a number of slightly different theoretical approaches within the NDM fraternity to studying decision making but they all share an interest in dynamic high pressure domains where experts are aiming for satisfactory rather than optimal decisions due to time and risk constraints.

A1.3 Recognition-Primed Decision Making (RPD)

Dr Gary Klein of Klein Associates, Ohio, conducts research into decision making by attempting to "get inside the head" of decision makers operating in many different domains. Klein's approach stemmed from his dissatisfaction with the applicability of traditional models of decision making to real life situations, particularly when the decisions could be lifesaving. He was interested in operational environments where experienced decision makers had to determine a course of action under conditions of high stakes, time pressures, dynamic settings, uncertainty, ambiguous information and multiple players.

Klein's research began with a study of urban fireground commanders who had to make decisions such as whether to initiate search and rescue, whether to begin an offensive attack or concentrate on defensive precautions and how to deploy their resources (Klein et al, 1986) They found that the fireground commanders' accounts of their decision making did not fit in to any conventional decision-tree framework.

"The fireground commanders argued that they were not "making choices", "considering alternatives", or "assessing probabilities". They saw themselves as acting and reacting on the basis of prior experience; they were generating, monitoring, and modifying plans to meet the needs of the situations. Rarely did the fireground commanders contrast even two options. We could see no way in which the concept of optimal choice might be applied. Moreover, it appeared that a search for an optimal choice could stall the fireground commanders long enough to lose control of the operation altogether. The fireground commanders were more interested in finding actions that were workable, timely, and cost-effective." (Klein et al, 1993, p139).

During post-incident interviews, they found that the commanders could describe other possible courses of action but they maintained that during the incident they had not spent any time deliberating about the advantages or disadvantages of these different options.

It appeared that these Incident Commanders had concentrated on assessing and classifying the situation in front of them. Once they recognised that they were dealing with a particular type of event, they usually also knew the typical response to tackle it. They would then quickly evaluate the feasibility of that course of action, imagining how they would implement it, to check whether anything important might go wrong. If they envisaged any problems, then the plan might be modified but only if they rejected it, would they consider another strategy.

Klein Associates have also studied other decision makers faced with similar demand characteristics (eg., tank platoon captains, naval warfare commanders, intensive care nurses) and found the same pattern of results. On the basis of these findings they developed a template of this strategy called the Recognition-Primed Decision Model. This describes how experienced decision makers can rapidly decide on the appropriate course of action in a high pressure situation.

...The Recognition Primed Decision Model (Klein,1996)

The model has evolved into three basic formats .

In the simplest version, ... Level 1, the decision maker recognizes the type of situation, knows the appropriate response and implements it.

If the situation is more complex and/ or the decision maker cannot so easily classify the type of problem faced, then ... in Level 2, there may be a more pronounced diagnosis (situation assessment) phase. This can involve a simple feature match where the decision maker thinks of several interpretations of the situation and uses key features to determine which interpretation provides the best match with the available cues. Alternatively, the decision maker may have to combine these features to construct a plausible explanation for the situation, this is called story building, an idea that was derived from legal research into juror decision making. Where the appropriate response is unambiguously associated with the situation assessment it is implemented as indicated in the Level 1 model.

In cases where the decision maker is less sure of the option, then the RPD model, Level 3 version indicates that before an action is implemented there is a brief mental evaluation to check whether there are likely to be any problems. This is called mental simulation or preplaying the course of action (an 'action replay' in reverse) and if it is deemed problematical then an attempt will be made to modify or adapt it before it is rejected. At that point the commander would re-examine the situation to generate a second course of action.

Key features of the RPD model are as follows:

- Focus on situation assessment
- Aim is to satisfice not optimize
- For experienced decision makers, first option is usually workable.
- Serial generation and evaluation of options (action plans)
- Check action plan will work using mental simulation
- Focus on elaborating and improving action plan
- Decision maker is primed to act

To the decision maker, the NDM type strategies (such as RPD) feel like an intuitive response rather than an analytic comparison or rational choice of alternative options. As 'intuition' is defined as, "the power of the mind by which it immediately perceives the truth of things without reasoning or analysis" then this may be an acceptable label for RPD which is rapid situation assessment to achieve pattern recognition and associated recall of a matched action plan from memory.

At present this appears to be one of the best models available to apply to the emergency situation whether the environment is civilian or military; onshore or offshore; aviation, industrial, or medical. In the USA, the RPD model is being widely adopted, it is being used at the National Fire Academy as well as in a number of military, medical, aviation and industrial settings (see Klein, 1998). The RPD model and associated research techniques have begun to generate a degree of interest in the UK, most notably by the Defence Research Agency and the Fire Service.

A1.4 Command roles and decision style

Obviously the RPD approach is not appropriate for all types of operational decisions and other NDM researchers have been developing taxonomies of the different types of decisions other emergency commanders, such as pilots, make in different situations .... The NASA Crew Factors researchers (Orasanu, 1995) have found that two key factors of the initial situation assessment are judgements of time and risk and that these may determine the appropriate decision strategy to use. The issue of dynamic risk analysis is a significant component of situation assessment on the fireground as discussed in Chapter 2 (see also *Fire Engineers Journal*, May, 1998).

If we consider the Orasanu model, the key skill is matching the correct decision style to the demands or allowances of the situation. For example not using the fast intuitive RPD style when there is time to evaluate options. Furthermore senior fire officers in strategic command roles may require special training to discourage them from using the fast RPD approach when a slower, analytical method would be more appropriate (Fredholm, 1997).

There are significant differences in the balance of cognitive skills required of commanders, depending on their role (rather than rank) in a given operation, ascending from operational or task level, to tactical command, to strategic command (Home Office, 1997). From studies of commanders' decision strategies (see Flin, 1996; Flin et al, 1997; Zsambok & Klein, 1997) these roles are briefly outlined below in terms of the decision skills required.

Strategic Command

This involves the overall policy of command and control, deciding the longer term priorities for tactical commanders and planning for contingencies depending on the enemy's response. The task also contains a strong analytical element, as co-ordination of multiple sources of information and resources demands an awareness that cannot be based on procedures alone.

...Decision process model for fixed wing pilots. (Orasanu.1995) Reprinted with permission of the Human Factors and Ergonomics Society

The decision making style assumed to be adopted for strategic decision making is creative or analytical, since the situations encountered will feature a number of novel elements or developments the strategic commander has not previously encountered.

Neither time pressure or high immediate risk should be influencing command at this level, where the aim if possible is to devise an optimal solution for the situation, taking into account the wider and longer term implications. The strategic commander is usually remote from the incident and will be supported throughout by a team of lower ranking officers.

Tactical Command

This refers to the planning and co-ordination of the actions determined at the strategic level.

Due to the higher time pressure at this level, decision making is based to a much greater extent on condition-action matching, or rule- based reasoning. This style is characterised by controlled actions derived from procedures stored in memory. Control of behaviour at this level is goal oriented and structured by feed- forward control' through a stored rule. Stored rules are of the type if (state) then (diagnosis) or if (state) then (remedial action).

The tactical decision maker is likely to be on-scene, with a remit to maintain a good mental model of the evolving plan and unfolding events. Situation assessment is expected to be a more significant component of tactical decision making than spending time choosing appropriate responses. However the tactical commander may have to create' time to engage in reflective thinking and when necessary to use more analytic decision strategies to evaluate alternative courses of action.

Kersholt (1997, p189) found from an interview study with battalion commanders of peace-keeping operations, that, decisions were mostly made analytically in the planning phase and intuitively during the execution of the mission. By analytic procedure we meant that several options were explicitly weighed against each other, whereas an intuitive decision meant that the commander immediately 'knew' which decision to take."

Operational Command

This involves front line or sector commanders who have to implement orders from the tactical level. They are operating in real time and have to react rapidly to situational demands. Decision making at this level is assumed to contain rule-based and intuitive elements. It is assumed that under time pressure and at high risk, they primarily make decisions based on pattern recognition (e.g. RPD) of the situations encountered. Ongoing situation awareness must remain very high as their performance depends on rapid identification of the situation and fast access to stored patterns of pre-programmed responses.

Only when time permits will they be able to engage in analytic decision making and option comparison. Striving to find optimal solutions runs the risk of stalling' their decision making, therefore their main objective is to find a satisfactory, workable course of action.

A1.5 Styles of Command Decision Making

From the above description of decision making techniques associated with particular command roles, there appear to be four main styles of decision making used by commanders: creative, analytical, procedural and intuitive.

The most sophisticated (and resource intensive) is creative problem solving which requires a diagnosis of an unfamiliar situation and the creation of a novel course of action. This is the most demanding of the four techniques, requires significant expertise and as Kersholt (1997) found, is more likely to be used in a planning phase rather than during an actual operation.

...Command decision styles

Analytical decision making also requires a full situation assessment, rigorous information search and then recall, critical comparison and assessment of alternative courses of action. Again with proper preparation, some of these option choices may already have been evaluated during exercises or planning meetings. These are the two most powerful decision techniques as they operate on large information sets, but consequently they require far greater cognitive processing. Thus, they take a longer time to accomplish, and for most individuals can only be used in situations of relative calm and minimal distraction.

In fast moving, high risk situations these styles are difficult if not impossible to use, and in order to maintain command and control, officers have to rely on procedural or intuitive styles which will produce a satisfactory, if not an optimal decision.

Procedural methods involve the identification of the problem faced and the retrieval from memory of the rule or taught method for dealing with this particular situation. Such decision methods (e.g. drills, routines and standard procedures) are frequently practised in training.

With experience, officers may also use the fastest style of decision making, intuitive or recognition-primed decision making described above. In this case there may not be a written rule or procedure but the commander rapidly recognises the type of situation and immediately recalls an appropriate course of action, on the basis of prior experience.

The evidence suggests that commanders use all four decision styles to a greater or lesser degree depending on the event characteristics and resulting task demands. For more senior commanders, distanced from the front line, the task characteristics change in terms of time frame, scale, scope and complexity, necessitating greater use of analytical and creative skills (Fredholm, 1997).

Studies of military and aviation commanders have shown that the following factors are of particular significance in determining decision style :

- available time.
- level of risk.
- situation complexity/familiarity, (or none at all).
- and availability of information.

The training implications of applying this new decision research to fire and rescue operations is first to determine the types of situations where experienced fire commanders use the intuitive RPD type of decision making. In these situations the critical focus will be on situation assessment. So the next stage is to discover the cues these experts use when quickly sizing up an incident and the responses they would choose to apply once they have assessed the situation.

Less experienced commanders need to be trained to recognize these key features or cues of different scenarios using simulated incidents with detailed feedback on their decision making. They need to develop a store of incident memories (from real events, simulator training (eg. VECTOR), case studies, expert accounts) which they can use to drive their search for the critical classifying information at a new incident.

The US Marines who favour the RPD model have developed a very useful volume of 15 decision exercises in *Mastering Tactics: A Tactical Decision Games Workbook* (Schmitt, 1994, see Klein, 1998). These are a series of tactical decision scenarios where a description of a problem is presented and officers are required to quickly work out and explain a solution to the problem which can then be discussed with the team and/or the trainer. This assists officers to learn the critical cues for given types of situations and to store methods of dealing with new situations.

In essence the basis of good command training must be a proper understanding of the decision making processes utilised by effective commanders.

Psychologists can offer a range of research techniques to begin to explore in a more scientific fashion the skills of incident command (eg. Burke, 1997; Flin et al, 1997). For instance, one of the most salient features of a fireground commander's decision task is the speed of fire development. Brehmer (1993) is particularly interested in this type of dynamic decision task, which he believes has four important characteristics: a series of decisions, which are interdependent, a problem which changes autonomously and as a result of the decision maker's actions, and a real time scenario.

He gives the following example, "Consider the decision problems facing a fire chief faced with the task of extinguishing forest fires. He receives information about fires from a spotter plane, and on the basis of this information, he then sends out commands to his firefighting units. These units then report back to him about their activities and locations as well as about the fire, and the fire chief uses this information (and whatever other information he may be able to get, e.g., from a personal visit to the fire and the fire fighting units) to issue new commands until the fire has been extinguished." (p1).

Brehmer and his colleagues have developed a computer programme (FIRE) based on a forest fire scenario which incorporates the four elements of dynamic decision making described above. The decision maker takes the role of the fire chief and using the grid map of the area shown on the computer screen, she or he has to make a series of decisions about the deployment of fire fighting resources with the goal of extinguishing the fire and protecting a control base.

The commander's actions are subject to feedback delays, that is time delay in actions being implemented or in the commander receiving status update information. Brehmer's studies have shown that decision makers frequently do not take such feedback delays into account, for example sending out too few firefighting units because they do not anticipate that the fire will have spread by the time they receive the status report.

He argues that the decision maker needs to have a good 'mental model' of the task in order to control a dynamic event, such as a forest fire, and his research has enabled him to identify several problems of model formation: dealing with complexity, balancing competing goals, feedback delays and taking into account possible side effects of actions. Brehmer (1993) uses control theory to encapsulate the dynamic decision process, "the decision maker must have clear goals, he must be able to ascertain the state of the system that he seeks to control, he must be able to change the system, and he must have a model of the system." (p10).

A1.6 Causes of Stress for Commanders

In fireground operations, stress may also have an impact on commanders' decision making and techniques for managing this need to be considered (see Flin 1996 for further details).

The effects of stress on commanders' thinking and decision making ability are of particular interest. Charlton (1992) who was responsible for the selection of future submarine commanders referred to the 'flight, fight or freeze' response manifested as problems in decision making, 'tunnel vision', misdirected aggression, withdrawl, and the 'butterfly syndrome' "where the individual flits from one aspect of the problem to another, without method solution or priority" (p54). He also mentions self delusion where the student commander denies the existence or magnitude of a problem, regression to more basic skills, and inability to prioritise.

Weiseath (1987) discussing the enhanced cognitive demands for leaders under stress describes reduced concentration, narrowing of perception, fixation, inability to perceive simultaneous problems, distraction, difficulty in prioritising and distorted time perception.

Brehmer (1993) argues that three 'pathologies of decision making' can occur, he calls these:

- thematic vagabonding- when the decision maker shifts from goal to goal,
- encystment - the decision maker focuses on only one goal that appears feasible, and as in i) fails to consider all relevant goals; and
- a refusal to make any decisions.

Not all researchers agree that the decision making of experienced Incident Commanders will be degraded by exposure to acute stressors. Klein (1998) points out that these effects are most typical when analytical decision strategies are used, in contrast, the recognition-primed type of decision strategy employed by experts under pressure may actually be reasonably stress proof.

A1.7 Leadership

Leadership ability is generally deemed to be a key attribute of an Incident Commander and to some extent may be regarded as an umbrella term for the required competencies which have to be trained. However, finding a precise specification of the required behaviours or the style of leadership is rather less frequently articulated.

Leadership within a military context embodies the concepts of command, control, organization and duty and there has been extensive military research into leadership much of which unfortunately, never sees the light of day outside the defence research community.

The dominant model of leadership for training in the British armed services, the emergency services and in lower level management, is Adair's (1988) Action Centred Leadership with its simple three circles model.

Adair developed his ideas from his experience as a British Army officer, and he maintained that the effective leader must focus on the needs of the individual, the task and the team. This functional model has not changed significantly since its initial exposition thirty years ago and continues to be taught in a wide range of management courses. While the three circles diagram and the associated advice to leaders is intuitively appealing, there has been little empirical work to test whether it can actually function as an explanatory theory of leadership in routine managerial duties or emergency command situations.

The managerial research literature on leadership is a progression from a long standing focus on leadership characteristics, to research in the 1960s on leader behaviours (e.g. autocratic vs democratic; team vs task), to an awareness that "one size fits all" recommendations of the best leadership style are unlikely to work. The contingency theories emphasised that leadership style cannot be considered in isolation. Thus, what is effective leadership behaviour is likely to be dependent on the leader's personality and skills, the situation, and the competence and motivation of the group being led. Thus the most effective leader needs to;

- be able to diagnose the situation (the task/problem, the mood, competence, motivation of the team),

- have a range of styles available (e.g. delegative, consultative, coaching, facilitating, directive),
- match her or his style to the situation (for example Hersey and Blanchard's (1988) model of situational leadership).

...Adairs Leadership Model.

In an emergency which has high time pressure and risk, then it is unlikely that a consultative leadership style would be totally appropriate and while the Incident Commander needs to solicit advice from available experts and to listen to the sector commanders, the appropriate style is likely to be closer to directive than democratic.

The need for a perceptible change in leadership style is very obvious when observing simulated emergency exercises when the time pressure and task demands are increased. Moreover, this sends a very important message to the rest of the team that the situation is serious and that they will also have to 'change gear' and sharpen their performance.

Within the business world, the current fashions in leadership style are the delegative, consultative styles, couched in notions of empowerment and transformational leadership. These approaches have not been developed with the Incident Commander in mind, and while it was argued above that a consultative style may be inappropriate, particularly in the opening stages of an incident, this does not mean that there should be no delegation to more junior commanders.

In a larger incident considerable authority has to be devolved to sector commanders who will be required to take critical decisions and who will not always have time or opportunity to seek the opinion of the Incident Commander. These individuals need to have the expertise and the confidence to make decisions as the need arises.

The essential point is that the commander should be comfortable with the style required and that the front-line commanders should have a clear understanding of their delegated authority and the Incident Commander's plan of action.

Finally the Incident Commander does not, and should not work alone. The need for effective team performance on the incident ground remains paramount. Recent advances in team training, known as Crew Resource Management (CRM) have been developed by the aviation industry and are now used in medicine and the energy industry. The focus is on non-technical skills relevant to incident command, such as leadership, situation awareness, decision making, team climate and communication (see Flin, 1995b; Salas et al, in press for further details). Fire officers who have studied this particular type of human factors training have argued that it has clear applications for the fire service (Bonney, 1995, Wynne, 1994).

REFERENCES

Brunacini, A. (1991) Command safety: A wake-up call. *National Fire Protection Association Journal*, January, 74-76.

Burke, E. (1997) Competence in command: Research and development in the London Fire Brigade. In R. Flin, E. Salas, M. Strub & L. Martin (Eds) *Decision Making under Stress*. Aldershot: Ashgate.

Driskell, J. & Salas, E. (1996) (Eds) *Stress and Human Performance*. Mahwah, NJ: LEA.

Flin, R. (1995a) Incident command: Decision making and team work. *Journal of the Fire Service College*, 1, 7-15.

Flin, R. (1995b) Crew Resource Management for teams in the offshore oil industry. *Journal of European Industrial Training*, 19,9, 23-27.

Flin, R. (1996) *Sitting in the Hot Seat: Leaders and Teams for Critical Incident Management*. Chichester: Wiley.

Flin, R., Salas, E., Strub, M. & Martin, L. (1997) (Eds) *Decision Making under Stress: Emerging Themes and Applications*. Aldershot: Ashgate.

Fredholm, L. (1997) Decision making patterns in major fire-fighting and rescue operations. In R. Flin, E. Salas, M. Strub & L. Martin (Eds) *Decision Making under Stress*. Aldershot: Ashgate.

Home Office (1997) *Dealing with Disaster*. Third edition. London: TSO

Klein, G. (1998) *Sources of Power. How People Make Decisions*. Cambridge, Mass: MIT Press.

Klein, G. (1997) The Recognition-Primed Decision (RPD) model: Looking back, looking forward. In C. Zsambok & G. Klein (Eds) *Naturalistic Decision Making*. Mahwah, NJ: Lawrence Erlbaum.

Klein, G., Calderwood, R., & Clinton-Cirocco, A. (1986) Rapid decision making on the fireground. In *Proceedings of the Human Factors Society 30th Annual Meeting*. San Diego: HFS.

Klein, G., Orasanu, J., Calderwood, R. & Zsambok, C. (1993). (Eds.) *Decision Making in Action*. New York: Ablex.

Murray, B. (1994) More guidance needed for senior commanders on the fireground. *Fire*, 87, June, 21-22.

Orasanu, J. & Fischer, U. (1997) Finding decisions in naturalistic environments: The view from the cockpit. In C. Zsambok & G. Klein (Eds) *Naturalistic Decision Making*. Mahwah, NJ: LEA.

Salas, B., Bowers, C & Edens, B. (in press) (eds.) *Applying Resource Management in Organisations*. New Jersey. LEA

Schmitt, J. (1994) *Mastering Tactics. Tactical Decision Game Workbook*. Quantico, Virginia. US Marine Corps Association.

Zsambok, C. & Klein, G. (1997) (Eds) *Naturalistic Decision Making.* Mahwah, NJ: LEA.

Legal Considerations in Command & Control: Legislative Requirements[3]

Whatever public safety endeavor one is involved in today, it is safe to assume that there will be legal considerations attached to one's decision making. The following legislative summation from Great Britain provides an insight into what manner of legal considerations should be evaluated and understood by Incident Commanders.

The responsibility placed on Incident Commanders by section 1 The Fire Services Act 1947 has increased by subsequent legislation or civil law cases since the 1947 Act.

Of major impact has been the introduction of the Health and Safety at Work Act 1974, hereto referred as H.A.S.A.W.A, and subordinate legislation such as the Management of Health and Safety at Work regs 1999.

In order to achieve effective and safe operations during any incident, the Incident Commander requires a knowledge of health and safety responsibilities and fireground systems to meet, so far as is reasonable, legal requirements.

A2.1 Fire Services Act 1947 (as amended)

Section 1

It shall be the duty of every Fire Authority in Great Britain to make provision for firefighting purposes, and in particular every Fire Authority shall secure:

- The services for their area of such a fire brigade and such as may be necessary to meet efficiently all normal requirements.

- The efficient training of the members of the fire brigade.

- Efficient arrangements for dealing with calls for the assistance of the fire brigade in case of fire and for summoning members of the brigade.

- Efficient arrangements for obtaining, by inspection or otherwise, information required for firefighting purposes with respect to the character of the building and other property in the area of the Fire Authority, the available water supplies and the means of access there to and other material local circumstances.

- Efficient arrangements for ensuring that reasonable steps are taken to prevent it mitigate damage to property resulting from measures taken in dealing with fires in the area of the Fire Authority.

- Efficient arrangements for the giving, when requested, of advice in respect of buildings and other property in the area of the Fire Authority as the fire prevention restricting the spread of fires and means of escape in case of fire.

Section 2 - Arrangements for Mutual Assistance

- It shall be the duty of every Fire Authority, so far as practicable to join in the making of schemes (hereafter in this section referred to as reinforcement schemes"). For securing the rendering of mutual assistance for the purpose of dealing with fires occurring in the areas of the Authorities participating in a reinforcement scheme where either:-

- It is necessary to supplement the services provided under the last foregoing section by the Authority in whose area the fire occurs.

- Reinforcements at any fire can be more readily obtained from the resources of other authorities participating in the scheme than from those of the authority in whose area the fire occurs.

- A Fire Authority may enter into arrangements with persons (not being other Fire Authorities) who maintain fire brigades to secure, on such terms as to payment or otherwise as may be provided by or under the arrangements, the provision by those persons of assistance for the purpose of dealing with fire occurring in the area of the Fire Authority where either:

- (a) It is necessary to supplement the services provided by the Authority under the last foregoing section,

- (b) Reinforcements at any fire occurring in the area of the Authority can be more readily obtained from the resources of the said persons than from the resources of the Authority.

Section 3 - Supplementary powers of Fire Authorities

- The powers of a Fire Authority shall include power:

- To employ the fire brigade maintained by them, or use any equipment so maintained, for purposes other than firefighting purposes for which it appears to the Authority to be suitable and, if they think fit, to make such change as they may determine for any services rendered in the course of such employment use.

Section 13 - Duty of Fire Authorities to ensure supply of water for firefighting.

A Fire Authority shall take all reasonable measures for ensuring the provision of an adequate supply of water and for securing that it will be available for use, in case of fire.

Section 30 - Powers of Firemen (Firefighters) and Police in Extinguishing Fires.

- Any member of the fire brigade maintained in pursuance of this Act who is on duty, any member of any other fire brigade who is acting in pursuance of any arrangements made under this Act, or any constable, may enter and if necessary break into any premises or place in which a fire has or is reasonably believed to have broken out, or any premises or place which it is necessary to enter for the purposes of extinguishing a fire or of

- protecting the premises or place from acts done for firefighting purposes, without the consent of the owner or occupier thereof, and may do all such things as he may deem necessary for extinguishing the fire or for protecting from fire, or from acts done as aforesaid, any such premises or place or for rescuing any person or property therein.

- At any fire the senior fire brigade officer present shall have the sole charge and control of all operations for the extinction of the fire, including the fixing of the positions of fire engines and apparatus, the attaching of hose to any water pipes or the use of any water supply, and the selection of the parts of the premises, object or place where the fire is, or of adjoining premises, objects or places, against which the water is to be directed.

- Any water undertakers shall, on being required by any such senior officer as is mentioned in the last preceding subsection to provide a greater supply and pressure of water for extinguishing a fire, take all necessary steps to enable them to comply with such requirement and may for that purpose shut off the water from the mains and pipes in any area; and no authority or person shall be liable to any penalty or claim by reason of the interruption of the supply of water occasioned only by compliance of the water undertakers with such a requirement.

- The senior officer of police present at any fire, or in the absence of any officer of police the senior fire brigade officer present, may close to traffic any street or may stop or regulate the traffic in any street whenever in the opinion that offer it is necessary or desirable to do so for firefighting purposes.

- In this section the expression senior fire brigade officer present", in relation to any fire, means the senior officer present of the fire brigade maintained in pursuance of this Act in the area in which the fire originates, or, if any arrangements or reinforcement scheme made under this Act provided that any other person shall have charge of the operations for the extinction of the fire, that other person.

Section 1 (1)

The Fire services act 1947 (amended) places a general duty on every fire authority, to make provision for fire fighting purposes. Within section 1 (1) (a) the Fire Authority are required to:

- *"Secure the services for their area of such a fire brigade and such equipment as may be necessary to meet efficiently all normal requirement's".*

In defining the term "normal requirements" and its relationship with incident command and legislation there is no statutory definition. However Halsbury's laws of England make the following observation:

- *It is thought that the reference to normal requirements does not imply that a fire brigade has no obligations in respect of abnormal fires. In considering however, the action to be taken in relation to a fire, regard must be had to all the factors, in particular e.g. the likelihood of other calls on the brigade, the danger to life and prop-*

erty and the value of the property concerned. Thus in certain circumstances, e.g. fire in a refuse dump, it might be best to allow the fire to burn itself out".

Whilst this is a general interpretation it clearly identifies that defensive fire fighting may in certain circumstances be a normal requirement at an incident.

A2.2 Health and Safety at Work Act 1974

General duties

The general duties in the 1974 Act relate to all persons at work and the protection of others who might be injured by the activities of persons at work. The aim is to ensure that all persons associated with the workplace, activities, articles and substances take responsibility for health and safety in relation to that workplace.

Employers duties to employees Employers duties to employees

Employers have responsibility for health and safety of employees.

Section 2 (1) imposes a general duty on the employer to ensure, so far as is reasonably practicable, the health, safety and welfare at work of all his employees.

This is the core provision of the Act and is expanded in section 2 (2). The duty is as concerned with human behaviour as with the physical environment

Section 2 (2) duties require safe plant, systems of work, substances, training and supervision, safe place of work, access and the working environment.

Duties of employer to non employees

Section 3 of the act imposes a general duty on employers for protection of persons other than their own employees. The general duty under 3(1) requires the employer;

- *To conduct his undertaking in such a way as to ensure so far as is reasonably practicable, that persons not in his employment who may be affected thereby are not thereby exposed to risks to their health and safety.*

The section places a duty on employers to have regard to safety of members of the public.

Duties of people in control of premises

Section 4 imposes a duty of care upon the controller (occupier) of premises to ensure, so far as is reasonably practicable that the premises, all means of access thereto or egress there from and any plant or substances in the premises, or provided for use there, is safe and without risks to health.

Duties regarding emissions into the atmosphere

Section 5 of the act imposes a general duty on persons in control of certain premises in relation to harmful emissions into the atmosphere. The Environment Act 1995 provides for the control of pollution responsibilities contained in part 1 of the Health and Safety at Work Act 1974 to be transferred to the Environment Agency as from 1st April 1996.

Supply of goods

Section 6 imposes duties on those supplying goods to the workplace.

Employees duty to take reasonable care

Section 7 of the Act imposes a duty upon each individual employee to take reasonable care while at work for the health and safety of him/herself and other persons.

It also imposes a duty for each employee to take regard of any duty or requirement imposed on his employer and to co-operate with the employer so far as is necessary to enable the duty or requirement to be performed or complied with.

Section 8 not intentionally or recklessly to interfere

Imposes a duty on all persons not intentionally or recklessly to interfere with or misuse anything provided in the interests of health, safety and welfare under the relevant statutory provisions.

This section could be invoked against anyone, not just employees.

Liability is for interference with something which has been provided specifically to comply with safety legislation.

Directors liability section 37

Section 37 enables liability to be imposed on senior management where the organisation is in breach of its duties. This section may only be invoked against the most senior management, who represent the alter ego (very brain) of the organisation. The section is particularly relevant to senior managers who have omitted to set up a safe system of work or to curb unsafe behaviour.

A2.3 Water Resources Act 1991

Sections 85-89 cover offences relating to polluting controlled waters.

Section 85 identifies contravention where a person causes or knowingly permits poisonous, noxious, polluting or solid matter to enter any controlled water. This includes trade effluent. The section also highlights contravention where a person permits any matter whatever to enter any inland freshwater so as to tend to impede the proper flow of the waters in a manner leading to, or likely to lead to a substantial aggravation of:

- pollution due to other causes;
- the consequences of such pollution

Section 89 provides defence as follows:

- A person shall not be guilty of an offence under section 85....; if:
- the entry is caused or permitted, or the discharge is made in any emergency in order to avoid danger to life or health;
- That person takes all steps as are reasonably practicable in the circumstances for minimizing the extent of the entry or discharge and of its polluting effects;

- Particulars of the entry or discharge are furnished to the Authority as soon as reasonably practicable after the entry occurs.

A2.4 The Environment Act 1995

Established the Environment Agency who are the authority referred to in the Water Resources Act 1991.

The Environment Agency has legal responsibilities enabling it to take remedial action in respect of controlled waters which have become polluted in an emergency situation, and addressing waste management matters brought about by the emergency situation. The Agency has responsibility relating to defined matters concerning air, land and water which impact the environment. The Agency responsibility does not extend to the effects of smoke on the environment other than when confined in a site regulated under the environment protection act.

It should be noted that in any action for pollution involving operations on a fire service incident ground, the Fire Authority, through the Incident Commander should be able to demonstrate that it took all practicable steps to minimize the pollution and that the balance at the time was one in which an emergency existed which involved danger to health or life which was greater than allowing discharge to occur.

A2.5 Regulations under the HSWA 1974 - Relevant to Incident Command

In 1992 a 'six pack' of Health and Safety regulations were introduced. These regulations made explicit those duties and responsibilities which were implied by the 1974 Act.

The following regulations are relevant to incident command because they are a major shift away from the prior safety legislation which looked at types of workplace or premises. The new regulations concentrate on combating a specific type of risk rather than the premises in which the risk occurred. As Risk assessment and management is a crucial part of incident command relevant factors are considered below.

The Management of Health and Safety Regulations 1992 have now been replaced by the Management of Health and Safety at Work Regulations 1999.

A2.6 Management of Health and Safety at Work Regs 1999

Regulation 3 Risk assessment

Every employer shall make a suitable and sufficient assessment of;

- The risks to the health and safety of his employees to which they are exposed whilst they are at work; and
- the risks to the health and safety of persons not in his employment arising out of or in connection with the conduct by him, of his undertaking.

For the purposes of identifying the measures he needs to take to comply with the requirements and prohibitions imposed upon him by or under the relevant statutory provisions.

Any assessment such as referred to in (a) or (b) above shall be reviewed by the employer if:

- There is reason to suspect that it no longer valid; or
- There has been a significant change in the matters to which it relates; and where as a result of any such review changes to an assessment are required, the employer shall make them.

Regulation 4 - Principles of prevention to be applied

Where an employer implements any preventive and protective measures he shall do so on the basis of the principles specified in Schedule 1 to these Regulations.

Regulation 5 - Health and safety arrangements.

Every employer shall make and give effect to such arrangements as are appropriate, having regard to the nature of his activities and the size of his undertaking, for the effective planning, organisation, control, monitoring and review of the preventative and protective measures.

Regulation 8 - Procedures for serious and imminent danger and for danger areas

Every employer shall;

- Establish and where necessary give effect to appropriate procedures to be followed in the event of serious and imminent danger to persons at work in his undertaking.
- Nominate a sufficient number of competent people to implement those procedures insofar as they relate to the evacuation of the premises of persons at work in his undertaking; and
- Ensure that none of his employees has access to any area occupied by him to which it is necessary to restrict access on the grounds of health and safety unless the employee has received adequate health and safety instruction.

Regulation 10 – Information for Employees

- Every employer shall provide his employees with comprehensible and relevant information on:
- the risk to their health and safety as identified by the assessments
- the preventable and protective measures
- the procedure referred to in regulation 8(1)(a)
- the identity of those persons nominated by him in accordance with regulation 8(1)(b)
- the risk notified to him in accordance with regulation 11(1)(c)

Regulation 11- Co-operation and co-ordination.

Where two or more employers share a workplace each such employer shall ;

- Co-operate with the other so far as is necessary to comply with the requirements and prohibitions imposed on them by or under the relevant statutory provisions

- (Taking into account the nature of his activities) take all reasonable steps to co- ordinate the measures he takes to comply with the requirements and prohibitions imposed upon him by or under the relevant statutory provisions with the measures the other employers concerned are taking to comply with the requirements and prohibitions imposed on them by or under the relevant statutory provisions.

Take all reasonable steps to inform the other employers concerned of the risks to their health and safety arising out of or in connection with the conduct by him of his undertaking

Regulation 12 (1) - Persons working in host employers undertaking.

Every employer and every self-employed person shall ensure that the employer of any employees from an outside undertaking who are working on the undertaking is provided with comprehensible information on;

- The risks to those employees health and safety arising out of or in connection with the undertaking ; and

- The measures taken by the main employer in compliance with the requirements and prohibitions imposed by or under the relevant statutory provisions.

- Every employer shall ensure that any person working in his undertaking who is not his employee is provided with appropriate instructions and comprehensible information regarding any risks to that persons health and safety which arise out of the conduct by that employer.

- Every employer shall also ensure that other employees are aware of emergency arrangements and receive sufficient information to enable them to identify persons nominated to implement evacuation procedures.

Regulation 14-Employees' Duties

Employees have a duty under section 7 of the Health and Safety at Work Act to take reasonable care for their own health and safety and that of others. Therefore, employees must use all work items provided by their employer correctly, in accordance with their training and the instructions they received to use them safety.

Employees should also notify any shortcomings in the health and safety arrangements, even when no apparent risk immediately exists, so the employer can take remedial action if needed.

A2.7 Workplace (Health Safety and Welfare Regs 1992)

Regulation 5(ACOP) Maintenance of workplace, and of equipment, devices and systems.

The workplace, and the equipment and devices mentioned in these regs should be maintained in an efficient state, in efficient working order and in good repair.

If a potentially dangerous defect is discovered, the defect should be rectified immediately or steps taken to protect anyone who may be put at risk, for example by preventing access until the work can be carried out or equipment replaced.

A2.8 Provision and use of work equipment regs 1998

Regulation 4 - Suitability of Work Equipment.

Every employer shall ensure that work equipment is so constructed or adapted to be suitable for the purpose for which it is to be used or provided.

In selecting work equipment, every employer shall have regard to the working conditions and to the risks to health and safety of persons which exist in the premises or undertaking in which that work equipment is to be used and any additional risks posed by the use of that work equipment.

Every employer shall ensure that work equipment is used only for operations for which, and under which, it is suitable

Regulation 7-Specific risks

Where the use of work equipment is likely to involve a specific risk to health and safety, every employer shall ensure that.

The use of that work equipment is restricted to those persons given the task of using it; The use of that work equipment is restricted to those persons given the task of using it;

The ACOP identifies that regulation 5 addresses the safety of work equipment from three aspects;

- its initial integrity
- the place where it will be used
- the purpose for which it will be used

This requires the employer to assess the location and take into account any risks that may arise from the particular circumstances. Employers should also take into account the fact that work equipment itself can sometimes cause risks to health and safety in particular locations which would otherwise be safe, ie generators in enclosed spaces.

Regulation 12-Protection Against Specific Hazards

One specified hazard particularly relevant to the incident ground is material falling from equipment, ie work at height or involving aerial appliances.

A2.9 Personal Protective Equipment at Work Regs 1992

Regulation 4 provision of personal protective equipment. Every employer shall ensure that suitable PPE is provided to his employees who may be exposed to a risk to their health or safety whilst at work except where and to the extent that such risk has been more adequately controlled by other means equally or more effective.

PPE shall not be suitable unless;

- It is appropriate for the risk or risks involved and the conditions at the place where exposure to the risk may occur.
- It takes account of ergonomic requirements and the state of health of the persons who may wear it.

So far as is practicable, it is effective to prevent or adequately control the risk or risks involved without increasing overall risk

Regulation 6-Assessment of PPE

The assessment shall include;

An assessment of any risk or risks to health or safety which have not been avoided by other means

Review of the assessment if there is reason to suspect it is no longer valid or there has been significant change in the matters to which it relates, and where as a result of any such review, changes are required, the employer shall make them.

Other regulations relevant to Incident command

It can be seen that regulations made since the HASAWA 1974, are based on the principles of risk assessment. The requirement for risk control measures such as information, instruction (both guidance and authoritative), training and supervision are also common themes. Therefore this subordinate legislation can provide the Fire Officer with guidance on the hazards and risks associated with specific activities and methods of control. ...

Incident Risk Management[4]

4.1 Health & Safety on the Incident Ground

The principal consideration of the Incident Commander is the safety of all personnel. This must be established by assessing the hazards that are present and the possible risks to the health and safety of those at the scene and adopting appropriate, safe systems of work.

The following summarises the philosophy of the fire service's approach to risk assessment:

- Firefighters will take some risk to save saveable lives;
- Firefighters will take a little risk to save saveable property;
- Firefighters will not take any risk at all to try to save lives or property that are already lost.

Regulation three of the Management of Health and Safety at Work Regulations 1999, requires that brigades carry out suitable and sufficient risk assessments of the risks to which operational personnel are exposed to.

The term "Dynamic Risk Assessment" is used to describe the continuing assessment of risk that is carried out in a rapidly changing environment.

The key elements of any assessment of risk are:

- Identification of the hazards;
- Assessment of the risks associated with the hazards;
- Identification of who is at risk;
- The effective application of measures that control the risk;

When considering what control measures to apply, the Incident and Sector Commanders need to maintain a balance between the safety of personnel and the operational needs of the incident. For example, whereas it may be considered appropriate to commit personnel into a hazardous environment for the purposes of saving life, it may be that purely defensive tactics are employed in a similar situation where there is no threat to life.

The Incident Commander must ensure that safe practices are followed and that, so far as is reasonably practicable under the circumstances, risks are eliminated or, if not, reduced to the minimum commensurate with the needs of the task. However, because personnel may be working in sectors or smaller teams, everyone must be constantly aware of their own safety as well as that of their colleagues and others who may be affected by the incident or work activity. Therefore all personnel should as a matter of course continually risk assess their position.

4.2 Dynamic Risk Assessment

"Dynamic Risk Assessment" is a process of risk assessment carried out in a changing environment, where what is being assessed is developing as the process itself is being undertaken. This is further complicated for the fire service commander in that, often, rescues have to be performed, exposures protected and stop jets placed before a complete appreciation of all material facts has been obtained.

It is nevertheless essential that an effective risk assessment is carried out at any scene of operations. However, in the circumstances of emergency incidents, it was less clear what methodology best served the need. Trials and experience have shown that it is impractical to expect the first arriving Incident Commander, in addition to the incident size-up and initial deployment and supervision of crews, to complete some kind of checklist or form.

Also, it is important that the outcome of a risk assessment is recorded, preferably in a way that is "time stamped" for later retrieval and analysis, such as would be achieved by transmission over the main scheme radio.

Although the dynamic management of risk is continuous throughout the incident, the focus of operational activity will change as the incident evolves. It is, therefore, useful to consider the process during three separate stages of an incident:

- The Initial Stage
- The Development Stage
- The Closing Stage

4.3 Initial Stage of Incident

There are 6 steps to the initial assessment of risk:

- Evaluate the situation, tasks and persons at risk
- Introduce and declare tactical mode
- Select safe systems of work
- Assess the chosen systems of work
- Introduce additional control measures
- Re-assess systems of work and additional control measures

Step 1. Evaluate the situation, tasks and persons at risk

On the arrival of the initial attendance the Incident Commander will need to gather information, evaluate the situation and then apply professional judgement to decide the most appropriate course of action. Hazards must be identified and the risks to firefighters, the public and the environment considered.

- In order to identify hazards the Incident Commander will initially need to consider:

- Operational intelligence information available from risk cards, fire safety plans etc.
- The nature of the tasks to be carried out
- The hazards involved in carrying out the tasks
- The risks involved to: firefighters, other emergency service personnel, the public and the environment
- The resources that are available e.g. experienced personnel, appliances and equipment, specialist advice.

Step 2. Introduce and declare tactical mode

The declaration of a tactical mode, which is the simple expression of whether it is appropriate to proceed to work in a hazard area or not, is a device to enable commanders of dynamic emergency incidents to comply with the principles of risk assessment and be seen to have done so. The detail of the process can be found in section 4.5 of this chapter. However, in simple terms, after a rapid appraisal of the situation the Incident Commander will either be comfortable in announcing 'offensive mode', which is the most usual mode of operation, or if not must announce 'defensive mode' until sufficient additional information has been gathered, control measures taken, etc. and eventually allow 'offensive' to be declared. This approach is commonly known as 'Default to Defensive'. See section 4.5 Tactical Mode)

Step 3. Select safe systems of work

The Incident Commander will then need to review the options available in terms of standard procedures. Incident Commanders will need to consider the possible systems of work and choose the most appropriate for the situation.

The starting point for consideration must be procedures that have been agreed in pre-planning and training and that personnel available at the incident have sufficient competence to carry out the tasks safely.

Step 4. Assess the chosen systems of work

Once a course of action, be it offensive or defensive, has been identified Incident Commanders need to make a judgement as to whether or not the risks involved are proportional to the potential benefits of the outcome. If YES proceed with the tasks after ensuring that:

- Goals, both individual and team are understood.
- Responsibilities have been clearly allocated.

- Safety measures and procedures are understood.

If NO then go back to step 3.

(see section 4.5 Tactical Mode)

Step 5. Introduce additional control measures

Incident commanders will need to eliminate, or reduce, any remaining risks to an acceptable level, if possible, by introducing additional control measures, such as:

- Use of Personnel Protective Equipment e.g. safety glasses, safety harnesses
- Use of BA
- Use of specialist equipment eg. HP, TL
- Use of Safety Officer(s)

Step 6. Re-assess systems of work and additional control measures

Even when safe systems of work are in place, there may well be residual risks. Where these risks remain, the Incident Commander should consider if the benefit gained from carrying out the tasks against the possible consequences if the risks are realised:

- If the benefits outweigh the risks, proceed with the tasks.
- If the risks outweigh the benefit do NOT proceed with the tasks, but consider viable alternatives.

Dynamic Risk Assessment Diagram

4.4 Development Stage of Incident

If an incident develops to the extent that sectors are designated, the Incident Commander will delegate the supervisory role to Sector Commanders. They will be responsible for the health and safety of all personnel within their sector.

Sector Commanders may feel that they can supervise safety within their own sectors. Alternatively, after consideration, the Sector Commander may feel it necessary to nominate a safety officer. This officer will be responsible to the Sector Commander.

As the incident develops changing circumstances may make the original course of action inappropriate, for example:

- Fire fighting tactics may change from defensive to offensive.
- New hazards and their associated risks may arise e.g. the effects of fire on

ON-SCENE COMPLICATION PLANNING 391

building stability.

- Existing hazards may present different risks.
- Personnel may become fatigued.

Both Incident and Sector Commanders, therefore, need to manage safety by constantly monitoring the situation and reviewing the effectiveness of existing control measures.

4.5 The Tactical Mode.

The Tactical Mode procedure assists the Incident Commander to manage an incident effectively without compromising the health and safety of personnel by:

- Ensuring that firefighting operations being carried out by a single crew, or sector, do not have adverse effects on the safety or effectiveness of firefighters in other crews or sectors. (For example, it will ensure that BA wearers inside a building are not subjected to an aerial monitor being opened up above them, or to the impact of a large jet through a window from another sector without warning).
- Generating a record of the outcome of the dynamic risk assessment process conducted by the Incident Commander.

On arrival at an emergency incident where immediate action is required, the Incident Commander will make an immediate judgement about whether it is safe to proceed with normal, offensive operations. Normally with usual procedures and control measure in place it will be, so 'offensive' can be announced. If the Incident Commander feels it is not safe enough, defensive tactics should be used until a suitably safe approach to deal with the incident can be decided upon. It is never acceptable not to be able to announce a tactical mode, if the Incident Commander is unsure, 'defensive' must be announced (ie. Default to defensive). As soon as the Incident Commander is able, a review of the risk assessment should be conducted (see 4.11).

The key to effective use of the Tactical Mode procedure is speed of application. The process is founded on the psychology of naturalistic decision-making, and specifically 'recognition primed decision making'. More details about these theories can be found in Appendix A1, but in application the principles are the same.

There are three Tactical Modes:

Offensive - This mode may apply to a sector, and/or the entire incident.

This is where the operation is being tackled aggressively. The Incident Commander will have established that the potential benefits outweighs the identified risks, so the Incident Commander will be committing crews into a relatively hazardous area, supported by appropriate equipment, procedures and training.

An offensive approach is appropriate when identified risks are managed by additional control measures (Risk Control);

- Elimination
- Isolation
- Substitution
- Control
- The correct level of PPE
- Appointing a "Safety Officer"

Offensive Mode is the normal mode of operation used at, for example, house fires, road traffic accidents and industrial premises to fight the fire, effect rescues, or close down plant, etc.

Examples;

- Committing BA crews to a smoke filled or toxic atmosphere to rescue persons or undertake firefighting action is an offensive action.
- Committing crews to a structural collapse to undertake rescues is an offensive action.
- Committing crews to an RTA rescue is an offensive action.
- Committing a crew to fight a field fire is an offensive action.

Defensive - This mode may apply to a sector, and/or the entire incident.

This is where the operation is being fought with a defensive approach. In defensive mode, the identified risks outweighs the potential benefits, so no matter how many additional control measures are put into place the risks are too great.

In these circumstances the Incident Commander would announce Defensive Mode, fight the fire with ground monitor jets and aerial jets, and protect exposure risks and adjoining property without committing crews into the hazard area.

Examples;

- Withdrawing a crew from a hazardous area because the risk has increased is a defensive action.
- Using jets outside a hazard area is a defensive action.
- Standing by awaiting expert advice, before committing crews is a defensive action.
- Standing by awaiting specialist equipment is a defensive action.
- RTA :- chemical tanker involved, the tanker is leaking a hazardous sub-

stance. No persons reported. Crews are standing by awaiting attendance of a specialist advisor and second tanker for decanting is a defensive action.

Transitional - This mode is never applied to a sector, but only to the whole incident.

Transitional should be declared where there is a combination of Offensive and Defensive modes in operation at the same incident, in two or more sectors. The main purpose of the announcement of 'Transitional Mode' is to keep commanders of sectors operating in defensive mode, using large jets and perhaps aerial monitors, aware that other personnel on the incident ground may be operating in areas of risk, which could be affected by their operations or tactics

An example of when a "Transitional Mode" would be adopted is; where a building fire being fought with the majority of sectors in Defensive Mode, has an annex that can be saved, safely, by using an Offensive Mode ie: by fighting the fire inside the annex. Here there may be, for example, three sectors in Defensive Mode and one in Offensive: the incident would be Transitional.

Before allowing a sector to operate in 'Offensive Mode' at an otherwise 'defensive' incident, which will cause the incident to become transitional, the Incident Commander must be satisfied that the actions of one sector will not adversely affect the safety of crews in any other sector. ...

4.6 The Application of Tactical Mode

A Tactical Mode should be decided upon and announced at all working incidents. As the incident grows and the Incident Commander's span of control increases, it is essential that all personnel are aware of the tactics on the incident ground and the prevailing Tactical Mode.

The first verbal message and further messages to brigade control should include a confirmation of the Tactical Mode for the information of oncoming appliances and officers.

A typical Informative Message might be 'Informative message from ADO Black at Green Street, Anytown; Factory premises, used for textile manufacturing, three floors, 20m x 20m. Ground and first floor well alight, three large jets in use, "WE ARE IN DEFENSIVE 'DELTA' MODE" (for easy of recognition over the radio, it has been found helpful to use the phonetic alphabet to suffix defensive 'Delta', or offensive 'Oscar').

This should then be updated by informing brigade control of which mode the incident is in at frequent intervals.

4.7 Adopting a Tactical Mode when Sectors are in use:

When the incident has been divided into sectors, the Incident Commander will retain responsibility for the Tactical Mode at all times.

There will be occasions when Sector Commanders wish to change the Tactical Mode in their sector from 'offensive' to 'defensive' quickly. For example; they may detect signs of collapse or obtain information about some previously unknown danger. In such circumstances, they must take the necessary action for the safety of the crews and then advise the Incident Commander of the developments.

However, if, the Sector Commander wishes to commit personnel internally in "Offensive Mode" when the prevailing mode is "Defensive", the permission of the Incident Commander must be sought and no change made until it is granted.

The Incident Commander will assess whether the Tactical Mode can change to Offensive in that sector, making the incident mode Transitional. This decision will be based on an understanding of the status of operations in all other sectors.

Sector Commanders must be involved in any intervention by the Incident Commander to change the Tactical Mode. Sector Commanders may then implement the change effectively and ensure that personnel under their command are aware of the prevailing Tactical Mode. However, it is more usual for the initiative to change mode to come from the Sector Commander.

4.8 Responsibilities for determining Tactical Mode.

i. Incident Commander

The Incident Commander should make an assessment of the incident and decide which Tactical Mode will be appropriate.

Any message sent should include which Tactical Mode is in operation at the incident. This should be repeated at regular and frequent intervals up until the time that the 'stop' message is sent and at appropriately regular intervals thereafter.

The Incident Commander should review and confirm the Tactical Mode on initial and all subsequent briefings to Crew and Sector Commanders.

ii. Sector Commanders

Sector Commanders should continually monitor conditions and operational priorities in the sector and ensure that the prevailing Tactical Mode continues to be appropriate.

They must immediately react to adverse changes, withdrawing personnel from risk areas without delay if necessary, and advise the Incident Commander of the change in conditions as soon as possible thereafter.

If the Sector commander considers it is appropriate to change Tactical Mode he/she must **seek the permission of the Incident Commander to do so.** (Before giving this permission, the Incident Commander will determine the status of all other operational sectors to ensure that nothing is in progress, or planned, in the other sectors, which would compromise the safety of personnel committed internally.)

It is appropriate to consider appointing a sector safety officer or officers, either for specific areas of concern (e.g. structure stability, dangerous terrain, etc) or for general support. Such safety officers report direct to the Sector Commander, even if a 'Safety Sector' has been designated, but must liaise with members of Safety Sector at every opportunity.

It is essential to update the tactical mode to the crews working in the sector at a suitably frequent interval.

iii. Crew Commanders

All Crew Commanders should continually monitor conditions in the risk area and draw the attention of the Sector Commander to significant developments, also react immediately to adverse changes and withdraw crew members from the risk area without delay where necessary.

iv. Safety Sector (if operating)

A safety sector may be established:

- To survey operational sectors, identifying hazards, and advise the Sector Commander as appropriate.
- To liaise with sector safety officers, if appointed, to support and exchange information.
- To confirm the validity of the initial risk assessment and record as appropriate.
- To act as an extra set of eyes and ears to the Sector Commanders in monitoring the safety of personnel.

4.9 Summary of the Procedure

- Arrive at incident
- Evaluate situation
- Carry out dynamic risk assessment and announce tactical mode
- Communicate tactical mode
- Commence operations
- Review tactical mode

There are only three Tactical Modes - **Offensive, Defensive**, or, **Transitional**.

Sectors can only be **Offensive** or **Defensive**.

The incident can be **Offensive, Defensive** or, if combinations of these two are in use, it will be **Transitional.**

The Incident Commander must adopt a Tactical Mode when operations are in progress.

When a Tactical Mode has been decided, the Incident Commander must ensure that everyone on the incident ground is aware of it.

Confirmation of the prevailing Tactical Mode must be maintained between Sector and Crew Commanders throughout the incident.

4.10 Examples of Application of Tactical Mode

Example 1
3-pump house fire. Ground floor well alight, persons reported, believed to be in a first floor bedroom.
Large jet to work through a front window to knock down the fire on the ground floor.

2 BA teams committed from the rear door up stairs to search the first floor. Incident is not sectorised

The incident is in Offensive Mode.

Later ...
Fire on the ground floor has been knocked down. BA team with hose reel enter ground floor to continue fire fighting.

The incident is in Offensive Mode.

Example 2
2-pump RTA persons trapped. Crews are working on the vehicles to effect rescues. Incident is not sectorised.

The incident is in Offensive Mode.

Example 3
2-pump grass fire railway embankment involved. Any firefighting operations being conducted are from a safe distance. Crews standing by awaiting confirmation of caution passed to rail operator. No personnel have been committed to the embankment. No other operations are under way.
Incident is not sectorised.

The incident is in Defensive Mode.

Later...
Caution has been confirmed and lookouts are in place. Crews are working on the embankment.

The incident is in Offensive Mode.

Example 4
2-pump RTA chemical tanker involved, the tanker is leaking a hazardous substance. No persons reported.

Road closed. Crews are standing by awaiting attendance of a specialist advisor and second tanker for decanting.

Incident is in Defensive Mode.

Later...
A crew has been committed in chemical protection suits to prevent the substance entering a drain. No operations at the crash scene.

The incident is in Offensive Mode.

Example 5
5 pump retail unit fire in a covered shopping mall. The retail unit is heavily involved in fire, all persons are accounted for. Smoke is issuing from the front of the unit into the shopping mall but is being contained and vented from a large atrium roof space. The smoke level is several metres above the mall floor and is stable. Operations in the mall are taking place in relatively fresh air and within easy reach of final exits. The back of the unit is outside the mall. Smoke is issuing from the unit's roof and from an open loading bay.

Crews are at work inside the mall with jets into the front of the retail unit. Crews are at work at the rear of the unit with jets through the loading bay. No crews have made an entry to the retail unit.

The incident is in Defensive Mode.

As a general guide in these circumstances, if conditions within a large building allow a sector or incident commander and associated staff to work within the building, then the risk assessments should be made on the basis of specific areas or compartments within the building rather than the whole building. Commanders and support staff should always work from an area of relative safety, so only crews committed beyond that area into a more hazardous environment could be considered as being committed offensively

5 pump retail unit fire in a covered shopping mall.

This is very similar to the principle of using a 'bridgehead' two floor below the fire floor of a multi-storey building for rigging and committing breathing apparatus wearers.

Example 6
Fire in multi-occupancy, single story range of premises. Crews in sector 1 are fighting a severe fire in a storage unit with two large jets and an aerial monitor, therefore in defensive mode. Crews in sector 2 and 4 (sector 3 is not in use) are conducting salvage operations in adjoining retail units using BA and are in offensive mode.

The incident is in Transitional Mode.

4.11 Risk Assessment Phase

Having carried out the dynamic risk assessment and established a tactical mode, the Incident Commander will be aware of the immediate hazards, the people at risk and the control methods necessary to protect those people.

Due to the changing nature of the environment at an incident, the Incident

Commander must ensure that as soon as resources permit, a more analytical form of risk assessment is carried out and, when necessary, new control measures implemented whenever the hazard or degree of risk demands it.

The outcome of the review of the risk assessment will either confirm that the dynamic risk assessment and chosen tactical mode was correct, or will result in a change of mode with the appropriate announcements and action occurring without delay.

At incidents that do not require sectorisation, responsibility for the completion of the analytical review of the risk assessment lies with the Incident Commander or any assistant if delegated.

At incidents that do require sectorisation, the responsibility for the risk assessment may be delegated to the Sector Commanders, who may in turn delegate to a deputy, as each sector is required to be assessed.

The review of the risk assessment should be completed at every incident as soon as practicable and reviewed when necessary. Any form used should be amended accordingly at the time of each review.

For incidents where a formal debrief may take place, the results of the risk assessment should be submitted to the Incident Commander for use at the debrief.

1. Analytical Risk Assessment Procedure

The following provides details of the procedure that is applied in West Yorkshire Fire Service.

For the purposes of this procedure,' Analytical Risk Assessment' includes the following elements:

- A formalised assessment of the hazards, who or what is at risk from those hazards, the likelihood and severity of risk.

- An assessment of existing control measures with additional control measures introduced as appropriate.

- Confirmation that the dynamic risk assessment and tactical mode was correct.

... Risk Management is a tool used on the incident ground. However it also can be used to feed relevant information from the incident ground, via the incident debrief, back into the risk assessment process at the systematic level, Thereby confirming or amending the brigades 'Generic Risk Assessment' or the 'Standard Operational Procedures'.

All pumping appliance carries a command support pack, containing a set of standard aid-memoirs and analytical risk assessment form. This pack is used to assist in the completion of the process.

2. Guide to completing the ' Conformation of Dynamic Risk Assessment' form

- Commence each risk assessment form for the incident by completing standard information;
- Address
- Is the assessment for an incident or for a sector
- Date and time
- Current Tactical Mode
- Use a hazard-spotting chart to identify the hazards. Treat each hazard separately; decide who is at risk and list the control measures in place.
- Use the eight-point grid to decide the SEVERITY and the LIKELIHOOD associated to each hazard.
- Multiply the severity and likelihood scores together and enter the total to calculate the risk rating; eg. tolerable, moderate, high, very high.
- Enter the total and the risk rating.
- Decide if existing control measures are adequate.
- If the answer is NO, list the control measures needed to reduce the risk.

When completed the form must be taken to the Command Unit for the Incident Commander to approve and put any recommendation in place, or to the Sector Commander if one is in place for approval and action prior to submitting to the Incident Commander.

Risk Assessment Process

Analytical Risk Assessment Form

4.12 Safety Responsibilities of Personnel at Incidents

All personnel on the incident ground MUST wear the personal protective equipment that has been provided.

This standard may only be varied by the Incident Commander, having considered the health and safety of all personnel and having taken all reasonable and practicable steps to minimise risks.

All personnel must be educated and trained in the procedures to be used at operational incidents and must be alert to the ever- changing environment at the scene of operations and the consequences of exposure to hazardous substances.

All personnel must be certain that they clearly understand the tasks that they are required to perform and must follow the instructions of the officer responsible for their area of work.

4.13 Safe Systems of Work - General

Operational procedures and practices are designed to promote safe operating systems (safe systems of work). To minimise the risk of injury Incident/Sector Commanders must ensure that recognised safe systems of work are being used so far as is reasonable and practicable.

Where possible, operational crews should work together in teams. Whenever practicable the teams should be made up of people who are familiar with each other and have trained together.

When necessary, safety briefings must be carried out and, as the incident develops, or where the risks of injury increase, those briefings must be more precise and appropriate precautions deployed.

4.13 The Closing Stage of Incident

The two key activities involved in the closing stages of an incident are:

Maintaining Control

Welfare

Incident debrief

Maintaining Control

The process of task and hazard identification, assessment of risk, planning, organisation, control, monitoring and review of the preventive and protect measures must continue until the last appliance leaves the incident ground.

There are usually fewer reasons for accepting risks at this stage, because there are fewer benefits to be gained from the tasks being carried out. Incident and Sector Commanders should therefore have no hesitation in halting work in order to maintain safety.

As the urgency of the situation diminishes, the Incident Commander may wish to nominate an officer to gather information for the post incident review. Whenever possible, this officer should debrief crews before they leave the incident, whilst events are still fresh in their minds.

Details of all near misses' i.e. occurrences that could have caused injury but did not in this instance, must be recorded because experience has shown that there are many near misses for every accident that causes harm. If, therefore, we fail to eradicate the causes of a near miss, we will probably fail to prevent injury or damage in the near future.

Welfare

The welfare of personnel is an important consideration. It must be given particular attention by the command team at arduous incidents or incidents that require a rapid turnover of personnel. The physical condition of crews must be continually monitored by supervisors.

Welfare includes provision of rest and feeding which should, where possible, be outside the immediate incident area and always away from any risk of direct or indirect contamination.

Incident Debrief

Following an incident any significant information gained, or lessons learned, must be fed back into the policy and procedures of the brigade. Relating to existing operational intelligence information, personal protective equipment, the provision and use of equipment, other systems of work, instruction, training, and levels of safety supervision etc.

It is important to highlight any unconventional system or procedure used which was successful or made the working environment safe.

It is equally important to highlight all equipment, systems or procedures which did NOT work satisfactorily, or which made the working environment unsafe **(see section 5.4)**

4.14 Summary of the Safety Function

- Identify safety issues.
- Initiate corrective action.
- Maintain safe systems of work.
- Ensure all personnel are wearing appropriate personal protection equipment.
- Observe the environment.
- Monitor physical condition of personnel.
- Regularly review.
- Sector Commanders or nominated Safety Officers should update the Incident Commander of any changing circumstances.

Case Study: ValuJet- May 11, 1996[5]

On Saturday, May, 11, 1996, Valujet flight 592 leaving Miami International Airport en route for Atlanta, Georgia crashed into the Florida Everglades shortly after take off. All 110 passengers and crew aboard were killed, making this one of the worst air disasters in South Florida's history. Our citizens had already experienced the destruction and aftermath left by Hurricane Andrew on August 24, 1992, and our community had survived the rebuilding efforts after several civil disturbances; so disaster was not something new to Miami-Dade County. Never the less the process and the impact of Valujet flight 592 recovery operations shall long be remembered and forever acknowledged, as an extraordinary undertaking that brought out the very best in the men and women who took part in the operation.

Aircraft Information

The aircraft was a McDonald Douglas DC-9-32 powered by 2 Pratt and Whitney JT8D-9A engines, configured with 113 passenger seats. Wing span is 93 feet 5 inches. The length if the plane is 119 feet 3 1/2 inches. Estimated maximum cruising speed was 564 mph. The craft was built in 1969. The tail registration read N904VJ.

The Operation Begins

As you will recognize, the handling of a disaster is similar to handling a major crime scene. The first priority in handling a major disaster is to identify, isolate, and secure the site as quickly as possible. The location and terrain are important in formulating a plan of operation. The location of the Valujet site allowed us the ability to easily restrict the private sector from the site.

We were in a helicopter over the site shortly after the incident, while the fire department was initiating their search and rescue operations we were taking photographs to start developing a plan of action. Initially the first flights over the area were difficult for us in locating the actual site. During the entire efforts of an operations different agencies will have different responsibilities and duties during ever-changing stages in the operations.

Terrain

The terrain is an area that is a dense marshland thick with razor-toothed sawgrass. The sawgrass grows nearly 12 feet tall, has serrated edges, and is rooted in a rich soil known as muck. The muck forms the bed of the everglades.

May is the height of the dry season for this area. The water table for most of its areas is 1 to 3 feet deep. During the summer rainy season starting in June the water can become as high as 5 feet in most of these areas.

The muck has been built from thousands of years by decaying vegetation. It lies over a porous lime rock known as Miami oolite. The muck is a sponge that makes the everglades possible. It keeps the roots and plants moist even in severe droughts, making it impossible to drain.

Evaluating the site

After establishing and securing the scene, evaluating the site is the next step in the process. The intensity of the scene will be dependent on the environment, location, known hazards, and conditions that might be present.

Flight 592 crashed at a site between 2-man made levees approximately 10 miles north of the Tamiami Trail (s.w. 8th street). The site was approximately 300 yards southeast of the L67 levee. The site was the size of approximately 2 football fields side by side. At the northwest edge of the site was a large black crater, which was the point of impact. The size of the crater was approximately 130 feet long by 40 wide. The crater was filled with jet fuel and hydraulic fluids used to power the plane and created zero visibility. It was due to the eerie feeling of the crater that the workers labeled this area, "the black hole".

Initial Start Up of Operations

The steps to the initial start up stage in a Disaster are to asses the scene/situation, identify the resources available to you, develop a command structure, and put your systems in place.

Some of the factors of selecting and establishing a command post site will be the distance to the scene, ease of access, the ability for security, and the availability for a press area.

What is being taught in critical incident management is a 3 tier perimeter. The outer perimeter is established as a larger border than the actual scene, to keep onlookers and nonessential personal from inhibiting the work area, an inner perimeter allowing for a command post and comfort area just outside the site, and the core or site itself.

Some of the leadership roles that were seen in place:

- A personnel officer to track who is on the site and decide the manpower needed for the operation.

- An equipment and supply officer to obtain and provide officers in the field with the equipment and supplies necessary to safely and expeditiously accomplish the operation.

- A provision officer to insure food, water, and other consumables are obtained and delivered to the workers.

- A facility officer to assign sites to responding units and agencies. Establish feeding and latrine areas. Coordinate and insure trash and waste pick ups.

- A liaison officer to work with other units and agencies as needed.

- An operations officer to coordinate and assign operations.

- A media relations officer to work with the vast amounts of media a large event will bring.

Accessibility to the scene

The first of many decisions of our operation began with transportation to and from the site for the workers. We were presented with several options. We had the long, narrow, bumpy, single roadway which was the L67 levee. The roadway ran south to north from s.w. 8th street in Miami-Dade County to Holiday park located in Broward County. Due to the structure and condition of the roadway it had to be traveled at a snails pace and took approximately 1 hour to 55 minutes.

During the first day's encounter with the site we used the aviation's helicopters for transportation of approximately 35 people. By the 3rd day of operations it was not feasible to use aviation due to our manpower force growing to over 150 people at the site daily.

The logical choice became transporting the crew and equipment by boat to and from the site facility each day. The 25 minute boat ride allowed us to safely and efficiently transport the workers and equipment. All supplies were continued to be air lifted in to the site with a minimal number of trips daily.

Compartmental-vs-Departmental Mobilization

A decision has to be made regarding staffing. A department will still have the daily operations which have to be attended to. Now you are faced with a manpower and staffing delima. Since the site was isolated and did not interfere with daily operations we were able to take only the entities and bureaus that were effectively needed and mobilize those units. For example, in crime scene we have 18 people assigned to the duties of major case incidents, covering 3-8 hour shifts. Our unit went to 2-12 hour work shifts to handle our daily operations. The 3rd remaining shift was reassigned to work the crash site. The crash site hours were regulated from sun up to sun down.

Constructing a city in the swamp

We were fortunate that the man-made levee had 2 large land areas constructed to off load boats into the water. This area was approximately 300 yards northwest of the crash site. This gave us an area to establish and construct an on-site work facility, our city in the swamp. The forward base as it was commonly referred to. This forward base was constructed over a period of several days. We were able to obtain tents from the fire department, medical examiners office, and the school board. These large tents provided us with shelter from the sun, allowed us a work place, and gave us a storage area for our equipment. The temperatures during most of this period was in the 90's, making work conditions practically unbearable.

The second off load area which was approximately 100 yards to the south, was designated and prepared as a work facility for the National Transportation Safety Board (NTSB) and later used by a private salvaging company brought in to salvage the wreckage.

Our facility had a small elevated area in the northwest corner which we were able to designate as a landing pad for the helicopters routinely delivering supplies for our operations. The plan was to make our operations base as comfortable as possible. Some of the obstacles that we were routinely exposed to were the saw grass, the extreme heat, muck, mosquitoes, decaying flesh, the fumes from the hydraulic and jet fuels, and frequent thunder storms.

We set up 2 large main tents with walls and doors. A couple of generators and portable air conditioners were brought in to power and cool the large tents. One of the large tents became our command center and housed our communications network, several computers, and equipment needed for our work.

The second large tent served as our mess hall and one section of it was cordoned off with cots for anyone needing rest from the fatigue and heat exhaustion brought on by the efforts.

Revisions in plans are a constant. In the middle of the second full week of operations we had quite a few heavy rain showers. Our levee quickly turned into standing puddles of water and muck. The fire department swiftly brought in wood and constructed platforms inside the tents to give our supplies and equipment a dry clean surface to sit on.

Safe Solid Structure

The Everglades is notorious for severe thunder storms which can come without warning. These storms have strong winds, heavy showers, and lightning. We needed a safe solid structure to escape to during these fierce storms. Two (2) large transit buses were brought in to use as a safe retreat during these storms. The buses also provided a cool safe shelter for the workers after suiting up for a mission that became delayed for whatever reason.

Bathroom & Wash Facilities

These types of operations will usually last for a long period of time. Portable latrines, wash stations, and a portable shower (trailer) were brought in for personal hygiene. Make sure that the facilities are separated and designated by gender.

Waste, garbage, & trash

There is a tremendous amount of trash and rubble accumulated daily during an operation of this magnitude. Arrangements will need to be made to empty the waste from the latrines and pick up the trash almost daily. These task have to be arranged to where they do not interfere with the other daily operations.

Hierarchy of authority

During the entire operations there will be different phases or stages to be dealt with. In the initial stage of the operation the fire department will have control and authority with other agencies assisting in the rescue and medical attention of survivors phase. Once all efforts have been exhausted for locating and assisting survivors, the authority is relinquished to the medical examiners office for the next stage involving the recovery of human remains. This will be done by the assistance of local Law Enforcement. The medical examiners office has the ultimate responsibility for identification of the remains and cause of death.

The National Transportation and Safety Board (NTSB) respond to all accidental disaster category events. The NTSB are responsible for the craft, vessel, or vehicle parts. They have the responsibility of answering 2 major questions of the incident. How the incident occurred? What can be done to prevent similar incidents in the future? Surprising to most, is that the NTSB is not a large organization. They are limited in number of staff as well as area offices spread throughout the United States. What they do have is the authority to mobilize and rely on local agencies to be their resources.

Media Attention

There were scores of media on the site within a few hours of this tragic incident. The time consuming and tedious efforts in a recovery operations can be a slow moving and painstaking news event. Human interest events relating to the workers and operational conditions were outlined daily to assist in media relations.

Briefing and sharing information

Each and every morning after sunrise we had a briefing to start the new days work. A safety presentation was repeated everyday during the briefings to allow new arrivals to our teams to be aware of any and all risk that would be encountered. Discussions on our plan of operations and any updated information would be shared. We also were given information by a member of our physiological services team in coping with the feelings and emotions that would be experience from our various exposures during our field operations.

Hazards at a scene

There are generally three types of hazards or exposures that may be encountered at the average response.

- Bio-hazards.
- Chemical hazards.
- Physical hazards.

During a major disaster or incident you can expect a combination of any of the three exposure conditions. There is the risk from the decaying flesh, the risk from the fluids and fuel used to control the craft, and the terrain, wildlife, and various other natural obstacles associated with the environment which we were dealing with.

Safety practices for chemical and bio-hazards are directed for exposure from inhalation, ingestion, and skin, eye, and mucous membrane contact. In hazard responses there are four threat levels of protective clothing.

The levels range from (level 4) the lowest level consisting of a coverall type garment with no respirator used for nuisance contamination's, to the highest fully encapsulated suit with a self contained breathing apparatus. Level 1 garments give the highest quality protection needed from exposures for the skin, respiratory system, and eye.

One of our large main tents housed our protective garments and supplies. Folding collapsible tables were set up along with folding metal chairs surrounding the tables. The backs of the chairs were positioned against the table with the bucket portion of the chair turned outward. The tables were stacked and organized (in an assembly line fashion) with our protective clothing. A team of workers were assigned to the tent. Their tasks were to keep the supplies stocked and assist the recovery teams in suiting up for an assigned mission, which we referred to as relays. It took each recovery team member approximately 15 minutes to prepare and suit up for a mission.

The suits are rated from 20 to 45 minutes of operations, depending on environment and temperatures. An important note for those controlling the logistics of these type operations; Order the largest sizes available in all of your protective garments. The need for the work being accomplished necessitates loose fitting garments. The garments can be altered or tailored with duct tape for smaller individuals. There is no way to expand the smaller size suits.

We learned a new term during this operation. "Rehydration" became your closest assets. We were constantly drinking water or energy type drinks. No carbonated type drinks were made available.

Decontamination Area

On all hazardous type scenes a cordoned off area away from all of the other work areas has to be established for a decontamination area. A decontamination area is a designated area established for the removal of dangerous goods from personnel. A three (3) series rinse station is established for this purpose. Our rinse stations were set up with a 1:20 dilution of fresh chlorine bleach and hoses for fresh water pumped in. This "Decon" station was organized with a team of workers assisting in the wash and rinse of the workers as they exited from their relays. The estimated time for the rinse station was approximately 15 minutes per worker. The disposable protective garments are shed and packaged in a clearly labeled biohazard bag for later incineration. The reusable garments are hung over a wooden rack constructed for a drying process. A gas powered pump with a long hose was used to bring fresh water to the decon area for the rinsing process.

Designated Biohazard area

A designated area needs to be established for securing the recovered remains. This cordoned off area will need limited access. When choosing this designated area consideration to other major access points of the forward operations base should be taken into account.

Practical search methods

There are five (5) search pattern techniques used in practical searching procedures for crime scenes. The strip or line, grid, circular or spiral, quadrant or zone, and the radius or wheel patterns. The organized method is usually chosen dependent on the scene and the number of searchers for the area to be searched. With the terrain, environment, and obstacles that we were faced with we chose an abbreviated combination of several of the search patterns. We needed to maintain the integrity of a line of searchers. With the terrain you could be walking ankle deep in muck and without warning a step could be taken sending the searcher chest deep in the muck. We used a long rope with flex cuffs tightened to the rope in two (2) foot intervals. Each diver was responsible to maintain his/her 2 foot section of ground. A long dowel rod with a flag at the top was put into the ground by the person at each end of the search line as the search was begun. At the completion of each search relay the flag process was repeated, for marking the area, to keep track of the areas searched.

Hopefully this article will assist you in some way. Its purpose is to give other geographic areas an idea of what is entailed to meet some of the challenges presented by a disaster or major incident. Is your department prepared? Do they have an organized plan of action in place just in case an unexpected event were to occur?

After completing most crime scenes the investigator or technician can walk away with the pride and sense of accomplishment in knowing that he has done his task to the best of his/her abilities. Each disaster or major incident has helped prepare us for further unexpected incidents. These events carry a roller coaster of emotions for the workers. The best the worker can settle for after completion of these types of incidents, is that he/she can walk away with the knowledge that the accomplishment has been in giving some type of closure to the families of the victim's.

Ethical Issues and Lessons from the SARS Outbreak[6]

The SARS outbreak posed a number of ethical challenges. Decision makers were required to balance individual freedoms against the common good, fear for personal safety against the duty to treat the sick, and economic losses against the need to contain the spread of a deadly disease. Decisions were often made with limited information and under short deadlines.

A working group of the University of Toronto Joint Centre for Bioethics undertook to draw the ethical lessons from the challenges of and responses to SARS in Toronto[2]. The working group identified five general categories of ethical issues arising from the SARS experience:

- Public health versus civil liberties: There are times when the interests of protecting public health override some individual rights, such as the freedom of movement. In public health, this takes its most extreme form with involuntary commitment to quarantine.

- Privacy of information and the public's need to know: While the individual has a right to privacy, the state may temporarily suspend this privacy right in case of serious public health risks, when revealing private medical information would help protect public health.

- Duty of care: Health care professionals have a duty to care for the sick while minimizing the possibility of transmitting diseases to the uninfected. Institutions in turn have a reciprocal duty to support and protect health care workers to help them cope with the situation, and to recognize their contributions.

- The problem of collateral damage: Restrictions on entry to SARS-affected hospitals meant that people were denied medical care, sometimes for severe illnesses. There were also restrictions on visits to patients in SARS-affected hospitals. Decision makers faced duties of equity and proportionality in making decisions that weighed the potential harm from these restrictions against benefits from containment of the spread of SARS through rapid and definitive intervention.

- Global interdependence: SARS underlines the increasing risk of emerging diseases and their rapid spread. It points to a duty to strengthen the global health system in the interests of all nations.

The Joint Centre working group suggests that an ethical framework be developed that would address the five issues noted above, and that would ensure that Canada is better prepared to deal with future health crises involving highly contagious diseases.

Four of these points bear brief elaboration.

Civil Liberties: During both SARS outbreaks, health care practitioners, patients and families were asked to place themselves under ten-day quarantines in their homes in order to reduce the risk of exposure of an infectious disease to the community. Other strategies used during SARS were widespread availability of disposable masks, self-surveillance and work-home quarantine (i.e., limiting contacts to those necessary for duties in the health care setting), and restrictions on assembly of groups. Although the *Health Protection and Promotion Act*[3] gives officials the power to force non-compliant individuals into quarantine, this was used only once during the outbreak.

Applying the principle of reciprocity, society has a duty to provide support and other alternatives to those whose rights have been infringed under quarantine. Intriguingly, after returning from quarantines, some health care practitioners reported feeling disconnected from the current state of the organization[4]. Focus groups with front-line workers also revealed that some in quarantine wished to continue participating in the battle against SARS by contacting patients and families to provide support and answer questions, or by helping with contact tracing.

Privacy: Disease reporting during an outbreak carries the risk of a breach of confidentiality. Boundaries of privacy vary from person to person. Some believe that there is a risk of privacy infringement only if confidentiality is not maintained and a social stigma or loss of employment ensues from the breach. The other view is that a privacy infringement is wrong regardless of whether any harm occurs as a result[5]. In either event, under the ethical value of proportionality, officials must use the least intrusive method to obtain their goal. Legislation such as the *Health Protection and Promotion Act* prohibits the release of personal information except in very specific circumstances where there is a public good to be served or added protection obtained by releasing an individual's name.

During SARS, Toronto Public Health named only two names - that of the deceased index case for Toronto and her deceased son, and this was done with the informed consent of the surviving family members, based on their understanding that this extraordinary step was necessary for the protection of public health. An unknown number of people had attended a funeral visitation at the home of the deceased index case, and public health authorities had no way of contacting these people individually to advise them that they had been exposed and to watch for symptoms and remain in isolation for ten days. Most of the remaining family members were already hospitalized and too ill to provide sufficient detail. Two probable SARS cases identified themselves to Toronto Public Health as a direct result of this announcement. Both were health care workers who could have spread the virus further with disastrous results. These details illustrate the knifeedge on which these decisions rest.

Duty of care: Health care providers constantly weighed serious health risks to themselves and their families against their obligation to care for patients with SARS. A substantial percentage of the probable SARS cases involved front-line providers. Nurses and physicians were at particular risk. Overall, it appears that 168 people or about 40% of those infected were health care workers. The Canadian Medical Association Code of Ethics calls on physicians to "consider first the well-being of the patient[6]," while the Canadian Nurses Association Code of Ethics for nursing stipulates that "nurses must provide care first and foremost toward the health and well-being of the person, family or community in their care[7]." Other health care professions in Canada have considered or adopted similar codes. SARS has taught us, however, that this ethical duty must be balanced by a countervailing duty: not to place others at risk by coming to work while ill and potentially contagious. What remains unclear are the limits to this duty: What is the point at which the duty of care is balanced by a right to refuse dangerous tasks? How is the duty of care modified by the occupational circumstances and professional obligations of different health care workers?

Just as health care practitioners have a duty to care for the sick, health care organizations clearly have a reciprocal duty to support and protect their workers. This meant providing the necessary safety equipment and appropriate education regarding the use of such equipment, providing information on risks and the need for precautionary measures and ensuring a safe working environment. Notwithstanding the enormous efforts that many institutions made with respect to internal communication and safeguards for health care workers, serious tensions arose with respect to occupational health and safety.

Many of these were avoidable, as they arose from directives around N95 masks and fit-testing which were either more stringent or interpreted more stringently, than necessary. Health care organizations did offer a variety of psychological supports to their staff, but many of these measures were instituted after SARS, rather than during the outbreak itself. What also emerged very clearly was that health care workers under siege in an emergency such as SARS greatly valued and deserved strong support from community and political leaders as well as co-workers and administrators.

Collateral effects: The ethical trade-offs posed by the collateral effects of caring for SARS patients were numerous. For example, the Catholic Health Association of Canada noted in its submission the serious impact on many patients, friends, and families from restricted visiting hours. Decision making was particularly challenging in critical care units[8]. The principle of equity required that decision makers balance controlling the spread of the disease on the one hand, and the rights of non-infected patients to access medical care, particularly urgent services on the other. The enormous human toll of the disruption to the system lies just beneath the statistics in Chapter 8. Countervailing this impact is the very real likelihood that the uncontained spread of SARS could have killed thousands. Such trade-offs make it very difficult to apply any ethical Procrustean bed in hindsight to the decisions made. However, an ethical framework of some type may be useful for future decision makers.

To this list the Committee would add two other issues.

First, the Canadian Association of Medical Microbiologists has noted the ethical challenges that arose in undertaking research during the SARS outbreak. Issues arose that cut across individual institutions and agencies, necessitating unprecedented coordination of expedited ethical reviews of research protocols and outbreak investigation proposals.

Second, scientific credit and collaboration also pose ethical challenges during an outbreak. For example, while many academic clinicians were fighting the SARS outbreak in Toronto, research scientists were testing the samples that were flooding the National Microbiology Laboratory in Winnipeg. They collaborated with the British Columbia Centre for Disease Control and genomics experts salaried by the British Columbia Cancer Agency to sequence the Toronto strain of the coronavirus. The University of British Columbia subsequently purchased a full-page advertisement during the outbreak to claim credit for the discovery. We thus had the situation where some academics were fighting a battle for all of Canada against a new infectious agent, and others were consumed with offering scientific advice to bring the outbreak under control, while others capitalized brilliantly on the availability of specimens and data to the benefit of all, winning scientific kudos in the process. How does one apportion a fair distribution of scientific credit in these difficult circumstances? Guidelines are needed to facilitate collaborative research and research publications during infectious disease outbreaks, particularly in a relatively small academic community such as that which exists in Canada.

A related ethical issue that arose from SARS is the seeking of patents on the SARS-associated coronavirus. Researchers in the United States, Canada and Hong Kong[9] have applied for patents on the coronavirus and its gene sequence. The US CDC and the British Columbia Cancer Agency publicly acknowledged taking this course of action to ensure that the virus and the sequence remain in the public domain (it is important to note that the sequences were published in *Science* magazine in early May, 2003)[10]. A news item in the June 20, 2003 edition of *The Lancet* reported that the US National Institute of Allergy and Infectious Diseases is making a SARS genome "chip" available to researchers around the world, free of charge, in an effort to spur research. The "chip" contains the 29,700 DNA base pairs of the SARS coronavirus designed from data from institutes in the US, Canada, and Asia that had sequenced the complete SARS coronavirus genome.

While this is a positive development, the patenting of organisms and genes such as SARS remains an issue and has raised myriad concerns[11,12]. The current patent system in Canada was not designed to address questions of DNA patenting and the commercialization of the human genome. Generally, raw products of nature are not patentable. However, a patent may be granted to the entire process of discovering and isolating, in the laboratory, strings of DNA that were not obvious before, rather than to a gene as it exists in nature. In order to patent a gene, a sequence or other similar material, the inventor must modify or identify the novel genetic sequences. The product of the sequence must be modified and the function in nature must be explained. These matters have been given point in Canada by the narrow decision (5-4) of the Supreme Court, in December 2002, to reject the patent of the Harvard 'Onco-mouse', not because of any primary principled objection to the concept, but because extant Canadian patent legislation did not contemplate such a claim. Patents had previously been granted in Canada for unicellular organisms; thus, there is ample precedent in Canada for patenting the genome of a virus. However, the ramifications of these practices are important, particularly where public funding or public health issues are concerned. This issue falls outside the Committee's mandate, but underscores the continuing uncertainty and concerns from a number of quarters about the patenting of organisms and genes in general. The Committee urges continued vigilance and debate concerning the application of the *Patent Act* and the corresponding frameworks surrounding the patent process to the unique challenges of patenting micro-organisms and other living entities.

9G. Recommendations

In light of the foregoing issues, the Committee recommends that:

- **9.1** The Government of Canada should embark on a time-limited intergovernmental initiative with a view to renewing the legislative framework for disease surveillance and outbreak management in Canada, as well as harmonizing emergency legislation as it bears on public health emergencies.

- **9.2** In the event that a coordinated system of rules for infectious disease surveillance and outbreak management cannot be established by the combined effects of the F/P/T Network for Communicable Disease Control, the Public Health Partnerships Program, and the above-referenced intergovernmental legislative review, the Government of Canada should initiate the drafting of default legislation to set up such a system of rules, clarifying F/P/T interactions as regards public health matters with specific reference to infectious diseases.

- **9.3** As part of Health Canada's legislative renewal process currently underway, the Government of Canada should consider incorporating in legislation a mechanism for dealing with health emergencies which would be activated in lockstep with provincial emergency acts in the event of a pan-Canadian health emergency.

- **9.4** The Government of Canada should launch an urgent and comprehensive review of the application of the *Protection of Information Privacy and Electronic Documents Act* to the health sector, with a view to setting out regulations that would clarify the applicability of this new law to the health sector, and/or creating new privacy legislation specific to health matters.

- **9.5** The Government of Canada should launch a comprehensive review of the treatment of personal health information under the *Privacy Act*, with a view to setting out regulations or legislation specific to the health sector.

- **9.6** The Canadian Agency for Public Health should create a Public Health Ethics Working Group to develop an ethical framework to guide public health systems and health care organizations during emergency public health situations such as infectious disease outbreaks. In addition to the usual ethical issues, the Working Group should develop guidelines for collaboration and co-authorship with fair apportioning of authorship and related credit to academic participants in outbreak investigation and related research, and develop templates for expedited ethics reviews of applied research protocols in the face of outbreaks and similar public health emergencies.

- **9.7** F/P/T departments/ministries of health should facilitate a dialogue with health care workers, their unions/associations, professional regulatory bodies, experts in employment law and ethics, and other pertinent government departments/ministries concerning duties of care toward persons with contagious illnesses and countervailing rights to refuse dangerous duties in health care settings.

Endnotes

[1] http://www.lacity.org/epd/pdf/eomppp/coverpg.pdf.

[2] http://www.upton.ma.us/html/psychology_of_command.html

[3] http://www.upton.ma.us/html/legal_considerations.html

[4] http://www.upton.ma.us/html/risk_management.html

[5] http://www.crime-scene-investigator.net/disaster.html

[6] http://www.hc-sc.gc.ca/english/protection/warnings/sars/learning/EngSe30_ch9.htm

Source: Fire Service Manual - Volume 2 - Incident Command
Crown Copyright 1999. Reproduced by West Yorkshire Fire and Civil Defence
Authority under licence from the Controller of Her Majesty's Stationery Office.

Source: Fire Service Manual - Volume 2 - Incident Command
Crown Copyright 1999. Reproduced by West Yorkshire Fire and Civil Defence
Authority under licence from the Controller of Her Majesty's Stationery Office.

A

NORTHERN VIRGINIA MASS CASUALTY INCIDENT PLAN

July 1, 2003 – June 30, 2004
NORTHERN VIRGINIA MCI PLAN
MASTER TABLE OF CONTENTS

MUTUAL AID OPERATIONS PLAN	SECTION 0
DIRECTIONS FOR USE AND COMMON RESPONSIBILITIES	SECTION 1
COMMUNICATION PROCEDURES	SECTION 2
ICS GLOSSARY	SECTION 3
ORGANIZATIONAL CHARTS	SECTION 4
UNIFIED COMMAND	SECTION 5
AREA COMMAND	SECTION 6
COMPLEX	SECTION 7
COMMAND	SECTION 8
OPERATIONS	SECTION 9
PLANNING	SECTION 10
LOGISTICS	SECTION 11
FINANCE/ADMINISTRATION	SECTION 12
RESOURCE TYPES AND MINIMUM STANDARDS	SECTION 13
IDENTIFICATION OF FUNCTIONAL AREAS AND PERSONNEL	SECTION 14
ACCOUNTABILITY	SECTION 15
EMERGENCY MEDICAL SERVICES	SECTION 16
STRUCTURE/HAZARDS AND SEARCH MARKING SYSTEMS	SECTION 17

HAZARDOUS MATERIALS	SECTION 18
DESIGNATION OF STRUCTURE AND GEOGRAPHICAL AREAS	SECTION 19
APPENDICES	
A – MASS CASUALTY SUPPORT UNIT INVENTORY	A-1
B – NORTHERN VIRGINIA HOSPITAL MOU	B-1
C – MEDCOMM HOSPITAL PROCEDURES	C-1

MUTUAL AID OPERATIONS PLAN

PREFACE	0-2
DISTRIBUTION LIST	0-3
CRITERIA AND PROCEDURE FOR REQUESTING ASSISTANCE	0-4
USE AND DEPLOYMENT OF PERSONNEL	0-5
COMMAND AND CONTROL	0-6
WITHDRAWAL OF ASSISTANCE	0-7
CHANGES TO THE OPERATIONAL PLAN	0-7

PREFACE

It is the intent of the Greater Metropolitan Washington Area Mutual Aid Operation Plan (MAOP) to create an environment where the fullest degree of understanding and cooperation among agencies who assist or require assistance under this plan is exercised.

This plan adopts the Incident Command System (ICS) as promulgated by the National Fire Academy under the Federal Emergency Management Agency's National Emergency Training Center. This provides a proven model that will allow the implementation of an incident management system for the following kinds of operations:

Single jurisdiction / Single-agency involvement,

Single-jurisdiction / Multi-agency involvement,

Multi-jurisdiction / Single-agency involvement, and

Multi-jurisdiction / Multi-agency involvement.

The ICS is designed to be used in response to emergencies caused by fire, floods, earthquakes, hurricanes, tornadoes, tidal waves, riots, hazardous materials, natural, or human-caused incidents, and is also applicable to incidents of peaceful mass gatherings that overwhelm normal resources.

It is the intent of this document to provide guidance rather than direct the operations of responding agencies.

It is therefore the responsibility of all signatory agencies to ensure that their respective local incident management plans used for day-to-day operations encompass all aspects of the ICS in structure and terminology.

This Operational Plan does not prevent any of the parties from entering into cooperative agreements with any other party for mutual cooperation during day-to-day operations; and in fact, all parties are encouraged to enter into such agreements in the interest of providing the most efficient, expedient, and effective public safety service to the general public. The adoption of the ICS ensures cooperative, systematic, congruent growth of incident management when the scope of the emergency grows above the capabilities of such agreements.

It is incumbent upon all signatory agencies to ensure that all personnel affected by this plan receive the training and have the qualifications necessary to perform the functions outlined within.

NORTHERN VIRGINIA JURISDICTIONS & AGENCIES SIGNING PLAN:

1) City of Alexandria

2) Arlington County

3) City of Fairfax

4) Fairfax County

5) Fort Belvoir

6) Loudoun County

7) Metropolitan Washington Airports Authority

8) Prince William County

OTHER METRO REGION JURISDICTIONS & AGENCIES SIGNING PLAN:

1) District of Columbia

2) Frederick County, Maryland

3) Montgomery County, Maryland

4) Prince George's County, Maryland

5) Naval District of Washington

CRITERIA AND PROCEDURE FOR REQUESTING ASSISTANCE

A. Criteria

1. An emergency shall exist or appear imminent.

2. The requesting jurisdiction shall have committed or shall have foreseen the need to commit its resources beyond the ability to sustain normal emergency response demand.

3. A jurisdiction lacks the resources needed to mitigate a specific type of incident.

B. Procedure

1. An official (or designee) of any signatory jurisdiction is authorized to determine the need for additional assistance when an emergency exists or appears imminent.

2. When it is determined by the designated official (or designee) of the affected jurisdiction that emergency assistance is required, they shall communicate this through the respective communications center(s) in accordance with the procedures set forth in Section 2 of the Field Operations Guide, Communication Procedures.

3. Requests for assistance shall include:

    a. The nature and location of the emergency;

    b. The number of personnel requested, whether specialized personnel are needed, and/ or the type of resource/equipment needed; and

    c. The location where the assisting units shall report and any special reporting instructions. Units responding shall follow procedures outlined in the Field Operations Guide as appropriate.

    d. The official receiving the request shall promptly advise the requesting jurisdiction of the extent to which the request has been fulfilled.

    e. No jurisdiction shall send assistance unless requested.

4. Local mutual aid agreements between jurisdictions shall supersede the procedures outlined within and will be followed accordingly.

THE USE AND DEPLOYMENT OF PERSONNEL

A. Use of Personnel

1. Assisting personnel shall be under the ultimate command of the Incident Commander.

2. Whenever possible, assisting personnel shall be deployed as integral units under their own supervisor. If such deployment is not possible, the assisting personnel shall be deployed as members of a team with officers of the requesting jurisdiction. If neither of these procedures is possible, the deployment shall be determined by the Incident Commander.

3. The nature of the incident should determine how the assisting personnel should be utilized.

B. Orders

1. Responsibility for the justification of refusal to obey any order rests with the refusing individual.

2. When any order conflicts with a previously issued order or directive, or with a departmental rule, regulation, or directive of the assisting personnel's agency, the conflict should be respectfully brought to the attention of the issuing officer. Attempts to resolve conflicts between orders should be made, when possible, between the parties involved, by summoning an individual higher in rank than both parties, to resolve the controversy. If the conflict cannot be resolved and the conflicting order is not rescinded, the order shall stand.

3. The responsibility for the conflicting order will rest with the issuing officer, and the assisting personnel shall not be answerable for disobeying any previously issued order, directive, rule, or regulation of the affected jurisdiction or of their own agency.

C. General

1. Evidence and confiscated/recovered property shall be processed in accordance with the established procedures of the controlling jurisdiction.

2. Responsibility for the investigation of any criminal act(s) related to the emergency shall rest with the affected jurisdiction unless state or federal law dictates otherwise. (This does not prevent the use of assisting personnel in conducting, or assisting in the conduct of, the investigation.)

3. Release of all information related to the incident shall be the responsibility of the Incident Commander by means of the assigned Public Information Officer.

4. Disposition of deceased persons shall be governed by the laws of the affected jurisdiction unless other disposition is dictated by state or federal law.

COMMAND AND CONTROL

A. Single-Jurisdictional Incidents

1. Single-jurisdictional incidents are defined as incidents in which the boundary of the operation does not cross legally determined jurisdictional boundaries and in which the determination of command and control is not an issue.

2. The jurisdiction in which the incident occurs shall have command and control authority. If the first responding unit is not from the affected jurisdiction, command and control authority shall be rapidly and expediently transferred to personnel of the affected jurisdiction unless otherwise determined by inter-jurisdictional response agreement. Upon arrival of the appropriate jurisdictional public safety official, the initial public safety individual assuming command and control shall automatically relinquish command of the incident unless otherwise directed by an official of the affected jurisdiction. Relinquishment of command shall be accomplished by:

a. Advising his/her communications network of the name, title, and agency of the appropriate relieving jurisdictional public safety official.

b. Appraising the jurisdictional public safety official of the nature and current status of the disaster, and of all actions taken prior to his/her arrival.

3. Upon assuming command of a disaster from the initial public safety individual, the relieving jurisdictional public safety official shall:

a Assume the role of the Incident Commander.

b. Announce his/her title, agency designation, and command post location.

4. Change of command as a result of the subsequent arrival of a senior jurisdictional public safety official, designated Incident Commander, or geographical change in command post location, shall be accomplished by announcing the name, title, and agency designation of the oncoming Incident Commander. Any subsequent changes of command shall follow the same procedure.

B. Multi-Jurisdictional Incidents

1. Multi-jurisdictional incidents are defined as incidents in which the boundary of the operation crosses legally determined boundaries and in which mitigation of the incident requires the cooperation of affected jurisdictions in order to produce an efficient and favorable outcome.

2. When jurisdictional boundaries are uncertain, two or more jurisdictions are affected, or doubt exists as to the ultimate responsibility for command, the first unit from one of the affected jurisdictions to arrive on the scene shall have command and control authority, with appropriate transfer of command contingent upon the arrival of senior official(s) from that same jurisdiction.

3. The senior or otherwise designated official whose units were first to arrive on the scene shall assume the role of Incident Commander with the ultimate responsibility of command and control authority until relieved or replaced as a result of the subsequent arrival of senior ranking official(s), determination of authority based on incident geographical location, or the decisions of appropriate higher government authority.

4. In multi-jurisdictional incidents, the senior official(s) who are on the scene from each affected jurisdiction shall be included in the Incident Management Team as determined by good practice and/or inter-jurisdictional response agreement(s). The Incident Commander shall head the Incident Management Team (IMT) and shall consult with team members in making decisions.

C. Final Determination of Incident Commander and IMT

1. As soon as practical after the onset of the incident, the appropriate public safety officials of the affected jurisdiction(s) shall confer, using the most

accessible means available, and make final determination of the Incident Commander and the IMT.

2. If issues related to command and control authority cannot be reconciled among chief public safety officials, the principal elected or appointed officials of the affected jurisdictions shall be responsible for their resolution.

WITHDRAWAL OF ASSISTANCE

Whenever possible, the assisting personnel and equipment shall be withdrawn pursuant to the mutual agreement of the requesting and assisting jurisdictions. If agreement is not possible, either the requesting or assisting jurisdiction may unilaterally withdraw the assisting personnel or equipment, after notifying the other(s) of the intended action.

CHANGES TO THE OPERATIONAL PLAN

1. Changes, including additions and deletions to the Operational Plan or any components, must be proposed in writing at a meeting of either the Fire Chiefs Committee or the Police Chiefs Committee of the Metropolitan Washington Council of Governments.

2. A sixty-day (60) period shall be provided for review by all signatories.

3. Each signatory shall transmit his/her agreement or disagreement with the proposed changes by the end of the sixty-day (60) review period.

4. All signatories must agree with the proposed changes in order to implement them.

SECTION 1
DIRECTIONS FOR USE AND COMMON RESPONSIBILITIES

CONTENTS	1-1
DIRECTIONS FOR USE	1-2
COMMON RESPONSIBILITIES	1-2
UNIT LEADER RESPONSIBILITIES	1-3

DIRECTIONS FOR USE

This Field Operations Guide (FOG) is designed for use when the incident exceeds the scope of inter jurisdictional response plans; when responding agencies/units are not operating under standing mutual aid operational agreements; or when the preceding is not applicable and its use is deemed appropriate by the host agency.

This guide has been adopted for use by emergency response agencies within the Metropolitan Washington Council of Governments (COG). All responding personnel should utilize this guide using the following steps:

1) Review Common Responsibilities.

2) Review the table of contents to become familiarized with Field Operations Guide layout.

3) Review section(s) on specific operations (Haz-Mat, EMS, etc..) where/when applicable.

COMMON RESPONSIBILITIES

The following is a checklist applicable to all ICS personnel:

    a. Receive assignment from your agency, including:

        1. Job assignment, e.g.., Strike Team designation, overhead position, etc..

        2. Reporting location.

        3. Reporting time.

        4. Travel instructions.

        5. Any special communications instructions, e.g.., frequency, channel.

    b. Review communication procedures

    c. Upon arrival at the incident, check in at designated location, such as:

        1. Incident Command Post.

        2. Base or Camps.

        3. Staging Areas.

        4. Helibases.

        5. If you are instructed to report directly to a line assignment, check in with the Division/Group Supervisor.

    d. Receive briefing from immediate supervisor.

    e. Acquire work materials.

    f. Organize and brief subordinates.

    g. Use clear text and ICS terminology (no codes) in all radio communications. All radio communications to the Incident Communications Center or Command Post will be addressed: "(Incident Name) Communications" or "(Incident Name) Command."

UNIT LEADER RESPONSIBILITIES

In ICS, a number of the Unit Leaders responsibilities are common to all units in all parts of the organization. Common responsibilities of Unit Leaders are listed below. These will not be repeated in Unit Leader Position Checklists in subsequent sections.

    a. Carry out assigned tactical orders.

    b. Participate in incident planning meetings, as required.

c. Determine current status of unit activities.

d. Confirm dispatch and estimated time of arrival of staff and supplies.

e. Assign specific duties to staff; supervise staff.

f. Develop and implement accountability, safety and security measures for personnel and resources.

g. Supervise demobilization of unit.

h. Provide Supply Unit Leader with a list of supplies to be replenished.

i. Maintain records, including log of unit activities.

SECTION 2
COMMUNICATION PROCEDURES

Washington Metropolitan Public Safety Mutual Aid Radio Network
Beginning in the year 2000, the Washington, D.C. area assembled a radio network, based on Motorola radio technology, allowing seamless interoperability as units provide mutual aid in the following participating jurisdictions:

Washington, D.C.:

Fire and EMS

Arlington County, Virginia:

Fire, EMS and Police

City of Alexandria, Virginia:

Fire, EMS and Police

Fairfax County, Virginia:

Fire, EMS, Police and Sheriffs

City of Fairfax, Virginia:

Fire, EMS and Police

Washington Metropolitan Airports Authority (MWAA):

Dulles International Airport

Fire, EMS, and Police

Reagan National Airport

Fire, EMS, and Police

Prince William County, Virginia	Fire and EMS
Loudoun County, Virginia:	Fire and EMS
Montgomery County, Maryland:	Fire and EMS
Prince George's County, Maryland:	Fire and EMS

All mobile and portable radios in these jurisdictions will have one zone for each of the participating jurisdictions programmed with 16 operational talk-groups. As units respond into adjacent jurisdictions, the host jurisdictions will identify an operational talk-group for all units to communicate. These talk-groups are uniquely identified by each jurisdiction by a number and alpha character.

Units outside the Washington, D.C. area responding to a catastrophic event may initiate communications either by calling the host jurisdiction dispatch center by cell phone, or attempt contact on one of the following call channels: National Public Safety Access Channel (NPSAC) Base 866.0125 MHz; Mobile 821.0125 MHz; private line tone 156.700 MHz; or Fire Mutual Aid Radio System (FMARS) 154.295 MHz.

Host jurisdictions will provide the following to incoming units: Location of the Incident Commander and Staging Area - At the staging area, a liaison with a portable radio will be assigned to each unit to assist with communications to the Incident Commander/Command Post, and direct units to the appropriate impact area to begin operations. The incoming unit commander will retain responsibility for personnel on his/her unit but will provide emergency service at the direction of the Incident Commander.

CITY OF ALEXANDRIA, VIRGINIA:

Emergency Communications Center:

Location: 900 Second Street

Alexandria, Virginia

Phone: 703-838-4660

ARLINGTON COUNTY, VIRGINIA:

Emergency Communications Center:

Location: 1400 North Uhle Street

Arlington, Virginia

Phone: 703-558-2222

FAIRFAX COUNTY, VIRGINIA:

Emergency Communications Center:

Location: 3911 Woodburn Road

Fairfax, Virginia

Phone: 703-691-FIRE (3473)

MONTGOMERY COUNTY, MARYLAND:

Emergency Communications Center:

Location: 120 Maryland Avenue

Rockville, Maryland

Phone: 240-777-0744

PRINCE GEORGE'S COUNTY, MARYLAND:

Emergency Communications Center:

Location: 7911 Anchor Street

Landover, Maryland

Phone: 301-499-8120

WASHINGTON, D.C.:

Emergency Communications Center:

Location: 300 McMillan Drive, NW

Washington, DC

Phone: 202-673-3266/3267

PRINCE WILLIAM COUNTY, VIRGINIA:

Emergency Communications Center:

Location: Three County Complex Court

Prince William, Virginia

Phone: 703-792-6500

METROPOLITAN WASHINGTON AIRPORTS AUTHORITY:

National Airport

Emergency Communications Center:

Phone: 703-417-8250

Dulles Airport

Emergency Communications Center:

Phone: 703-572-2970

LOUDOUN COUNTY, VIRGINIA:

Emergency Communications Center:

Location: 16600 Courage Court

Leesburg, Virginia

Phone: 703-777-0637

SECTION 3

ICS GLOSSARY

This glossary contains definitions of terms frequently used in ICS documentation which are, for the most part, not defined elsewhere in this guide.

Agency Executive or Administrator. Chief executive officer (or designee) of the agency or jurisdiction that has responsibility for the incident.

Agency Representative. Individual assigned to an incident from an assisting or cooperating agency who has been delegated full authority to make decisions on all matters affecting that agency's participation at the incident. Agency representatives report to the Incident Liaison Officer.

Allocated Resources. Resources dispatched to an incident that have not yet checked-in with the Incident Communications Center.

Area Command. Area Command is an expansion of the incident command function, primarily designed to manage a very large incident that has multiple incident management teams assigned. However, an Area Command can be established at any time that incidents are close enough that oversight direction is required among incident management teams to ensure conflicts do not arise.

Assigned Resources. Resources checked-in and assigned work tasks on an incident.

Assistant. Title for subordinates of the Command Staff positions. The title indicates a level of technical capability, qualifications, and responsibility subordinate to the primary positions. Assistants may also be used to supervise unit activities at camps.

Assisting Agency. An agency directly contributing suppression, rescue, support, or service resources to another agency.

Available Resources. Resources assigned to an incident and available for an assignment.

Base. That location at which the primary logistics functions are coordinated and administered. (Incident name or other designator will be added to the term Base.) The Incident Command Post may be collocated with the base. There is only one base per incident.

Branch. That organizational level having functional or geographic responsibility for major parts of incident operations. The Branch level is organizationally between Section and Division/Group in the Operations Section, and between Section and Units in the Logistics Section.

Camp. A geographical site, within the general incident area, separate from the Base, equipped and staffed to provide food, water, and sanitary services to incident personnel.

Chief. ICS title for individuals responsible for command of the functional Sections: Operations, Planning, Logistics and Finance/Administration.

Clear Text. The use of plain English in radio communications transmissions. No Ten Codes, or agency specific codes are used when using Clear Text.

Command. The act of directing, ordering and/or controlling resources by virtue of explicit legal, agency, or delegated authority.

Command Officer. An Officer who is not a part of the staffing of a single resource.

Command Post (CP). That location at which primary Command functions are executed; usually collocated with the Incident Base. Also referred to as the Incident Command Post (ICP).

Command Staff. The Command Staff consists of the Information Officer, Safety Officer, and Liaison Officer, who report directly to the Incident Commander.

Communications Officer. Responsible for the handling of voice and data communications for the Incident Commander.

Communications Unit. Functional Unit within the Service Branch of the Logistics Section. This unit is responsible for the incident communications plan, the installation and repair of communications equipment, and operation of the Incident Communications Center. Also may refer to a vehicle used to provide the major part of an Incident Communications Center.

Company. Any mobile piece of equipment having a minimum complement of personnel as determined by the assisting jurisdiction.

Company Officer/Commander. The individual responsible for command of a Company. This designation is not specific to any particular rank.

Compensation/Claims Unit. Functional Unit within the Finance/Administrative Section. Responsible for financial concerns resulting from injuries or fatalities at an incident.

Complex. A complex is two or more individual incidents located in the same general proximity, which are assigned to a single Incident Commander or Unified Command to facilitate management.

Cooperating Agency. An agency supplying assistance other than direct suppression, rescue, support, orservice functions to the incident control effort (e.g., Red Cross, law enforcement agency, telephone company, etc.).

Coordination. The process of systematically analyzing a situation, developing relevant information, and informing appropriate command authority (for its decision) of viable alternatives for selection of the most effective combination of available resources to meet specific objectives. The coordination process (which can be either intra- or interagency) does not in and of itself involve command dispatch actions. However, personnel responsible for coordination may perform command or dispatch functions within limits as established by specific agency delegations, procedures, legal authority, etc..

Coordination Center. Term used to describe any facility that is used for the coordination of agency or jurisdictional resources in support of one or more incidents.

Cost Sharing Agreements. Agreements between agencies or jurisdictions to share designated costs related to incidents. Cost sharing agreements are normally written but may also be verbal between authorized agency or jurisdictional representatives at the incident.

Cost Unit. Functional Unit within the Finance/Administration Section. Responsible for tracking costs, analyzing cost data, making cost estimates and recommending cost-saving measures.

Crew. A specific number of personnel assembled for an assignment such as search, ventilation, or hose line deployment and operations. A crew operates under the direct supervision of a Crew Leader.

Demobilization Unit. Functional Unit within the Planning Section. Responsible for assuring orderly, safe, efficient demobilization of resources committed to the incident.

Deputy. A fully qualified individual who, in the absence of a superior, could be delegated the authority to manage a functional operation or perform a specific task. In some cases, a Deputy could act as relief for a superior and therefore must be fully qualified in the position. Deputies can be assigned to the Incident Commander, General Staff, and Branch Directors.

Director. ICS title for individuals responsible for command of a Branch.

Disaster. Any event of unusual or severe effect, threatening or causing extensive damage to life and/or property and requiring extraordinary measures to protect lives, meet human needs and achieve recovery. A disaster will demand resources beyond local capabilities and require extensive mutual aid and support needs.

Dispatch. The implementation of a command decision to move a resource(s) from one place to another.

Dispatch Center. A facility from which resources are directly assigned to an incident.

Division. That organization level having responsibility for operations within a defined geographic area or with functional responsibility. The Division level is organizationally between the Strike Team and the Branch. (See also Group.)

Documentation Unit. Functional Unit within the Planning Section. Responsible for recording/protecting all documents relevant to the incident.

Emergency. A condition of disaster or of extreme peril to the safety of persons and property.

Facilities Unit. Functional Unit within the Support Branch of the Logistics Section. Provides fixed facilities for the incident. These facilities may include the Incident Base, feeding areas, sleeping areas, sanitary facilities, and a formal Command Post.

Finance/Administration Section. Responsible for all costs and financial actions of the incident. Includes the Time Unit, Procurement Unit, Compensation/Claims Unit, and the Cost Unit.

Food Unit. Functional Unit within the Service Branch of the Logistics Section. Responsible for providing meals for personnel involved with the incident.

General Staff. The group of incident management personnel composed of: The Incident Commander The Operations Section Chief The Planning Section Chief The Logistics Section Chief The Finance/Administration Section Chief

Ground Support Unit. Functional Unit within the Support Branch of the Logistics Section. Responsible for fueling/maintaining/repairing vehicles and the transportation of personnel and supplies.

Group. That organizational level having responsibility for a specified functional assignment at an incident (ventilation, salvage, water supply, etc.).

Helibase. A location within the general incident area for parking, fueling, maintenance, and loading of helicopters.

Helispot. A designated location where a helicopter can safely take off and land. Also referred to as Landing Zone (LZ).

Incident. An occurrence or event, either human-caused or caused by natural phenomena, that requires action by emergency response personnel to prevent or minimize loss of life or damage to property and/or natural resources.

Incident Action Plan. The Incident Action Plan contains general control objectives reflecting the overall incident strategy, and specific action plans for the given operational period.

Incident Command Post (ICP). That location at which the primary command functions are executed and usually collocated with the incident base.

Incident Command System (ICS). The combination of facilities, equipment, personnel, procedures, and communications operating within a common organizational structure with responsibility for the management of assigned resources to effectively accomplish stated objectives pertaining to an incident.

Incident Commander (IC). The individual responsible for the management of all incident operations.

Incident Objectives. Statements of guidance and direction necessary for the selection of appropriate strategy(ies), and the tactical direction of resources. Incident objectives are based on realistic expectations of what can be accomplished when all allocated resources have been effectively deployed. Incident objectives must be achievable and measurable, yet flexible enough to allow for strategic and tactical alternatives.

Initial Response. Resources initially committed to an incident.

Investigation/Intelligence Officer. Responsible for the investigation and intelligence function concerning the disaster.

Jurisdiction. A single geographical area defined by boundaries with its own elected body and public safety agencies with resources to respond to an incident/disaster.

Jurisdictional Agency. The agency having jurisdiction and responsibility for a specific geographical area.

Landing Zone. A designated location where a helicopter can safely take off and land. Also referred to as Helispot.

Leader. The individual responsible for command of a Task Force, Strike Team, or Functional Unit.

Liaison Officer. The point of contact for assisting or coordinating agencies. Member of the Command Staff.

Logistics Section. Responsible for providing facilities, services and materials for the incident.

Mass Gathering. An assemblage of people of such magnitude to adversely affect the ability of a jurisdiction to provide normal response to emergency incidents.

Mayday. Term used by personnel to signal that they are in immediate danger and in need of assistance.

Medical Unit. Functional Unit within the Service Branch of the Logistics Section. Responsible for providing emergency medical treatment of emergency personnel. This unit does not provide treatment for civilians.

Message Center. The Message Center is part of the Communications Center and is collocated or placed adjacent to it. It receives, records, and routes information about resources reporting to the incident, resource status, and administration and tactical traffic.

Mobilization Center. An off incident location at which emergency service personnel and equipment are temporarily located pending assignment, release, or reassignment.

Officer. ICS title for the Command Staff positions of Safety, Liaison, and Information. Also used when a single individual performs a unit function within Planning, Logistics, or Finance.

Operational Period. The period of time scheduled for execution of a given set of operational actions as specified in the Incident Action Plan.

Operations Section. Responsible for all tactical operations at the incident.

Out-of-Service Resources. Resources assigned to an incident but unable to respond for mechanical, rest, or personnel reasons.

Overhead Personnel. Personnel who are assigned to supervisory positions, which include Incident Commander, Command Staff, General Staff, Directors, Supervisors and Unit Leaders.

Planning Meeting. A meeting, held as needed throughout the duration of an incident, to select specific strategies and tactics for incident control operations and for service and support planning.

Planning Section. Responsible for the collection, evaluation, dissemination and use of information about the development of the incident and the status of resources.

Procurement Unit. A functional unit within the Finance/Administration Section. Responsible for financial matters involving vendors.

Property Control Officer. Responsible for the receipt, documentation, custody and control of personal property and items having evidential value recovered from the disaster site.

Public Information Officer. Responsible for the dissemination of factual and timely reports to the news media and the interface with the media or other appropriate agencies requiring information direct from the incident scene. Member of the Command Staff.

Public Safety Chaplain. Person responsible for the establishment and direction of incident religious support. The Public Safety Chaplain is a functional unit within the Logistics Section.

Public Safety Official. Any public safety individual of appropriate rank or any civilian designated by proper authority.

Rapid Intervention Team. A team consisting of a minimum of one Officer and two firefighters equipped to assist other firefighters in need of rapid rescue.

Recorder. Person assigned to record information. May be utilized by any ICS position having need.

Regional Area. The geographical area encompassed by, but not limited to, the members of the Metropolitan Washington Council of Governments.

Reinforced Response. Those resources requested in addition to the initial response.

Reporting Locations. Any facility(ies)/ location(s) where incident assigned resources may check in. The locations are: Incident Command Post - Resources Unit, Base, Camp, Staging Area, Helibase or Division/ Group Supervisor for direct line assignments. (Check-in at one location only.)

Resources. All personnel and major items of equipment available, or potentially available, for assignment to incident tasks on which status is maintained.

Resource Status Unit (RESTAT). Functional Unit within the Planning Section. Responsible for recording the status and accounting of resources committed to incident. Also responsible for the evaluation of resources currently committed to the incident, the impact that additional responding resources will have on the incident, and anticipated resource needs.

Rehabilitation. Also known as Rehab; rest and treatment of incident personnel who are suffering from the effects of strenuous work and/or extreme conditions.

Safety Officer. Responsible for monitoring and assessing safety hazards, unsafe situations, and developing measures for ensuring personnel safety. Member of the Command Staff.

Section. That organization level having functional responsibility for primary segments of incident operations such as Operations, Planning, Logistics, and Finance/Administration. The Section level is organizationally between Branch and Incident Commander.

Section Chief. Title referring to a member of the General Staff. Service Branch. A Branch within the Logistics Section. Responsible for service activities at the incident. Components include the Communications, Medical, and Food Units.

Single Resource. An individual, a piece of equipment and its personnel complement, or a crew or team of individuals with an identified work supervisor that can be used on an incident.

Situation Status Unit (SITSTAT). Functional Unit within the Planning Section. Responsible for analysis of the situation as it progresses. Reports to the Planning Section Chief.

Staging Area. A location near the incident where incident personnel and equipment are assigned on a three-minute available status.

Strategy. The general plan or direction selected to accomplish incident objectives.

Strike Team. A group (typically five) of the same kind and type of resources, with common communications and a leader.

Supervisor. ICS title for individuals responsible for command of a Division or a Group.

Supply Unit. Functional Unit within the Support Branch of the Logistics Section. Responsible for ordering equipment/supplies required for incident operations.

Support Branch. A Branch within the Logistics Section. Responsible for providing personnel, equipment, and supplies to support incident operations. Components include Supply, Facilities, and Ground Support Units.

Tactics. Deploying and directing resources on an incident to accomplish the objectives designated by strategy.

Task Force. A group of any type and kind of resources, with common communications and a leader, temporarily assembled for a specific mission, not to exceed five (5) resources.

Technical Specialists. Personnel with special skills who are activated only when needed. Technical Specialists may be needed in areas such as water resources, environmental concerns, resource use, training areas, etc.

Time Unit. Functional Unit within the Finance Section. Responsible for record keeping of time for personnel working at an incident.

Unified Command. In ICS, Unified Command is a unified team effort, which allows all agencies with responsibility for the incident, either geographical or functional, to manage an incident by establishing a common set of incident objectives and strategies. This is accomplished without losing or abdicating agency authority, responsibility, or accountability.

Unit. That organization element having functional responsibility for a specific incident planning, logistic, or finance activity.

SECTION 4

ORGANIZATIONAL CHARTS

CONTENTS	4-1
ICS ORGANIZATIONAL HIERARCHY	4-2
ICS ORGANIZATION CHART	4-3
INITIAL RESPONSE ORGANIZATION CHART	4-4
REINFORCED RESPONSE ORGANIZATION CHART	4-5
MULTI-DIVISION ORGANIZATION CHART	4-6
MULTI-BRANCH ORGANIZATION CHART	4-7

INCIDENT COMMAND SYSTEM

ORGANIZATION CHART

4-4

INITIAL RESPONSE ORGANIZATION CHART

Initial response resources are managed by the initial response Incident Commander,

who will perform all Command and General Staff functions.

4-5

REINFORCED RESPONSE ORGANIZATION CHART

STRUCTURE FIRE EXAMPLE

In the extended/reinforced response situation, the Incident Commander continues to directly manage all resources. The IC has now designated a Safety Officer, Staging Area, an Attack Group, a Search Group, and an EMS Group.

4-6

MULTI-DIVISION ORGANIZATION CHART

The Incident Commander has filled several Command and General Staff positions. Some Units in the Planning and Logistics Sections have been established. The Operations Section has established two divisions and an Air Tactical Group Supervisor position.

4-7

MULTI-BRANCH ORGANIZATION CHART

All Command and General Staff positions have been filled as well as many of the Units. The Operations Section has now established a two-Branch organization and an Air Operations organization.

5-1

SECTION 5

UNIFIED COMMAND

CONTENTS	5-1
UNIFIED COMMAND	5-2
IMT FUNCTIONS	5-2
POSITION CHECKLISTS	5-3
IMT GROUP COORDINATOR	5-3
IMT GROUP AGENCY REPRESENTATIVES	5-3
SITUATION ASSESSMENT UNIT	5-4
RESOURCES UNIT	5-4
PUBLIC INFORMATION UNIT	5-4

5-2

UNIFIED COMMAND

A key component of effective incident management is the concept of Unified Command. In ICS, Unified Command is a unified, cooperative, team effort that allows all agencies with responsibility for the incident, either geographical or functional, to manage an incident by establishing a common set of incident objectives and strategies. This is accomplished without losing or abdicating agency authority, responsibility, or accountability. The following are examples of when Unified Command is applied:

- Incidents that impact more than one political jurisdiction.
- Incidents involving multiple agencies (or departments) within the same political jurisdiction.
- Incidents that impact (or involve) several political and functional agencies.

In order to facilitate the process of Unified Command during such incidents, top management personnel from responsible agencies/jurisdictions and those heavily supporting the effort and/or significantly impacted by use of local resources will convene to form an Incident Management Team (IMT). The Incident Management Team is supported by personnel, facilities, equipment, procedures, and communications integrated into a common system with responsibility for coordination of assisting agency resources and support to agency emergency operations.

IMT FUNCTIONS

a. Evaluate new incidents.

b. Prioritize incidents.

   i. Life threatening situation

   ii. Real property threatened

   iii. High damage potential

   iv. Incident complexity

c. Ensure agency resource situation status is current.

d. Determine specific agency resource requirements.

e. Determine agency resources availability (available for out-of-jurisdiction assignment at this time).

f. Determine need and designate regional mobilization centers.

g. Allocate resources to incidents based on priorities.

h. Anticipate future agency/regional resource needs.

i. Communicate IMT decisions back to agencies/incidents.

j. Review policies/agreements for regional resource allocations.

k. Review need for other agency involvement in the IMT.

l. Provide necessary liaison with out-of-region facilities and agencies as appropriate.

POSITION CHECKLISTS

IMT GROUP COORDINATOR

The IMT Group Coordinator serves as a facilitator in organizing and accomplishing the mission, goals and direction of the IMT Group. The Coordinator will:

a. Facilitate the IMT Group decision process by obtaining, developing and displaying situation information.

b. Fill and supervise necessary unit and support positions within the IMT.

c. Acquire and manage facilities and equipment necessary to carry out the IMT Group functions.

d. Implement the decisions made by the IMT Group.

IMT GROUP AGENCY REPRESENTATIVES

The IMT Group is made up of top management personnel from responsible agencies/jurisdictions and those heavily supporting the effort and/or significantly impacted by use of local resources. Agency representatives involved in an IMT Group must be fully authorized to represent their agency. Their functions can include the following:

a. Ensure that current situation and resource status is provided by their agency.
b. Prioritize incidents by an agreed upon set of criteria.
c. Determine specific resource requirements by agency.
d. Determine resource availability for out-of-jurisdiction assignments and the need to provide resources in mobilization centers.
e. As needed, designate area or regional mobilization and demobilization centers within their jurisdictions.
f. Collectively allocate scarce, limited resources to incidents based on priorities.
g. Anticipate and identify future resource needs.
h. Review and coordinate policies, procedures and agreements as necessary.
i. Consider legal/fiscal implications.
j. Review need for participation by other agencies.
k. Provide liaison with out-of-area facilities and agencies as appropriate.
l. Critique and recommend improvements to the IMT and IMT Group operations.
m. Provide personnel cadre and transition to emergency or disaster recovery as necessary.

SITUATION ASSESSMENT UNIT

The Situation Assessment Unit (this is also referred to in some agencies and EOCs as the Intelligence Unit) in an IMT is responsible for the collection and organization of incident status and situation information. The unit evaluates, analyzes and displays information for use by the IMT Group. Functions include the following:

a. Maintain incident situation status including location, type, size, potential for damage, control problems and any other significant information.
b. Maintain information on environmental issues, cultural and historic resources or sensitive populations and areas.
c. Maintain information on meteorological conditions and forecast conditions that may have an effect on incident operations.
d Request/obtain resource status information from the Resources Unit or agency dispatch sources.
e Combine, summarize and display data for all appropriate incidents according to established criteria.

f Collect information on accidents, injuries, deaths and any other significant occurrences.

g Develop projections of future incident activity.

RESOURCES UNIT

The Resources Unit, if activated in an IMT, maintains summary information by agency on critical equipment and personnel committed and available within the IMT area of responsibility. Status is kept on the overall numbers of critical resources rather than on individual units. Functions can include the following:

a Maintain current information on the numbers of personnel and major items of equipment committed and/or available for assignment.

b Identify both essential and excess resources.

c Provide resource summary information to the Situation Assessment Unit as requested.

PUBLIC INFORMATION UNIT

The Public Information Unit is designed to satisfy the need for regional information gathering. The unit will operate an information center to serve the print and broadcast media and other governmental agencies. It will provide summary information from agency/incident information officers and identify local agency sources for additional information to the media and other government agencies. Functions are to:

a. Prepare and release summary information to the news media and participating agencies.

b Assist news media visiting the IMT fatality and provide information on its function. Stress joint agency involvement.

c Assist in scheduling media conferences and briefings. Assist in preparing information materials, etc., when requested by the IMT Group Coordinator.

d Coordinate all matters related to public affairs (VIP tours, etc.).

e Act as escort for facilitated agency tours of incident areas, as appropriate.

SECTION 6

AREA COMMAND

CONTENTS	6-1
AREA COMMAND DESCRIPTION	6-2
POSITION CHECKLISTS	6-3
AREA COMMANDER	6-3
AREA COMMAND PLANNING CHIEF	6-4
AREA COMMAND LOGISTICS CHIEF	6-4

AREA COMMAND

Area Command is an expansion of the incident command function primarily designed to manage a very large incident that has multiple incident management teams assigned. However, an Area Command can be established at any time that incidents are close enough that oversight direction is required among incident management teams to ensure conflicts do not arise.

The functions of Area Command are to coordinate the determination of:

1. Incident objectives
2. Incident strategies
3. Priorities for the use of critical resources allocated to the incident assigned to the Area Command.

The organization is normally small with personnel assigned to Command, Planning and Logistics. Depending on the complexity of the interface between the incidents, specialists in other areas such as aviation may also be assigned to Area Command.

POSITION CHECKLISTS

AREA COMMANDER (SINGLE/UNIFIED AREA COMMAND)

The Area Commander is responsible for the overall direction of incident management teams assigned to the same incident or to incidents in close proximity. This responsibility includes ensuring that conflicts are esolved, incident objectives are established and strategies are selected for the use of critical resources.

Area Command also has the responsibility to coordinate with local, state, federal and volunteer assisting and/ or cooperating organizations.

These actions will generally be conducted in the order listed.

a Obtain briefing from the agency executives(s) on agency expectations, concerns and constraints.

b Obtain and carry out delegation of authority from the agency executive for overall management and direction of the incidents within the designated Area Command.

c If operating as a Unified Area Command, develop working agreement for how Area Commanders will function together.

d Delegate authority to Incident Commanders based on agency expectations, concerns and constraints.

e Establish an Area Command schedule and time line.

f Resolve conflicts between incident realities and agency executive wants.

g Establish appropriate location for the Area Command facilities.

h Determine and implement an appropriate Area Command organization. Keep it manageable.

I Determine need for technical specialists to support Area Command.

j Obtain incident briefing and Incident Action Plans from Incident Commanders (as appropriate).

k Assess incident situations prior to strategy meetings.

l Conduct a joint meeting with all Incident Commanders.

m Review objectives and strategies for each incident.

n Periodically review critical resource needs.

o Maintain a close coordination with the agency executive.

p Establish priority use for critical resources.

q Review procedures for interaction within the Area Command.

r Approve Incident Commanders requests for and release of critical resources.

s Coordinate and approve demobilization plans.

t Maintain log of major actions/decisions.

AREA COMMAND PLANNING CHIEF

The Area Command Planning Chief is responsible for collecting information from incident management teams in order to assess and evaluate potential conflicts in establishing incident objectives, strategies and the priority use of critical resources.

a Obtain briefing from Area Commander.

b Assemble information on individual incident objectives and begin to identify potential conflicts and/or ways for incidents to develop compatible operations.

c Recommend the priorities for allocation of critical resources to incidents.

d Maintain status on critical resource totals (not detailed status).

e Ensure that advance planning beyond the next operational period is being accomplished.

f Prepare and distribute Area Commander's decisions or orders.

g Prepare recommendations for the reassignment of critical resources as they become available.

h Ensure demobilization plans are coordinated between incident management teams and agency dispatchers.

I Schedule strategy meeting with Incident Commanders to conform with their planning processes.

j Prepare Area Command briefings as requested or needed.

k Maintain log of major actions/decisions.

AREA COMMAND LOGISTICS CHIEF

The Area Command Logistics Chief is responsible for providing facilities, services and material at the Area Command level, and for ensuring effective use of critical resources and supplies among the incident management teams.

    a Obtain briefing from the Area Commander.

    b Provide facilities, services and materials for the Area Command organization.

    c Ensure coordinated airspace and temporary flight restrictions are in place and understood.

    d Ensure coordinated communication links are in place.

    e Assist in the preparation of Area Command decisions.

    f Ensure the continued effective and priority use of critical resources among the incident management teams.

    g Maintain log of major actions/decisions.

SECTION 7

COMPLEX

A complex is two or more individual incidents located in the same general geographic proximity that are assigned to a single Incident Commander or Unified Command to facilitate management.

The diagram at the right illustrates a number of incidents in the same general proximity.

Management responsibility for all of these incidents has been assigned to a single incident management team. A single incident may be complex but it is not referred to as a "complex." A complex may be in place with or without the use of Unified and/or Area Command.

A typical organization would be as follows:

SECTION 8

COMMAND

CONTENTS	8-1
ORGANIZATION CHART	8-2
POSITION CHECKLISTS	8-2
INCIDENT COMMANDER	8-2
PUBLIC INFORMATION OFFICER	8-3
LIAISON OFFICER	8-3
AGENCY REPRESENTATIVES	.8-4
SAFETY OFFICER	8-5

ORGANIZATION CHART
POSITION CHECKLISTS
INCIDENT COMMANDER

Te Incident Commander's responsibility is the overall management of the incident. On most incidents the command activity is carried out by a single Incident Commander. The Incident Commander is selected by qualifications and experience.

The Incident Commander may have a deputy, who may be from the same agency or from an assisting agency.

Deputies may also be used at section and branch levels of the ICS organization. Deputies must have the same qualifications as the person for whom they work, as they must be ready to take over that position at any time.

   a Review Common Responsibilities.

   b Assess the situation and/or obtain a briefing from the prior Incident Commander.

   c Determine incident objectives and strategy.

   d Establish the immediate priorities.

   e Establish an Incident Command Post.

   f Establish an appropriate organization.

   g Ensure planning meetings are scheduled as required.

   h Approve and authorize the implementation of an Incident Action Plan.

   I Ensure that adequate safety measures are in place.

   j Coordinate activity for all Command and General Staff.

   k Coordinate with key people and officials.

   l Approve requests for additional resources or for the release of resources.

   m Keep agency administrator informed of incident status.

   n Approve the use of trainees, civilian, and auxiliary personnel.

   o Authorize release of information to the news media.

   p Order the demobilization of the incident when appropriate.

PUBLIC INFORMATION OFFICER

Te Public Information Officer is responsible for developing and releasing information about the incident to the news media, to incident personnel, and to other appropriate agencies and organizations.

Only one Public Information Officer will be assigned for each incident, including incidents operating under Unified Command and multi-jurisdiction incidents. The Public Information Officer may have assistants as necessary, and the assistants may also represent assisting agencies or jurisdictions.

Agencies have different policies and procedures relative to the handling of public information. The following are the major responsibilities of the Public Information Officer that would generally apply on any incident:

    a Review Common Responsibilities.

    b Determine from the Incident Commander if there are any limits on information release.

    c Develop material for use in media briefings.

    d Obtain Incident Commander's approval of media releases.

    e Inform media and conduct media briefings.

    f Arrange for tours and other interviews or briefings that may be required.

    g Obtain media information that may be useful to incident planning.

    h Maintain current information summaries and/or displays on the incident and provide information on status of incident to assigned personnel.

    I Maintain log of unit activity.

LIAISON OFFICER

Incidents that are multi-jurisdictional, or have several agencies involved, may require the establishment of the Liaison Officer position on the Command Staff.

Only one Liaison Officer will be assigned for each incident, including incidents operating under Unified Command and multi-jurisdiction incidents. The Liaison Officer may have assistants as necessary, and the assistants may also represent assisting agencies or jurisdictions.

The Liaison Officer is the contact for the personnel assigned to the incident by assisting or cooperating agencies. These are personnel other than those on direct tactical assignments or those involved in a Unified Command.

    a Review Common Responsibilities.

    b Be a contact point for Agency Representatives.

    c Maintain a list of assisting and cooperating agencies and Agency Representatives.

    d Assist in establishing and coordinating interagency contacts.

    e Keep agencies supporting the incident aware of incident status.

    f Monitor incident operations to identify current or potential inter-organizational problems.

    g Participate in planning meetings, providing current resource status including limitations and capability of assisting agency resources.

    h Maintain log of unit activity.

AGENCY REPRESENTATIVES

In many multi-jurisdiction incidents, an agency or jurisdiction will send a representative to assist in coordination efforts.

An Agency Representative is an individual assigned to an incident from an assisting or cooperating agency who has been delegated authority to make decisions on matters affecting that agency's participation at the incident. Agency Representatives report to the Liaison Officer, or to the Incident Commander in the absence of a Liaison Officer.

a Review Common Responsibilities.

b Ensure that all agency resources are properly checked-in at the incident.

c Obtain briefing from the Liaison Officer or Incident Commander.

d Inform assisting or cooperating agency personnel on the incident that the Agency Representative position for that agency has been filled.

e Attend briefings and planning meetings as required.

f pvide input on the use of agency resources unless resource technical specialists are assigned from the agency.

g Cooperate fully with the Incident Commander and the General Staff on agency involvement at the incident.

h Ensure the well-being of agency personnel assigned to the incident.

I Advise the Liaison Officer of any special agency needs or requirements.

j Report to home agency dispatch or headquarters on a prearranged schedule.

k Ensure that all agency personnel and equipment are properly accounted for and released prior to departure.

l Ensure that all required agency forms, reports and documents are complete prior to departure.

m Have a debriefing session with the Liaison Officer or Incident Commander prior to departure.

SAFETY OFFICER

The Safety Officer function is to develop and recommend measures for assuring personnel safety, and to assess and/or anticipate hazardous and unsafe situations.

Only one Safety Officer will be assigned for each incident. The Safety Officer will report to the Incident Commander. The Safety Officer may have assistants as necessary, and the assistants may also represent assisting agencies or jurisdictions. Safety assistants may have specific responsibilities such as air operations, hazardous materials, etc.

a. Review Common Responsibilities.
b. Participate in planning meetings.
c. Identify hazardous situations associated with the incident.
d. Review the Incident Action Plan for safety implications.
e. Exercise emergency authority to stop and prevent unsafe acts.
f. Ensure accountability procedures are in place. (See Section 15.)
g. Size up need for and effectiveness of:
  1. Acountability plans/procedures;
  2. Rapid Intervention plans/procedures;
  3. Protective clothing needs of personnel and assistants;
  4. Scene security measures;
  5. Safety zones; and
  6. Avenues of access/egress.
h. Organize, assign and brief assistants as needed.
i. Review and approve the medical plan and ensure that adequate re-hab for all personnel is established.
j. Develop hazardous materials site safety plan as required.
k. Maintain log of unit activity.

SECTION 9

OPERATIONS

CONTENTS	9-1
ORGANIZATION CHART	9-2
POSITION CHECKLISTS	9-2
OPERATIONS SECTION CHIEF	9-2
BRANCH DIRECTOR	9-3
DIVISION/GROUP SUPERVISOR	9-3
STRIKE TEAM/TASK FORCE LEADER	9-4
SINGLE RESOURCE	9-4
STAGING AREA MANAGER	9-5
AIR OPERATIONS BRANCH DIRECTOR	9-5
AIR TACTICAL GROUP SUPERVISOR	9-6
HELICOPTER COORDINATOR	9-7

AIR SUPPORT GROUP SUPERVISOR	9-7
HELIBASE MANAGER	9-8
HELISPOT (LANDING ZONE) MANAGER	9-9

ORGANIZATION CHART

POSITION CHECKLISTS

OPERATIONS SECTION CHIEF

The Operations Section Chief, a member of the General Staff, is responsible for the management of all operations directly applicable to the primary mission. The Operations Chief activates and supervises organization elements in accordance with the Incident Action Plan and directs its execution. The Operations Chief also directs the preparation of unit operational plans, requests or releases resources, makes expedient changes to the Incident Action Plan as necessary, and reports such changes to the Incident Commander.

a. Review Common Responsibilities.

b. Develop operations portion of Incident Action Plan.

c. Brief and assign operations personnel in accordance with Incident Action

d. Plan. Supervise operations.

e. Determine need and request additional resources.

f. Review suggested list of resources to be released and initiate recommendation for release of resources.

g. Assemble and disassemble strike teams assigned to Operations Section.

h. Report information about special activities, events, and occurrences to Incident Commander.

i. Ensure that adequate safety measures and accountability procedures are in place.

j. Maintain log of unit activity.

BRANCH DIRECTOR

The Branch Directors when activated, are under the direction of the Operations Section Chief, and are responsible for the implementation of the portion of the Incident Action Plan appropriate to the Branches.

a. Review Common Responsibilities.

b. Develop with subordinates alternatives for Branch control operations.

c. Attend planning meetings at the request of the Operations Chief.

d. Review Division/Group assignments for Divisions/Groups within Branch. Modify based on effectiveness of current operations.

e. Assign specific work tasks to Division/Group Supervisors.

f. Supervise Branch operations.

   g. Ensure that adequate safety measures and accountability procedures are in place.

   h. Resolve logistic problems reported by subordinates.

   i. Report to Operations Chief when Incident Action Plan is to be modified; additional resources are needed; surplus resources are available; or hazardous situations or significant events occur.

   j. Maintain log of unit activity.

DIVISION/GROUP SUPERVISOR

The Division/Group Supervisor reports to the Operations Section Chief (or Branch Director when activated). The Supervisor is responsible for the implementation of the assigned portion of the Incident Action Plan, assignment of resources within the Division/Group, and reporting on the progress of control operations and status of resources within the Division/Group.

   a. Review Common Responsibilities.

   b. Implement Incident Action Plan for Division/Group.

   c. Provide Incident Action Plan to Strike Team Leaders, when available.

   d. Review Division/Group assignments and incident activities with subordinates and assign tasks.

   e. Ensure that Incident Communications and/or Resources Unit is advised of all changes in status of resources assigned to the Division/Group.

   f. Coordinate activities with adjacent Divisions.

   g. Determine need for assistance on assigned tasks.

   h. Ensure that adequate safety measures and accountability procedures are in place.

   i. Submit situation and resources status information to Branch Director or Operations Chief.

   j. Report hazardous situations, special occurrences, or significant events (e.g., accidents, sickness) to immediate supervisor.

   k. Ensure that assigned personnel and equipment get to and from assignments in a timely and orderly manner.

   l. Resolve logistics problems within the Division/Group.

   m. Participate in the development of Branch plans for next operational period.

   n. Maintain log of unit activity.

STRIKE TEAM/TASK FORCE LEADER

The Strike Team/Task Force Leader reports to a Division/Group Supervisor and is responsible for performing tactical assignments assigned to the Strike Team or Task Force. The Leader reports work progress, resources status, and other important information to a Division/Group Supervisor, and maintains work records on assigned personnel.

 a. Review Common Responsibilities.

 b. Review assignments with subordinates and assign tasks.

 c. Monitor work progress and make changes when necessary.

 d. Coordinate activities with adjacent Strike Team, Task Forces, and Single Resources.

 e. Travel to and from active assignment area with assigned resources.

 f. Retain control of assigned resources while in available or out-of-service status.

 g. Ensure that adequate safety measures and accountability procedures are in place.

 h. Submit situation and resource status information to Division/Group Supervisor.

 i. Maintain log of unit activity.

SINGLE RESOURCE

The person in charge of a single tactical resource will carry the unit designation of the resource.

 a. Review Common Responsibilities.

 b. Review assignments.

 c. Obtain necessary equipment/supplies.

 d. Review weather/environmental conditions for assignment area.

 e. Ensure that adequate safety measures and accountability procedures are in place.

 f. Monitor work progress.

 g. Ensure adequate communications with supervisor and subordinates.

 h. Keep supervisor informed of progress and any changes.

 i. Inform supervisor of problems with assigned resources.

 j. Brief relief personnel, and advise them of any change in conditions.

 k. Return equipment and supplies to appropriate unit.

STAGING AREA MANAGER

The Staging Area Manager is responsible for managing all activities within a Staging Area.

a. Review Common Responsibilities.

b. Proceed to Staging Area.

c. Establish Staging Area layout.

d. Determine any support needs for equipment, feeding, sanitation and security.

e. Establish check-in function as appropriate.

f. Post areas for identification and traffic control.

g. Respond to request for resource assignments. (Note: This may be direct from Operations or via the incident Communications Center).

h. Determine required resource levels from the Operations Section Chief.

i. Advise the Operations Section Chief when reserve levels reach minimums.

j. Maintain and provide status to Resource Unit of all resources in Staging Area.

k. Maintain Staging Area in orderly condition.

l. Demobilize Staging Area in accordance with Incident Demobilization Plan.

m. Maintain log of unit activity.

n. Ensure that proper coordination is maintained with air/ground ambulance coordinator(s).

AIR OPERATIONS BRANCH DIRECTOR

The Air Operations Branch Director, who is ground based, is primarily responsible for preparing the air operations portion of the Incident Action Plan. The plan will reflect agency restrictions that have an impact on the operational capability or utilization of resources (e.g., night flying, hours per pilot). After the plan is approved, Air Operations is responsible for implementing its strategic aspects-those that relate to the overall incident strategy as opposed to those that pertain to tactical operations (specific target selection).

Additionally, the Air Operations Branch Director is responsible for providing logistical support to helicopters operating at the incident. Specific tactical activities (target selection, suggested modifications to specific tactical actions in the Incident Action Plan) are normally performed by the Air Tactical Group Supervisor working with ground and air resources.

a. Review Common Responsibilities.

b. Organize preliminary air operations.

c. Request declaration (or cancellation) of restricted air space area (FAA Regulation 91.137).

d. Participate in preparation of the Incident Action Plan through Operation Section Chief. Ensure that the Air Operations portion of the Incident Action Plan takes into consideration the Air Traffic Control requirements of assigned aircraft.

e. Perform operational planning for air operations.

f. Determine coordination procedures for use by air organization with ground Branches, Divisions or Groups.

g. Coordinate with appropriate Operations Section personnel.

h. Supervise all Air Operations activities associated with the incident.

i. Evaluate helibase locations.

j. Establish procedures for emergency reassignment of aircraft.

k. Inform the Air Tactical Group Supervisor of the air traffic situation external to the incident.

l. Consider requests for non-tactical use of incident aircraft.

m. Resolve conflicts concerning non-incident aircraft.

n. Coordinate with Federal Aviation Administration (FAA).

o. Update Air Operations Plans.

p. Report to the Operations Section Chief on air operations activities.

q. Report special incidents/accidents.

r. Arrange for an accident investigation team when warranted.

s. Maintain log of unit activity.

AIR TACTICAL GROUP SUPERVISOR

The Air Tactical Group Supervisor is primarily responsible for the coordination of aircraft operations when fixed and/or rotary-wing aircraft are operating at an incident. These coordination activities are performed by the Air Tactical Group Supervisor while airborne. The Air Tactical Group Supervisor reports to the Air Operations Branch Director.

a. Review Common Responsibilities.

b. Determine what aircraft are operating within area of assignment.

c. Manage air tactical activities based upon Incident Action Plan

d. Establish and maintain communications and Air Traffic Control with pilots, Air Operations, Helicopter Coordinator, and Air Support Group (usually Helibase Manager).

e. Coordinate approved flights of non-incident aircraft or non-tactical flights in restricted air space area.

f. Receive reports of non-incident aircraft violating restricted airspace area.

g. Make tactical recommendations to approved ground contact (Operations Section Chief, Branch Director, or Division Supervisor).

h. Inform Air Operations Branch Director of tactical recommendations affecting the air operations portion of the Incident Action Plan.

i. Report on Air Operations activities to the Air Operations Branch Director. Advise Air Operations immediately if aircraft mission assignments are causing conflicts in the Air Traffic Control System.

j. Maintain log of unit activity.

HELICOPTER COORDINATOR

The Helicopter Coordinator is primarily responsible for coordinating tactical or logistical helicopter mission(s) at the incident. The Helicopter Coordinator can be airborne or on the ground, operating from a high vantage point. The Helicopter Coordinator reports to the Air Tactical Group Supervisor. Activation of this position is contingent upon the complexity of the incident and the number of helicopters assigned. There may be more than one Helicopter Coordinator assigned to an incident.

a. Review Common Responsibilities.

b. Determine what aircraft are operating within incident area of assignment.

c. Survey assigned incident area to determine situation, aircraft hazards, and other potential problems.

d. Coordinate Air Traffic Control with pilots, Air Operations Branch Director, Air Tactical Group Supervisor, and the Air Support Group (usually Helibase Manager) as the situation dictates.

e. Coordinate the use of assigned ground-to-air and air-to-air communications frequencies with the Air Tactical Group Supervisor, Communications Unit, or local agency dispatch center.

f. Ensure that all assigned helicopters know appropriate operating frequencies.

g. Coordinate geographical areas for helicopter operations with Air Tactical Group Supervisor and make assignments.

h. Determine and implement air safety requirements and procedures.

i. Ensure that approved night flying procedures are in operation.

j. Receive assignments, brief pilots, assign missions, and supervise helicopter activities.

k. Coordinate activities with Air Tactical Group Supervisor, Air Support Group, and ground personnel.

l. Maintain continuous observation of assigned helicopter operating area and inform Air Tactical Group Supervisor of incident conditions including any aircraft malfunction or maintenance difficulties and anything that may affect the incident.

m. Inform Air Tactical Group Supervisor when mission is completed and reassign helicopter(s) as directed.

n. Request assistance or equipment as required.

o. Report incidents or accidents to Air Operations Director and Air Tactical Group Supervisor immediately.

p. Maintain log of unit activity.

AIR SUPPORT GROUP SUPERVISOR

The Air Support Group Supervisor is primarily responsible for supporting and managing helibase and helispot operations. This includes providing 1) fuel and other supplies; 2) maintenance and repair of helicopters; 3) keeping records of helicopter activity; and 4) providing enforcement of safety regulations. These major functions are performed at helibases and helispots. During landing and take-off and while on the ground, helicopters are under the control of the Air Support Groups, Helibase, or Helispot Managers. The Air Support Group Supervisor reports to the Air Operations Director.

a. Review Common Responsibilities.

b. Obtain copy of the Incident Action Plan from the Air Operations Branch Director.

c. Participate in Air Operations Branch Director planning activities.

d. Inform Air Operations Branch Director of group activities.

e. Identify resources/supplies dispatched for Air Support Group.

f. Request special air support items from appropriate sources through Logistics Section.

g. Identify helibase and helispot locations (from Incident Action Plan) or from Air Operations Branch Director.

h. Determine need for assignment of personnel and equipment at each helibase and helispot.

i. Coordinate special requests for air logistics.

j. Maintain coordination with airbases supporting the incident.

k. Coordinate activities with Air Operations Branch Director.

l. Obtain assigned ground to air frequency for helibase operations.

m. Inform Air Operations Branch Director of capability to provide night flying service.

n. Ensure compliance with each agency's operations checklist for day and night operations.

o. Ensure dust abatement procedures are implemented at helibase and helispots.

p. Provide crash-rescue service for helibases and helispots.

q. Ensure that Air Traffic Control procedures are established between helibase, helispots, Air Tactical Group Supervisor, and Helicopter Coordinator.

r. Maintain log of unit activity.

HELIBASE MANAGER

a. Review Common Responsibilities.

b. Obtain Incident Action Plan.

c. Participate in Air Support Group planning activities. d. Inform Air Support Supervisor of helibase activities.

e. Report to assigned helibase. Brief pilots and assigned personnel.

f. Manage resources/supplies dispatched to helibase.

g. Ensure helibase is posted and cordoned.

h. Coordinate helibase Air Traffic Control with pilots, Air Support Group Supervisor, Air Tactical Group Supervisor, and Helicopter Coordinator.

i. Ensure helicopter fueling, maintenance, and repair services are provided.

j. Supervise manifests for and loading of personnel and cargo.

k. Ensure dust abatement techniques are provided and used at helibases and helispots.

I. Ensure security is provided at each helibase and helispot.

m. Ensure crash-rescue services are provided for the helibase.

n. Request special air support items from the Air Support Group Supervisor.

o. Receive and respond to special requests for air logistics.

p. Coordinate activities with Air Support Group Supervisor.

q. Display organization and work schedule at each helibase, including helispot organization and assigned radio frequencies.

r. Solicit pilot input concerning selection and adequacy of helispots, communications, AirTraffic Control, operational difficulties, and safety problems.

s. Maintain log of unit activity.

HELISPOT (LANDING ZONE) MANAGER

a. Review Common Responsibilities.

b. Obtain Incident Action Plan.

c. Report to assigned helispot.

d. Coordinate activities with Helibase Manager.

e. Inform Helibase Manager of helispot activities.

f. Manage resources and supplies dispatched to helispot.

g. Request special air support items from Helibase Manager.

h. Coordinate Air Traffic Control and communications with pilots, Helibase Manager, Helicopter Coordinator, and Air Tactical Group Supervisor when appropriate.

i. Ensure crash-rescue services are available.

j. Ensure that dust control is adequate, that debris cannot blow into the rotor system, that touchdown zone slope is not excessive, and that rotor clearance is sufficient.

k. Do manifests for and perform loading of personnel and cargo.

l. Coordinate with pilots for proper loading and unloading and safety problems.

m. Maintain log of unit activity.

SECTION 10
PLANNING SECTION

CONTENTS	10-1
ORGANIZATION CHART	10-2
POSITION CHECKLISTS	10-2
PLANNING SECTION CHIEF	10-2
RESOURCES UNIT LEADER	10-3
CHECK-IN RECORDER	10-3
SITUATION UNIT LEADER	10-4
TECHNICAL SPECIALISTS	10-4
WATER RESOURCES SPECIALIST	10-4
RESOURCE USE SPECIALIST	10-5

ORGANIZATION CHART

*May be assigned wherever their services are required.

POSITION CHECKLISTS

PLANNING SECTION CHIEF

The Planning Section Chief, a member of the Incident Commander's General Staff, is responsible for the collection, evaluation, dissemination and use of information about the development of the incident and status of resources. Information is needed to 1) understand the current situation, 2) predict the probable course of incident events, and 3) prepare alternative strategies and control operations for the incident.

   a. Review Common Responsibilities.

   b. Collect and process situation information about the incident.

   c. Supervise preparation of the Incident Action Plan.

   d. Provide input to the Incident Commander and Operations Section Chief in preparing the Incident Action Plan.

   e. Assign and/or re-assign personnel resources either on-site or reporting for duty to ICS organizational positions as required.

   f. Establish information requirements and reporting schedules for Planning Section units (e.g., Resources, Situation Units).

   g. Determine need for any specialized resources in support of the incident.

   h. If requested, assemble and disassemble strike teams and task forces not assigned to operations.

   i. Establish special information collection activities as necessary, e.g., weather, environmental, toxins, etc.

   j. Assemble information on alternative strategies.

   k. Provide periodic predictions on incident potential.

   l. Report any significant changes in incident status.

   m. Compile and display incident status information.

   n. Oversee preparation and implementation of Incident Demobilization Plan.

   o. Incorporate plans, (e.g., Traffic, Medical, Communications) into the Incident Action Plan.

   p. Maintain log of unit activity.

RESOURCES UNIT LEADER

The Resources Unit Leader is responsible for maintaining the status of all assigned resources (primary and support) at an incident. This is achieved by overseeing the check-in of all resources, maintaining a status-keeping system indicating current location and status of all resources, and maintenance of a master list of all resources (e.g., key supervisory personnel, primary and support resources, etc.).

a. Review Common Responsibilities.

b. Review Unit Leader Responsibilities.

c. Establish check-in function at incident locations.

d. Maintain and post the current status and location of all resources.

e. Maintain master roster of all resources checked in at the incident.

f. A Check-in Recorder reports to the Resources Unit Leader and is responsible for accounting for all resources assigned to an incident.

CHECK-IN RECORDER

Check-in recorders are needed at each check-in location to ensure that all resources assigned to an incident are accounted for.

a. Review Common Responsibilities.

b. Establish communications with the Communication Center.

c. Post signs so that arriving resources can easily find the check-in locations.

d. Transmit check-in information to Resources Unit on a regular pre-arranged schedule.

SITUATION UNIT LEADER

The collection, processing and organizing of all incident information takes place within the Situation Unit. The Situation Unit may prepare future projections of incident growth, maps and intelligence information.

a. Review Common Responsibilities.

b. Begin collection and analysis of incident data as soon as possible.

c. Prepare, post, or disseminate resource and situation status information as required, including special requests.

d. Prepare periodic predictions or as requested.

e. Provide photographic services and maps if required.

f. Prepare appropriate directories (e.g., maps, instructions, etc.) for inclusion in the demobilization plan.

g. Distribute demobilization plan (on and off-site).

h. Ensure that all Sections/Units understand their specific demobilization responsibilities.

i. Supervise execution of the Incident Demobilization Plan.

j. Brief Planning Section Chief on demobilization progress.

TECHNICAL SPECIALISTS

Certain incidents or events may require the use of Technical Specialists who have specialized knowledge and expertise. Technical Specialists may function within the Planning Section, or be assigned wherever their services are required.

WATER RESOURCES SPECIALIST

a. Review Common Responsibilities.

b. Participate in the development of the Incident Action Plan and review general control objectives including alternative strategies presently in effect.

c. Collect and validate water resource information within the incident area.
d. Prepare information on available water resources.

e. Establish water requirements needed to support fire suppression actions.

f. Compare incident control objectives as stated in the Plan, with available water resources and report inadequacies or problems to Planning Section Chief.

g. Participate in the preparation of Incident Action Plan when requested.

h. Respond to requests for water information.

i. Collect and transmit records and logs to Documentation Unit at the end of each operational period.

j. Maintain log of unit activity.

RESOURCE USE SPECIALIST

a. Review Common Responsibilities.

b. Participate in the development of the Incident Action Plan and review general control objectives including alternative strategies as requested.

c. Collect information on incident resources as needed.

d. Respond to requests for information about limitations and capabilities of resources.

e. Collect and transmit records and logs to Documentation Unit at the end of each operational period.

f. Maintain log of unit activity.

SECTION 11

LOGISTICS

CONTENTS	11-1
ORGANIZATION CHART	11-2
POSITION CHECKLISTS	11-3
LOGISTICS SECTION CHIEF	11-3
SERVICE BRANCH DIRECTOR	11-3
COMMUNICATIONS UNIT LEADER	11-4

MEDICAL UNIT LEADER	11-4
REHABILITATION MANAGER	11-5
FOOD UNIT LEADER	11-5
SUPPORT BRANCH DIRECTOR	1-5
SUPPLY UNIT LEADER	11-6
RECEIVING AND DISTRIBUTION MANAGER	11-6
FACILITIES UNIT LEADER	11-7
FACILITY MAINTENANCE SPECIALIST	11-7
SECURITY MANAGER	11-7
BASE MANAGER	11-8
CAMP MANAGER	11-8
GROUND SUPPORT UNIT LEADER	11-9
EQUIPMENT MANAGER	11-9
PUBLIC SAFETY CHAPLAIN	11-10

LOGISTICS ORGANIZATION CHART
POSITION CHECKLISTS

LOGISTICS SECTION CHIEF

The Logistics Section Chief, a member of the General Staff, is responsible for providing facilities, services, and material in support of the incident. The Section Chief participates in development and implementation of the Incident Action Plan and activates and supervises the Branches and Units within the Logistics Section.

    a. Review Common Responsibilities.

    b. Plan organization of Logistics Section.

    c. Assign work locations and preliminary work tasks to Section personnel.

    d. Notify Resources Unit of Logistics Section units activated including names and locations of assigned personnel.

    e. Assemble and brief Branch Directors and Unit Leaders.

    f. Participate in preparation of Incident Action Plan.

    g. Identify service and support requirements for planned and expected operations.

    h. Provide input to and review Communications Plan, Medical Plan and Traffic Plan.

    I Coordinate and process requests for additional resources.

j. Review Incident Action Plan and estimate Section needs for next operational period.

k. Advise on current service and support capabilities.

l. Prepare service and support elements of the Incident Action Plan.

m. Estimate future service and support requirements.

n. Receive Demobilization Plan from Planning Section.

o. Recommend release of unit resources in conformity with Demobilization Plan.

p. Ensure general welfare and safety of Logistics Section personnel.

q. Maintain log of unit activity.

SERVICE BRANCH DIRECTOR

The Service Branch Director, when activated, is under the supervision of the Logistics Section Chief, and is responsible for the management of all service activities at the incident. The Branch Director supervises the operations of the Communications, Medical, and Food Units.

a. Review Common Responsibilities.

b. Obtain working materials.

c. Determine level of service required to support operations.

d. Confirm dispatch of Branch personnel.

e. Participate in planning meetings of Logistics Section personnel. f. Review Incident Action Plan.

g. Organize and prepare assignments for Service Branch personnel.

h. Coordinate activities of Branch Units.

i. Inform Logistics Chief of Branch activities.

j. Resolve Service Branch problems.

k. Maintain log of unit activity.

COMMUNICATIONS UNIT LEADER

The Communications Unit Leader, under the direction of the Service Branch Director or Logistics Section Chief, is responsible for developing plans for the effective use of incident communications equipment and facilities; installing and testing of communications equipment; supervision of the Incident Communications Center; distribution of communications equipment to incident personnel; and the maintenance and repair of communications equipment.

a. Review Common Responsibilities.

b. Review Unit Leader Responsibilities.

c. Determine unit personnel needs.

d. Prepare and implement a Communications Plan.

e. Ensure the Incident Communications Center and Message Center are established.

f. Ensure communications systems are installed and tested.

g. Ensure an equipment accountability system is established.

h. Provide technical information as required on:

   1. Adequacy of communications systems currently in operation;

   2. Geographic limitations on communications systems;

   3. Equipment capabilities/limitations;

   4. Amount and types of equipment available; and

   5. Anticipated problems in the use of communications equipment.

i. Supervise Communications Unit activities.

j. Maintain records on all communications equipment as appropriate.

k. Recover equipment from relieved or released units.

MEDICAL UNIT LEADER

The Medical Unit Leader, under the direction of the Service Branch Director or Logistics Section Chief, is primarily responsible for the development of the Medical Plan, obtaining medical aid and transportation for injured and ill incident personnel, and preparation of reports and records.

a. Review Common Responsibilities.

b. Review Unit Leader Responsibilities.

c. Participate in Logistics Section/Service Branch planning activities.

d. Establish Medical Unit.

e. Prepare the Medical Plan.

f. Prepare procedures for major medical emergency.

g. Declare major medical emergency as appropriate.

h. Respond to requests for medical aid, medical transportation, and medical supplies.

i. Prepare and submit necessary documentation.

REHABILITATION MANAGER

The Rehabilitation Manager reports to the Medical Unit Leader and is responsible for the rehabilitation of incident personnel who are suffering from the effects of strenuous work and/or extreme conditions.

a. Review Common Responsibilities.

b. Designate responder rehabilitation location(s) and have location(s) announced on radio with radio designation "Rehab."

c. Request necessary medical personnel to evaluate medical condition of personnel being rehabilitated.

d. Request necessary resources for rehabilitation of personnel, e.g., water, juice, personnel.

e. Request through Food Unit or Logistics Section Chief food as necessary for personnel being rehabilitated.

f. Release rehabilitated personnel to Planning Section for reassignment.

g. Maintain appropriate records and documentation.

FOOD UNIT LEADER

The Food Unit Leader is responsible for supplying the food needs for the entire incident, including all remote locations (e.g., Camps, Staging Areas), as well as providing food for personnel unable to leave tactical field assignments.

a. Review Common Responsibilities.

b. Review Unit Leader Responsibilities.

c. Determine food and water requirements.

d. Determine method of feeding to best fit each facility or situation.

e. Obtain necessary equipment and supplies and establish cooking facilities.

f. Ensure that well-balanced menus are provided.

g. Order sufficient food and potable water from the Supply Unit.

h. Maintain an inventory of food and water.

i. Maintain food service areas, ensuring that all appropriate health and safety measures are being followed.

j. Supervise caterers, cooks, and other Food Unit personnel as appropriate.

SUPPORT BRANCH DIRECTOR

The Support Branch Director, when activated, is under the direction of the Logistics Section Chief, and is responsible for development and implementation of logistics plans in support of the Incident Action Plan. The Support Branch Director supervises the operations of the Supply, Facilities and Ground Support Units.

a. Review Common Responsibilities.

b. Obtain work materials.

c. Identify Support Branch personnel dispatched to the incident.

d. Determine initial support operations in coordination with Logistics Section Chief and Service Branch Director.

e. Prepare initial organization and assignments for support operations.

f. Assemble and brief Support Branch personnel.

g. Determine if assigned Branch resources are sufficient.

h. Maintain surveillance of assigned units' work progress and inform Section Chief of activities.

i. Resolve problems associated with requests from Operations Section.

j. Maintain log of unit activity.

SUPPLY UNIT LEADER

The Supply Unit Leader is primarily responsible for ordering personnel, equipment and supplies; receiving, and storing all supplies for the incident; maintaining an inventory of supplies; and servicing non-expendable supplies and equipment.

a. Review Common Responsibilities.

b. Review Unit Leader Responsibilities.

c. Participate in Logistics Section/Support Branch planning activities.

d. Determine the type and amount of supplies en route.

e. Review Incident Action Plan for information on operations of the Supply Unit.

f. Develop and implement safety and security requirements.

g. Order, receive, distribute, and store supplies and equipment.

h. Receive and respond to requests for personnel, supplies and equipment.

i. Maintain inventory of supplies and equipment.

j. Service reusable equipment.

k. Submit reports to the Support Branch Director.

RECEIVING AND DISTRIBUTION MANAGER

The Receiving and Distribution Manager is responsible for receipt and distribution of all supplies and equipment (other than primary resources) and the service and repair of tools and equipment. The Receiving and Distribution Manager reports to the Supply Unit Leader.

a. Review Common Responsibilities.

b. Order required personnel to operate supply area.

c. Organize physical layout of supply area.

d. Establish procedures for operating supply area.

e. Set up filing system for receiving and distribution of supplies and equipment.

F. Maintain inventory of supplies and equipment.

g. Develop security requirement for supply area.

h. Establish procedures for receiving supplies and equipment.

i. Submit necessary reports to Supply Unit Leader.

j. Notify Ordering Manager of supplies and equipment received.

k. Provide necessary supply records to Supply Unit Leader.

FACILITIES UNIT LEADER

The Facilities Unit Leader is primarily responsible for the layout and activation of incident facilities, e.g., Base, Camp(s) and Incident Command Post. The Unit provides sleeping and sanitation facilities for incident personnel and manages Base and Camp(s) operations. Each facility (Base, Camp) is assigned a manager who reports to the Facilities Unit Leader and is responsible for managing the operation of the facility. The basic functions or activities of the Base and Camp Managers are to provide security service and general maintenance. The Facility Unit Leader reports to the Support Branch Director.

a. review Common Responsibilities.

b. Review Unit Leader Responsibilities.

c. Receive a copy of the Incident Action Plan.

d. Participate in Logistics Section/Support Branch planning activities.

e. Determine requirements for each facility.

f. Prepare layouts of incident facilities.

g. Notify Unit Leaders of facility layout.

h. Activate incident facilities.

i. Provide Base and Camp Managers.

j. Provide sleeping facilities.

k. Provide security services.

l. Provide facility maintenance services: sanitation, lighting, clean-up.

FACILITY MAINTENANCE SPECIALIST

The Facility Maintenance Specialist is responsible for ensuring proper sleeping and sanitation facilities are maintained; providing shower facilities; providing and maintaining lights and other electrical equipment; and maintaining the Base, Camp and Incident Command Post facilities in a clean and orderly manner.

a. Review Common Responsibilities.

b. Request required maintenance support personnel and assign duties.

c. Obtain supplies, tools, and equipment.

d. Supervise/perform assigned work activities.

e. Ensure that all facilities are maintained in a safe condition.

f. Disassemble temporary facilities when no longer required.

g. Restore area to pre-incident condition.

SECURITY MANAGER

The Security Manager is responsible for providing safeguards needed to protect personnel and property from loss or damage.

a. Review Common Responsibilities.

b. Establish contacts with local law enforcement agencies as required.

c. Contact the Resource Use Specialist for crews or Agency Representatives to discuss any special custodial requirements that may affect operations.

d. Request required personnel support to accomplish work assignments.

e. Ensure that support personnel are qualified to manage security problems.

f. Develop Security

Endnotes

1
http://www.northern.vaems.org/Documents/No%20Va%20MCI%20Plan.pdf#search='jurisdiction,%20multi%20agency,%20response'

B

FEDERAL EMERGENCY MANAGEMENT AGENCY (FEMA): ROLES AND RESPONSIBILITIES IN A TERRORIST INCIDENT

LAW ENFORCEMENT

CHECKLIST OF CONSIDERATIONS

List all those items the Law Enforcement Representative should consider during a terrorist incident. The Law Enforcement Representative and Emergency Management Coordinator should determine what level of detail should be included in this annex. Examples of possible tasks include:

- ❏ Maintain the integrity of the crime scene.
- ❏ Provide security at the following:

_____ Shelters

_____ Emergency Operations Center (EOC)

_____ Command Post

_____ Disaster Site

_____ Hospitals

_____ Temporary Morgue

_____ Jail

_____ Joint Information Center (JIC)

_____ Joint Operations Center (JOC)

_____ Other Medical Care Centers

- ❏ Secure impassable roads. Fire Services and Public Works may provide support for this task.
- ❏ Request necessary assistance from Public Works to identify routes that need barricades and signs.
- ❏ Coordinate with the Road Commission or Department of Public Works in rerouting traffic and putting the appropriate signs in place.

- ❏ Ensure that security passes are issued to appropriate personnel who have authority to enter secured areas.
- ❏ Implement any curfews ordered by the Governor or Chief Executive Official. Describe how curfew will be enforced (through citations or arrest, etc.)
- ❏ Enforce quarantine controls, if applicable.
- ❏ Develop a method and a location for a "lost and found" service. Inform the Public Information Officer (PIO) of the details of how the public can access this service.
- ❏ Ensure that vehicles blocking evacuation routes and routes to health care centers are removed. If necessary, request that Public Works or Road Commission trucks move vehicles off the road.
- ❏ Maintain records of where vehicles are being taken. Inform the PIO of the details of how the public can reclaim their vehicles.
- ❏ Ensure that prisons and jails are notified of the potential threat, and determine whether proper safety and security precautions are being taken.
- ❏ Ensure that staff are not working more than X hours.
- ❏ Activate, or request activation of, mutual aid agreements.
- ❏ Assist the warning agency, as needed, in notifying the public of an impending emergency.

FIRE SERVICES
CHECKLIST OF CONSIDERATIONS

List all those items that the Fire Service agencies should consider in a terrorist incident. The Fire Services Representative and Emergency Management Coordinator should determine what level of detail should be included in this annex. Examples of possible tasks include:

- ❏ Maintain incident site safety.
- ❏ Decontaminate victims/rescuers (in consultation with public health officials).
- ❏ Activate, or request activation of, search and rescue teams, as needed.
- ❏ Provide communications and other logistical supplies, as needed.
- ❏ Assist building inspectors in performing fire safety inspections at facilities designated as shelters.
- ❏ Provide trained personnel to inspect damaged buildings before occupancy, after repairs have been done.
- ❏ Notify Public Works of the gas valves turned off so that the return of gas service can be coordinated.

- Activate Radiological Monitoring Teams, as needed.

- Coordinate the fire department's role in providing emergency medical services, if appropriate.

- Report disaster-related damage information to the Emergency Management Coordinator, Damage Assessment Representative, etc. as it is encountered.

- Assist in traffic control by providing personnel to direct traffic at certain intersections, as requested by the law enforcement organization.

- Assist in warning the population, if assigned.

- Participate in the Joint Information Center at the scene. Coordinate the release of information with the Public Information Officer.

- Determine the locations of the different staging areas. Notify appropriate EOC staff of their locations.

- Keep emergency service organizations informed of existing dangers associated with the incident.

PUBLIC WORKS

CHECKLIST OF CONSIDERATIONS

List all those items that the Public Works Director should consider during a terrorist incident. The Public Works Director and Emergency Management Coordinator should determine what level of detail should be included in this annex. The Public Works Representative will need to work closely with all of the listed agencies on how to contact them and request equipment and expertise. Examples of possible tasks include:

- Provide barricades and signs for road closures and boundary identification. Ensure that there are adequate barricades and activate, or request activation of, appropriate mutual aid agreements, if necessary.

- Assist in identifying boundaries of areas in which access must be controlled.

- Provide vehicles and personnel to transport essential goods such as food, medical supplies, and other needed items.

- Notify law enforcement of the location(s) of vehicles being towed.

- Contact the appropriate Department of Transportation official to request travel restrictions on State highways, if necessary.

- Determine the extent and cause(s) of damage and outages faced by local utilities. Report this information to EOC staff.

- Coordinate with utility companies in the restoration of essential services. Provide appropriate assistance, such as debris clearance, to expedite restoration.

- Provide engineering expertise to inspect public structures to determine whether they are safe to use. Develop teams to inspect roads, bridges, buildings, infrastructure, etc. (These teams may be called upon to assist in assessing damage for public assistance grants from the Federal government, if applicable.)
- Ensure that Public Works crews report damage information to the Emergency Management Coordinator, supervisor, damage assessment representative, etc. Note: This includes damage to public facilities, debris clearance requirements, emergency protective measures, and other damage information, as appropriate.
- Prioritize and coordinate the use of generators and fuel supplies.
- Prioritize and coordinate the use of emergency lighting.
- Assist in identifying and obtaining the appropriate construction equipment to support response and recovery within the jurisdiction.
- Determine where debris should be piled initially, then determine a permanent location for debris. If necessary, coordinate security of debris sites with law enforcement personnel.
- Determine what support Public Works crews can provide during a terrorist incident.

EMERGENCY MEDICAL SERVICES
CHECKLIST OF CONSIDERATIONS

List all those items that emergency medical services should consider during a terrorist incident. The Emergency Medical Representative and Emergency Management Coordinator should determine what level of detail should be included in this annex. Examples of possible tasks include the following:

- Ensure that responding emergency medical teams coordinate with the unified command.
- Ensure that personal protection protocols have been implemented.
- If necessary, establish a triage area in close proximity to but outside of the hot zone.
- Ensure that the triage areas has adequate medical supplies.
- Provide for a medical supply inventory to determine what, if any, supplies are needed, including appropriate antidotes and antibiotics, and the number of ambulances needed and being used.
- Prepare to augment medical supplies and resources. (Augmenting emergency medical supplies and equipment is a critical pre-disaster planning consideration!)
- Ensure that each ambulance unit, as well as paramedic units, are tracking resources used during the response.

- Determine what, if any, medical resources and systems need augmenting on the scene. (How would you utilize mutual aid, hospital staff, etc.?)
- Augment universal precaution supplies.
- Ensure that a casualty tracking system is established.
- Direct on-scene volunteers to a volunteer registration area.
- Maintain a liaison with the Human Services Representative to request additional medical personnel, when necessary.
- Coordinate security at triage centers, CCPs, etc. with law enforcement personnel.
- Establish and maintain field communications and coordination with the command post and other responding emergency teams, as well as telephone or radio communications with hospitals.
- Appoint someone to serve as a liaison to the unified command and the EOC.
- Implement hazardous materials procedures, as needed.

PUBLIC HEALTH SERVICES
CHECKLIST OF CONSIDERATIONS

List all those items that the Public Health Service agencies should consider during a terrorist incident. The Public Health Service Representative and the Emergency Management Coordinator should decide what level of detail should be included in this annex. Examples of possible tasks include:

- Coordinate with hospitals and other health/medical care facilities in the investigation of a bioterrorist event.
- Implement assessment and surveillance procedures to assess the numbers of persons and area affected to determine the potential public health impact.
- Provide technical assistance and guidance for the monitoring of private citizens and emergency workers for exposure to chemical, radiological, or biological contaminants. (Note: Most local public health or emergency management groups will not have the capability to monitor for all chemical, radiological, and biological agents.)
- Provide for administration of preventive measures, such as vaccines and antibiotics.
- Coordinate information sharing with all Federal, State, and local public health and medical officials, and with EOC personnel.
- Provide advice and guidance on the monitoring of public and private water sources, and request the issuance of appropriate public health warnings, if necessary.

- Provide advice and guidance on the monitoring of public and private sewage disposal systems. Request issuance of appropriate public health warnings, if necessary.
- Provide for the inspection of food service establishments or those temporarily established for emergency workers or disaster victims to ensure the safety of food products prior to distribution and consumption.
- Work with the PIO to issue advisories on food preservation, disposal of contaminated or spoiled products, or consumption of homegrown and other potentially contaminated products.
- Work with waste haulers to arrange for special pickup and disposal of waste items to minimize prolonged exposure of potential health and safety hazards.
- Prioritize and coordinate enforcement of nuisance abatement ordinances to keep debris from becoming a health hazard. Advise the EOC of the need for such emergency ordinances, if necessary. (The Section Representative may wish to list existing local ordinances.)
- Ensure that proper vector control and pest management activities are in place. (The Section Representative may want to list possible measures that may be ordered and implemented.)
- Provide advice and guidance to the local animal control unit to protect public health.
- Ensure that emergency workers are aware of the availability of crisis counseling.
- Ensure that medical care facilities are capable of relocating patients in the event that an evacuation is ordered.
- Provide appropriate protection, prophylaxis and treatment for citizens and emergency workers.
- Provide advice and guidance for monitoring exposed individuals for health concerns.
- Provide for injury and illness assessment and surveillance activities.
- Establish a registry system to provide for the ongoing monitoring and follow up of exposed persons, including emergency workers.
- Notify health service institutions of special mass casualty treatment requirements.
- Monitor patient care capacity of casualty-receiving hospitals, and provide for forward transportation of overflow patients to surrounding regions, using the NDMS structure.
- Provide supervision for decontamination and other exposure reduction methods. (Refer to the appropriate SOP for specific information according to the type of incident.)

- ❏ Work with the Medical Examiner in providing for a mass fatality mortuary service.
- ❏ Assist with site identification for a temporary morgue.
- ❏ Implement Mass Casualty Standard Operating Procedures detailing the identification of the deceased, release of remains to next of kin, collection and storage of personal property, etc.

UNIFIED COMMAND CHECKLIST

Instructions: The checklist below presents the minimum requirements for all Incident Commanders. Note that some activities are one-time actions, while others are ongoing or repetitive for the duration of an incident.

COMPLETED/ NOT APPLICABLE	TASKS
❏	Obtain an incident briefing and Incident Briefing Form (ICS Form 201) from the prior Incident Commander.
❏	Assess the incident situation.
❏	Determine incident goals and strategic objectives.
❏	Establish the immediate priorities.
❏	Establish an Incident Command Post.
❏	Conduct the initial briefing.
❏	Activate elements of the Incident Command System, as required.
❏	Brief the command staff and section chiefs.
❏	Ensure that planning meetings are conducted.
❏	Approve and authorize the implementation of the incident action plan.
❏	Ensure that adequate safety measures are in place.
❏	Determine information needs and inform command personnel.
❏	Coordinate staff activity.
❏	Coordinate with key people and officials, including the EOC and JOC.
❏	Manage incident operations.
❏	Approve requests for additional resources and requests for release of resources.
❏	Approve the use of trainees at the incident.

COMPLETED/ NOT APPLICABLE	TASKS
❏	Authorize release of information to the news media.
❏	Ensure that the Incident Status Summary (ICS Form 209 or local form) is completed and forwarded to the dispatch center(s).
❏	Approve a plan for demobilization.
❏	Release resources and supplies.

MAJOR RESPONSIBILITIES AND TASKS

The major responsibilities of the Incident Commander are listed below. Following each are tasks for implementing the responsibility.

RESPONSIBILITY	TASKS
Conduct Initial Briefing	❏ Obtain and review the Incident Briefing Form (ICS Form 201 or local form) with the Incident Commander.
	❏ Meet with the prior Incident Commander (as appropriate) and selected staff available at that time.
	❏ Review and/or prepare plans for the use of on-scene and allocated resources scheduled to arrive before the next planning meeting.
Set Up Required Organization Elements	❏ Confirm the dispatch and/or arrival of requested organizational elements.
	❏ Hold a briefing and assign work tasks to general and command staffs. This briefing should include:
	■ The contents of the Incident Briefing Form.
	■ A summary of the incident organization.
	■ A review of current incident activities.
	■ A summary of resources already dispatched.
	■ The time and location of the first planning meeting.
	■ Special instructions, including specific delegation of authority to carry out particular functions.

RESPONSIBILITY	TASKS
	❑ Reassign the prior Incident Commander to a position within the incident organization (as appropriate).
	❑ Request required additional resources through normal dispatch channels.
	❑ Notify the Resources Unit of the command and general staff organizational elements activated, including the name of the person assigned to each position.

Ensure Planning Meetings are Conducted	❑ Schedule a meeting time and location.
	❑ Notify the attendees, including:
	   ■ Prior Incident Commander (required at first general planning meeting).
	   ■ Command and general staffs.
	   ■ Others as desired (e.g., communications, resources, and Situation Unit and Operations Branch Directors).
	❑ Develop the general objectives for the incident action plan.
	❑ Participate in the development of the incident action plan for the next operational period.
	❑ Participate in the preparation of logistics services and support requirements associated with the incident action plan (e.g., the communications plan).
	❑ Review safety considerations with the Safety Officer.
	❑ Summarize the decisions made about the:
	   ■ General strategy selected.
	   ■ Control objectives selected for the next operational period.
	   ■ Resources required.
	   ■ Service and support requirements.

RESPONSIBILITY	TASKS
Approve and Authorize Implementation of the Incident Action Plan	Note: In some instances, these tasks may be done orally. ❏ Review the incident action plan for completeness and accuracy. ❏ Make any required changes and authorize the release of the plan.
Determine Information Needs from Staff	❏ Identify any special information desired from each section chief. ❏ Prepare information item lists for each section and command staff element (as appropriate). ❏ Provide lists to appropriate personnel or facility. (Note: This may be done orally in some situations.)
Manage Incident Operations	❏ Review information concerning significant changes in the status of the situation, predicted incident behavior, weather, or status of resources. ❏ Review modification to the current incident action plan received from the Operations Section Chief. ❏ Identify any major changes to incident operations which are required immediately.
Approve Requests for Additional Resources	❏ Review requests for additional resources. ❏ Determine the condition and advisability of activating out-of-service resources. ❏ Have the Planning Section Chief provide a list of resources for reassignment if out-of-service resources are to be activated. Include the time needed, reporting location, and to whom to report. ❏ To obtain additional resources from off the incident, direct the Logistics Section Chief to forward the request through normal channels.

RESPONSIBILITY	TASKS
Authorize Information Release	❏ Review materials submitted by the Information Officer for release to the news media.
	❏ Check information release policies and constraints with involved jurisdiction officials.
	❏ Authorize the release of the final copy.
Report Incident Status	❏ Have the Incident Status Summary Report (ICS Form 209 or local form) prepared.
	❏ Ensure that the incident status summary is submitted to local agency dispatch centers, as required.
Approve Demobilization Planning	❏ Review recommendations for the release of resources and supplies from the Demobilization Unit.
	❏ Schedule a demobilization planning meeting.
	❏ Ensure that current and future resource and supply requirements have been closely estimated.
	❏ Establish general service and support requirements.
	❏ Modify specific work assignments for general and command staff, as required.
	❏ Summarize the actions to be taken.
	❏ Have the Planning Section Chief document the demobilization plan.
Coordinate Staff Activity	❏ Periodically check the progress on assigned tasks of Logistics, Planning, Operations, and Finance/Administration Sections, as well as command staff personnel.
	❏ Ensure that the general welfare and safety of personnel is adequate.
	❏ Notify the Resources Unit of changes to the command or general staff organization, including the name of the person assigned to each position.

RESPONSIBILITY	TASKS
Release Resources and Supplies	❑ Review recommendations for any release of resources and supplies from the general staff.
	❑ Approve release recommendations.
	❑ Ensure that local agency dispatch centers are notified of the intended release.
	❑ Direct the Planning Section Chief to prepare an assignment list for the release of resources.
	❑ Direct the Logistics Section Chief to release supplies.

INFORMATION OFFICER CHECKLIST

RESPONSIBILITIES

The Information Officer, a member of the command staff, is responsible for the collection and release of information about the incident to the news media and other appropriate agencies and organizations. The Information Officer reports to the Incident Commander.

Instructions: The checklist below presents the minimum requirements for Information Officers. Note that some items are one-time actions, while others are ongoing or repetitive throughout the incident.

COMPLETED/ NOT APPLICABLE	TASKS
❑	Contact the appropriate agency to coordinate public information activities.
❑	Establish a Joint Information Center (JIC), whenever possible.
❑	Determine from the Incident Commander if there are any limits on information release.
❑	Arrange for necessary work space, materials, telephones, and staffing.
❑	Obtain copies of the Incident Commander's Situation Status Summary Report (ICS Form 209 or local form).
❑	Prepare an initial information summary as soon as possible after arrival.
❑	Observe constraints on the release of information imposed by the Incident Commander.

COMPLETED/ NOT APPLICABLE	TASKS
❏	Obtain approval for information release from the Incident Commander.
❏	Release news to the media and post information at the Incident Command Post and other appropriate location(s).
❏	Attend meetings between the media and incident personnel.
❏	Arrange for meetings between the media and incident personnel.
❏	Provide escort service to the media and VIPs.
❏	Provide protective clothing for the media and VIPs (as appropriate).
❏	Respond to special requests for information.
❏	Maintain the unit log (ICS Form 214 or local form).

The major responsibilities of the Information Officer are listed below. Following each are tasks for implementing the responsibility.

RESPONSIBILITY	TASKS
Identify Information Officer Activities	❏ Contact the jurisdiction's responsible agency to determine what other external public information activities are being performed for this incident.
	❏ Establish the coordination of information acquisition and dissemination.
	❏ Compile the information, and maintain records.
Establish an Information Center as Required	❏ Establish an information center adjacent to the Incident Command Post area where it will not interfere with Incident Command Post activities.
	❏ Contact the Facilities Unit for any support required to set up the information center.

RESPONSIBILITY	TASKS
Prepare a Press Briefing	❏ Obtain from the Incident Commander any constraints on the release of information.
	❏ Select the information to be released (e.g., the size of the incident, the agencies involved, etc.).
	❏ Prepare the material for release (obtained from the Incident Briefing [ICS Form 201 or local form], Situation Unit status reports, etc.).
	❏ Obtain the Incident Commander's approval for release. (Note: The Incident Commander may give blanket release authority.)
	❏ Release the information for distribution to the news media.
	❏ Release the information to press representatives at the joint information center (JIC).
	❏ Post a copy of all information summaries in the Incident Command Post area and at other appropriate incident locations (e.g., base, camps, etc.).
Collect and Assemble Incident Information	❏ Obtain the latest situation status and fire behavior prediction information from the appropriate Situation Unit Leader.
	❏ Observe incident operations.
	❏ Hold discussions with incident personnel.
	❏ Identify special event information (e.g., evacuations, injuries, etc.).
	❏ Contact external agencies for additional information.
	❏ Review the current incident action plan (ICS Form 202 or local form).
	❏ Repeat the above procedures as necessary to satisfy media needs.

RESPONSIBILITY	TASKS
Provide Liaison between Media and Incident Personnel	
	❏ Receive requests from the media to meet with incident personnel and vice versa.
	❏ Identify the parties involved in the request (e.g., the Incident Commander for TV interviewers, etc.).
	❏ Determine if policies have been established to handle requests, and, if so, proceed accordingly.
	❏ Obtain any required permission to satisfy a request (i.e., the Incident Commander's).
	❏ Fulfill the request or advise the requesting party of the inability to do so, as the case may be.
	❏ Coordinate as necessary with the Incident Commander for news media flights into the incident area.
Respond to Special Requests for Information	❏ Receive request for information.
	❏ Determine if the requested information is currently available, and, if so, provide it to the requesting party.
	❏ Determine if currently unavailable information can be reasonably obtained by contacting incident personnel.
	❏ Assemble the desired and/or available information, and provide it to the requesting party.
Maintain the Unit Log	❏ Record the Information Officer's actions on the unit log (ICS Form 214 or local form).
	❏ Collect and transmit information summaries and unit logs to the Documentation Unit at the end of each operational period.

SAFETY OFFICER CHECKLIST

RESPONSIBILITIES

The Safety Officer, a member of the command staff, is responsible for monitoring and assessing hazardous and unsafe situations and developing measures for assuring personnel safety. The Safety Officer will correct unsafe acts or conditions through the regular line of authority, although he or she may exercise emergency authority to stop or prevent unsafe acts when immediate action is required. The Safety Officer maintains an awareness of active and developing situations, approves the medical plan, and includes safety messages in each incident action plan. The Safety Officer reports to the Incident Commander.

Instructions: The checklist below presents the minimum requirements for Safety Officers. Note that some items are one-time actions, while others are ongoing or repetitive throughout the incident.

COMPLETED/ NOT APPLICABLE	TASKS
❑	Obtain an incoming briefing from the Incident Commander.
❑	Identify hazardous situations associated with the incident.
❑	Participate in planning meetings.
❑	Review the incident action plan.
❑	Identify potentially unsafe situations.
❑	Exercise emergency authority to stop and prevent unsafe acts.
❑	Investigate accidents that have occurred within the incident area.
❑	Assign assistants as needed.
❑	Review and approve the medical plan (ICS Form 206 or local form).
❑	Maintain the unit log (ICS Form 214 or local form).

The major responsibilities of the Safety Officer are listed below. Following each are tasks for implementing the responsibility.

RESPONSIBILITY	TASKS
Obtain a Briefing from the Incident Commander	❑ Receive a briefing from the Incident Commander to obtain: ■ Relieved Incident Commander's Incident Briefing (ICS Form 201 or local form). ■ Summary of the incident organization. ■ Special instructions. ❑ Obtain a copy of the incident action plan from the Incident Commander.

RESPONSIBILITY	TASKS
Identify Hazardous Situations Associated with the Incident Environment Prior to First Planning Meeting	❑ Identify and resolve unsafe situations in the incident area (e.g., unsafe sleeping areas, absence of protective clothing etc.). ❑ Compile and record hazardous and potentially hazardous situations for presentation at the planning meeting.
Attend the Planning Meeting to Advise on Safety Matters	❑ Review the suggested strategy and control operations as presented at the planning meeting. ❑ Identify potentially hazardous situations associated with the proposed plans and/or strategies. ❑ Advise the general staff of such situations.
Identify Potentially Unsafe Situations	❑ Review the incident action plan. ❑ Receive reports from incident personnel concerning safety matters. ❑ Review reports to identify hazardous environmental and operational situations. ❑ Personally survey the incident environment and operations, as appropriate. ❑ Obtain and review Situation Unit information to identify unsafe situations.
Advise Incident Personnel in Matters Affecting Personnel Safety	❑ Identify potentially hazardous situations. (See previous tasks.) ❑ Determine the appropriate actions to ensure personnel safety. ❑ Coordinate with incident supervisory personnel, as required. ❑ Advise incident personnel as to the appropriate action.

RESPONSIBILITY	TASKS
Exercise Emergency Authority to Prevent or Stop Unsafe Acts	❏ Identify potentially hazardous situations. (See previous tasks.) ❏ Determine the severity of the situation. ❏ Determine if the situation requires the use of emergency authority, and, if so, exercise that authority to prevent or stop the act. ❏ Coordinate with the appropriate supervisory personnel.
Investigate (or Coordinate Investigation of) Accidents that Occur within the Incident Area	❏ Receive notification of the accident. ❏ Obtain information concerning the accident by: ■ Interviewing personnel. ■ Visiting the scene of the accident. ■ Photographing the scene (if appropriate). ■ Collecting evidence (if appropriate). ■ Collecting reports prepared by involved personnel. ❏ Reconstruct the accident events. ❏ Identify the cause of the accident (if possible). ❏ Recommend corrective action. ❏ Prepare the accident report and submit it to the Incident Commander.
Review the Medical Plan	❏ Coordinate with the Medical Unit Leader on the preparation of the medical plan (ICS Form 206 or local form). ❏ Review the plan for completeness. ❏ Discuss areas of concern with the Medical Unit Leader and provide instructions for correction.

RESPONSIBILITY	TASKS
Maintain the Unit Log	❑ Record the Safety Officer's actions on the unit log (ICS Form 214 or local form).
	❑ Collect and transmit required records and logs to the Documentation Unit at the end of each operational period.

LIAISON OFFICER CHECKLIST

RESPONSIBILITIES

The Liaison Officer is responsible for interacting (by providing a point of contact) with the assisting and cooperating agencies, including fire agencies, the American Red Cross, law enforcement, public works and engineering organizations, and others. When agencies assign agency representatives to the incident, the Liaison Officer will coordinate their activities. As a member of the command staff, the Liaison Officer reports to the Incident Commander.

Instructions: The checklist below presents the minimum requirements for Liaison Officers. Note that some items are one-time actions, while others are ongoing throughout the incident.

COMPLETED/ NOT APPLICABLE	TASKS
❑	Obtain a briefing from Incident Commander.
❑	Provide a point of contact for assisting and/or coordinating with agency representatives.
❑	Identify representatives from each involved agency, including a communications link and his or her location.
❑	Keep agencies supporting the incident aware of incident status.
❑	Respond to requests from incident personnel for interorganizational contacts.
❑	Monitor incident operations to identify current or potential inter-organizational contacts.
❑	Participate in planning meetings, providing current resource status, including limitations and capability of assisting agency resources.
❑	Maintain the unit log (ICS Form 214 or local form).

The major responsibilities of the Liaison Officer are listed below. Following each are tasks for implementing the responsibility.

RESPONSIBILITY	TASKS
Obtain a Briefing	❏ Receive a briefing from the Incident Commander and obtain the: ■ Incident Briefing Report (ICS Form 201 or local form). ■ Summary of the incident organization. ■ Names of agencies currently involved in the incident. ■ Special instructions from the Incident Commander. ❏ Obtain the incident action plan, when available.
Provide Point of Contact for Assisting and/or Cooperating Agencies	❏ Identify assisting and cooperating agencies from: ■ The Incident Briefing Report (ICS Form 201 or local form). ■ Local dispatchers. ❏ Determine if assisting and cooperating agencies have assigned agency representatives. If so, obtain their names, locations, and communication channels by contacting: ■ The agencies. ■ The Incident Commander. ■ The agencies' senior officers at the scene. ❏ Receive requests for contacts between incident personnel and agency personnel. ❏ Identify the appropriate personnel to contact (either incident or agency personnel). ❏ Establish contact with the appropriate personnel. ❏ Take the necessary action to satisfy requests. ❏ Notify concerned personnel.

RESPONSIBILITY	TASKS
Identify Current or Potential Interagency Problems	❑ Receive complaints pertaining to matters such as a lack of logistics, inadequate communications, and personnel problems.
	❑ Personally observe incident operations to identify current or potential interagency problems.
	❑ Notify the appropriate personnel of current or potential problems.
Maintain the Unit Log	❑ Record key actions on the unit log (ICS Form 201 or local form).
	❑ Collect and transmit the required records and logs to the Documentation Unit at the end of each operational period.

AGENCY REPRESENTATIVE CHECKLIST

RESPONSIBILITIES

An Agency Representative is assigned to an incident from an assisting or cooperating agency with full authority to make decisions on all matters affecting that agency's participation at the incident. Agency Representatives report to the Liaison Officer, if that position has been filled. If there is no Liaison Officer, Agency Representatives report to the Incident Commander. There will be only one Agency Representative from each agency assigned to the incident.

Instructions: The checklist below presents the minimum requirements for Agency Representatives. Note that some of the activities are one-time actions, while others are ongoing throughout the incident.

COMPLETED/ NOT APPLICABLE	TASKS
❑	Check in at the Incident Command Post. Complete the check-in list (ICS Form 211 or local form). Ensure that all agency resources have completed check-in.
❑	Obtain a briefing from the Liaison Officer or Incident Commander.
❑	Establish a working location. Advise agency personnel at the incident that the agency representative position has been filled.
❑	Attend planning meetings, as required.
❑	Provide input on the use of agency resources if no

resource technical specialists are assigned.

NOT APPLICABLE	TASKS
❏	Cooperate fully with the Incident Commander and general staff on the agency's involvement at the incident.
❏	Oversee the well-being and safety of agency personnel assigned to the incident.
❏	Advise the Liaison Officer of any special agency needs or requirements.
❏	Determine if any special reports or documents are required.
❏	Report to agency dispatch or headquarters on a pre-arranged schedule.
❏	Ensure that all agency personnel and/or equipment are properly accounted for and released prior to your departure.
❏	Ensure that all required agency forms, reports, and documents are complete prior to your departure.
❏	Hold a debriefing session with the Liaison Officer or Incident Commander prior to departure.

PLANNING SECTION CHIEF CHECKLIST

RESPONSIBILITIES

The Planning Section Chief, a member of the Incident Commander's general staff, is responsible for the collection, evaluation, dissemination, and use of information regarding the development of the incident and status of resources. Information is needed to:

- Understand the current situation.
- Predict the probable course of incident events.
- Prepare alternative strategies and control operations for the incident.

The Planning Section Chief reports directly to the Incident Commander. The Planning Section Chief may have a deputy. The deputy's responsibilities will be as delegated by the Planning Section Chief. Unit functions may be combined if workload permits.

Instructions: The checklist below presents the minimum requirements for Planning Section Chiefs. Note that some activities are one-time actions, while

FEDERAL EMERGENCY MANAGEMENT AGENCY (FEMA)

others are ongoing and repetitive throughout the incident.

COMPLETED/ NOT APPLICABLE	TASKS
☐	Obtain a briefing from the Incident Commander.
☐	Activate Planning Section units.
☐	Collect and process situation information about the incident.
☐	Reassign initial response personnel to incident positions, as appropriate.
☐	Establish information requirements and reporting schedules for all ICS organizational elements for use in preparing the Incident action plan.
☐	Notify the Resources Unit of the Planning Section units which have been activated, including the names and locations of assigned personnel.
☐	Establish a weather data collection system, when necessary.
☐	Supervise the preparation of the Incident action plan (see planning process checklist).
☐	Assemble information on alternative strategies.
☐	Assemble and disassemble strike teams not assigned to operations.
☐	Identify the need for use of specialized resource(s).
☐	Perform operational planning for the Planning Section.
☐	Provide periodic predictions on incident potential.
☐	Compile and display the staff incident status summary information.
☐	Advise the general staff of any significant changes in incident status.
☐	Provide the incident traffic plan.
☐	If requested, assemble and disassemble strike teams and task forces not assigned to operations.
☐	Supervise the Planning Section units.
☐	Prepare and distribute the Incident Commander's orders.
☐	Instruct the Planning Section units on how to distrib-

ute incident information.

COMPLETED/ NOT APPLICABLE	TASKS
❏	Ensure that normal agency information collection and reporting requirements are being met.
❏	Oversee preparation of incident demobilization plan.
❏	Prepare recommendations for the release of resources (to be submitted to the Incident Commander).

The major responsibilities of the Planning Section Chief are stated below. Following each responsibility are procedures for implementing the activity.

RESPONSIBILITY	TASKS
Obtain Briefing from Incident Commander	❏ Receive briefing from the Incident Commander and obtain: ■ Incident Commander's Incident Briefing (ICS Form 201 or local form). ■ Summary of resources dispatched to the incident. ■ Initial restrictions concerning work activities.
Activate Planning Section Units	❏ Determine from the Incident Commander's briefing what Planning Section personnel have been dispatched. ❏ Confirm dispatch of Planning Section personnel. ❏ Plan preliminary organization of Planning Section. ❏ Identify units to be activated. ❏ Estimate personnel required. ❏ Compare preliminary plan with personnel dispatched, as appropriate. ❏ Establish time intervals at which data are to be supplied by Planning Section units. ❏ Assign work locations and work tasks to Planning Section personnel. ❏ Request additional personnel as required. ❏ Notify Resources Unit of Planning Section units activated, including names and locations of

assigned personnel.

RESPONSIBILITY	TASKS
Reassign Initial Attack Personnel to Incident Positions	❏ Review the situation to identify the need for personnel familiar with the incident area. ❏ Identify personnel who are most familiar with the incident area. ❏ Arrange for reassignment of these personnel to incident positions. ❏ Ensure adequate Planning Section personnel are available to complete the Operational Planning Worksheet (ICS Form 215 or local form).

Supervise Preparation of Incident Action Plan	❏ Establish information requirements and reporting schedules for all ICS organizational elements to use in preparing the incident action plan and attachments.
	❏ Present general incident control objectives, including alternatives.
	❏ Participate in a discussion of specific operations being considered, and provide detailed information concerning:
	  ■ Resource availability.
	  ■ Situation status.
	  ■ Situation predictions.
	  ■ Weather.
	  ■ Communication capabilities.
	  ■ Environmental impact and cost of resources use information.
	❏ Participate in selection of operational objectives for the next operational period.
	❏ Assemble appropriate material for inclusion in the incident action plan.

❏ Ensure that all operations support and service needs are coordinated with the Logistics Section prior to release of the incident action plan.

RESPONSIBILITY	TASKS
	❏ Document and distribute the incident action plan to the Incident Commander, section chiefs, branch directors, unit leaders, division/group supervisors, incident command staff, and strike team/task force leaders. ❏ Receive notification of incident action plan changes from the Operations Section Chief. ❏ Distribute incident action plan changes to recipients of the plan. Note: The Planning Section Chief may include in the meeting those Planning Section technical specialists deemed necessary.
Assemble Information on Alternative Strategies	❏ Review the current situation status, resource status, weather, and prediction reports for the current incident status. ❏ Develop alternative strategies using technical specialists and operations personnel, as appropriate. ❏ Identify resources required to implement the alternative operational objectives. ❏ Contact the involved agency dispatch center to identify resource availability for the incident. ❏ Document alternatives for presentation to the Incident Commander and his or her staff.
Assemble Strike Teams/Task Forces not Assigned to Operations	❏ Prior to each planning meeting, identify individual resources not assigned to the Operations Section. ❏ Periodically review operations activity to determine the need to assemble additional strike teams/task forces from individual resources. ❏ Determine strike teams/task forces to be assembled by type, location, and strike team leader.

❑ Request the Resources Unit to select specific resources to assign to each strike team/task force and assign a designator.

RESPONSIBILITY	TASKS
	❑ Request the Resources Unit to notify strike team/task force leaders and resources to assemble into assigned strike teams/task forces by preparing a list of assignments and submitting the assignment list to the communications center for assignment.
	Note: The specifications for each kind/type of strike team/task force must be followed and all units must have a common communications link. If needed, arrange for additional radios through the Communications Unit.
Disassembling Strike Teams	Note: Strike teams are not disassembled unless there is a need for a specific resource or fewer resources than in a strike team and/or it would be inappropriate to use a full strike team. When strike teams are disassembled at the incident, the individual units must be identified and carried by the resources unit. ❑ Disassemble strike teams (or task forces) for demobilization. ❑ Reassemble strike teams that have been disassembled for purposes other than demobilization at the earliest possible time. ❑ Review alternative operational objectives to determine the need for the use of individual resources versus task forces. ❑ Identify individual resources and strike teams that are not assigned to the Operations Section (including their leaders). ❑ Determine if there is an adequate number of individual resources to meet the needs of the incident. ❑ Determine the strike teams that can be disassembled. ❑ Request the Resources Unit to prepare reassignment of strike team leaders to manage task forces. ❑ Request the Resources Unit to reassign resources by designating resources to a specific mission or to other units in staging areas, the base, or camps.

- ❑ Request the Resources Unit to notify strike team/task force leaders of disassembly and reassignment of resources (as required).

RESPONSIBILITY	TASKS
Identify Need for use of Specialized Resources	❑ As part of the planning function, identify the need for technical specialists. ❑ Request personnel with required special knowledge/ experience to be assigned to the Planning Section. Note: Some specialists may be assigned temporarily or for a short duration.
Perform Operational Planning for Planning Section	❑ Review the incident action plan with the Planning Section Chief. ❑ Plan the organization of the Planning Section by identifying units to be activated and estimating the number of personnel required. ❑ Request needed additional personnel from the Resources Unit. ❑ If personnel are not available from the Resources Unit, request them directly from the Logistics Section Chief. ❑ Give specific work tasks including work locations to the Planning Section staff.
Provide Periodic Predictions on Incident Potential	❑ Obtain the latest incident prediction information and incident action plan. ❑ Obtain the current situation status summary from the Situation Unit. ❑ Identify risks and possible hazards. ❑ Estimate work accomplishment for the prediction period. ❑ Document predictions on the course of the incident.

- Present predictions at the planning meeting and display in the Incident Command Post area.
- Repeat procedures at the intervals specified by the Incident Commander or upon occurrence of significant events.

RESPONSIBILITY	TASKS
	- If the prediction indicates a significant change in the course of the incident, immediately notify the Incident Commander and the Operations Section Chief.
Compile and Display Incident Status Information	- Display incident status summary information at a common location in the Incident Command Post area, including multiple overlays, if needed. - Receive information from the Situation Unit, Resources Unit, and the incident prediction and review information for completeness. Specify location and method of display. - Ensure that all reports are displayed. - Repeat these procedures at intervals specified by the Incident Commander or upon occurrence of significant events.
Advise General Staff of any Significant Changes in Incident Status	- Reported significant changes in incident status to the general staff immediately. - Receive requests for incident status information from the general staff. - Obtain incident status information from appropriate sources. - Assemble and summarize the requested information in an appropriate form. - Supply the information to the general staff.
Prepare and Provide Incident Traffic Plan information:	Note. The traffic plan will include the following

- Specified routes to reporting locations for resources dispatched to the incident.
- Specified routes inside general incident area.
- Traffic flow inside ICS facilities.

RESPONSIBILITY	TASKS
To prepare the traffic plan:	❏ Review control operations to determine the locations of planned operations activities and the locations of all incident facilities.
	❏ Review the information obtained from the Situation Unit and/or agency dispatch center(s) to determine existing roadways and their characteristics and capabilities.
	❏ Establish traffic routing factors and coordinate traffic flow plans with appropriate agency representatives.
	❏ Document the traffic plan and attach in to the incident action plan.
	❏ Ensure that Ground Support Unit receives a copy of the traffic plan.
Supervise Planning Section Units	❏ Maintain communications with Planning Section personnel.
	❏ Coordinate the activities of all Planning Section units.
	❏ Ensure the general safety and welfare of Planning Section personnel.
Prepare and Distribute Incident Commander's Orders	❏ Identify orders being issued in the name of the Incident Commander.
	❏ Document all formal operational orders given by the Incident Commander.
	❏ Identify the organizational elements responsible for executing the orders.
	❏ Distribute the orders in accordance with local policy.
Instruct Planning Section Units on	

Distribution of Incident Information	❏ Contact section chiefs and command staff to determine major information categories they want to receive automatically from the Planning Section. ❏ Consolidate the information and prepare a list for each unit.

RESPONSIBILITY	TASKS
Prepare Recommendations for Release of Resources	❏ Identify the number of out-of-service resources and/or individuals by reviewing the current resource status information. ❏ Review the latest situation status and incident prediction information. ❏ Estimate current and future requirements for resources. ❏ Identify and list any potentially surplus resources. ❏ Review the surplus resource list with Operations Section personnel and the Logistics Section Chief. ❏ Modify the surplus resource list as necessary. ❏ Upon approval of the Operations Section Chief and the Logistics Chief, present the list of resources recommended for release to the Incident Commander. ❏ Document the approved demobilization plan. ❏ Prepare an assignment list specifying resources to be released and submit it to the Logistics Section Chief for notification of the involved resources. ❏ Distribute the demobilization plan to the general staff, incident command staff, and agency dispatch centers.
Submit Documentation to Documentation Unit	❏ Submit all documentation to Documentation Unit at the end of each operational period.

SITUATION UNIT LEADER CHECKLIST

RESPONSIBILITIES

The Situation Unit is primarily responsible for the collection and organization of incident status and situation information, and the evaluation, analysis, and display of that information for use by ICS personnel.

Instructions: The checklist below presents the minimum requirements for Situation Unit Leaders. Note that some items are one-time actions, while others are ongoing and repetitive throughout the incident.

COMPLETED/ NOT APPLICABLE	TASKS
❑	Report to and receive a briefing and special instructions from the person in charge of planning activities when you arrive.
❑	Prepare and maintain the Incident Command Post display.
❑	Assign duties to situation status personnel.
❑	Confirm the dispatch and estimated time of arrival of requested Situation Unit personnel and request additional personnel (or release excess personnel).
❑	Collect all incident-related data at the earliest possible opportunity, and continue to do so throughout incident.
❑	Post data on unit work displays and Incident Command Post displays at scheduled intervals or as requested by command post personnel.
❑	Participate in incident planning meetings, as required by the Incident Commander.
❑	Develop and implement accountability, safety and security measures for personnel and resources.
❑	Prepare the Incident Summary Form (ICS Form 209 or local form) before each planning meeting.
❑	Provide photographic services and maps, as necessary.
❑	Provide resources and situation status information in response to specific requests.
❑	Maintain the Situation Unit records.
❑	Receive the order to demobilize the Situation Unit.

❏	Dismantle the Situation Unit displays and place them in storage.
❏	List the expendable supplies that need replenishing and file the list with the Supply Unit.
❏	Maintain the unit log (ICS Form 214 or local form).

RESOURCES UNIT LEADER CHECKLIST
RESPONSIBILITIES

The Resources Unit is primarily responsible for:

- ❏ Seeing that incident resources are properly checked in.
- ❏ The preparation and processing of resource status change information.
- ❏ The preparation and maintenance of displays, charts, and lists which reflect the current status and location of operational resources, transportation, and support vehicles.
- ❏ Maintaining a file or check-in list of resources assigned to the incident.

Instructions: The checklist below presents the minimum requirements for Resources Unit Leaders. Note that some items are one-time actions, while others are ongoing or repetitive throughout the incident.

COMPLETED/ NOT APPLICABLE	TASKS
❏	Report to and obtain a briefing and special instructions from the Planning Section Chief.
❏	Establish check-in procedures at specified incident locations.
❏	Using the Incident Briefing Form (ICS Form 201 or local form), prepare and maintain the Incident Command Post display (organizational chart and resource allocation and deployment sections).
❏	Assign duties to resource unit personnel.
❏	Confirm the dispatch of and estimated time of arrival for ordered Resources Unit personnel. (Request additional personnel or release excess personnel.)
❏	Establish contacts with incident facilities by telephone or through the communications center, and begin maintenance of resource status.

☐	Participate in Planning Section meetings, as required by the Planning Section Chief.
☐	Gather, post, and maintain incident resource status.
☐	Gather, post, and maintain resources status of transportation and support vehicles and personnel.
☐	Maintain a master list of all resources checked at the incident.

COMPLETED/ NOT APPLICABLE	TASKS
☐	Prepare the organization Assignment List (ICS Form 203 or local form) and Organization Chart (ICS Form 204 or local form).
☐	Prepare the appropriate parts of Division Assignment Lists (ICS Form 204 or local form).
☐	Provide resource summary information to the Situation Unit, as requested.
☐	Receive the order to demobilize the Resources Unit.
☐	List the expendable supplies that need replenishing and file with the Supply Unit Leader.
☐	Maintain the unit log (ICS Form 214 or local form).

DOCUMENTATION UNIT LEADER CHECKLIST
RESPONSIBILITIES

The Documentation Unit is responsible for:

- Maintaining accurate and complete incident files.
- Providing duplication service to incident personnel.
- Pack and store incident files for legal, analytical, and historical purposes.

Instructions: The checklist below presents the minimum requirements for Documentation Unit Leaders. Note that some activities are one-time actions, while others are ongoing throughout the incident.

COMPLETED/ NOT APPLICABLE	TASKS
☐	Obtain a briefing from the Planning Section Chief.
☐	Establish a work area.
☐	Establish and organize incident files.

COMPLETED/ NOT APPLICABLE	TASKS
☐	Establish a duplication service and respond to requests.
☐	Retain and file duplicate copies of official forms and reports.
☐	Accept and file reports and forms submitted by ICS units.
☐	Check on the accuracy and completeness of records submitted for files.

COMPLETED/ NOT APPLICABLE	TASKS
☐	Correct errors or omissions by contacting the appropriate ICS units.
☐	Provide duplicates of forms and reports to authorized requestors.
☐	Prepare incident documentation for the Planning Section Chief when requested.
☐	Maintain, retain, and store incident files for after incident use.
☐	Maintain the unit log (ICS Form 214 or local form).

DEMOBILIZATION UNIT LEADER CHECKLIST
RESPONSIBILITIES

The demobilization of the resources and personnel from a major incident is a team effort involving all elements of the incident command organization. The Demobilization Unit develops the demobilization plan and coordinates and supports the implementation of that plan throughout the incident command organization. Several units of the incident command organization-primarily in logistics-are responsible for assisting in the demobilization effort. These units also should participate in the preparation of the plan.

The Demobilization Unit Leader is responsible for the preparation of the demobilization plan and assisting sections and/or units in ensuring that an orderly, safe, and cost-effective movement of personnel and equipment is accomplished from the incident.

Individual agencies and/or contractors may have additional specific procedures to follow in the process of incident demobilization.

Instructions: The checklist below presents the minimum requirements for Demobilization Unit Leaders. Note that some activities are one-time actions, and others are ongoing or repetitive throughout the incident.

COMPLETED/ NOT APPLICABLE	TASKS
☐	Follow ICS general instructions.
☐	Obtain a briefing from the Planning Section Chief.
☐	Review the incident resource records (ICS Forms 201, 211, 219 or local forms) to determine the probable size of the demobilization effort.
☐	Assess and fill unit needs for additional personnel, workspace, and supplies.

COMPLETED/ NOT APPLICABLE	TASKS
☐	Obtain objectives, priorities, and constraints on demobilization from the Planning Section Chief, agency representatives, and contractors, as applicable.
☐	Meet with agency representatives to determine:
	■ Personnel rest and safety issues.
	■ Coordination procedure with cooperating and/or assisting agencies.
☐	Be aware of ongoing Operations Section resource needs.
☐	Obtain identification and description of surplus resources and probable release times.
☐	Coordinate with the Planning Section to arrange shifts to assure priority resources are available for release.
☐	Develop release procedures in coordination with other sections and/or units and agency dispatch center(s).
☐	Coordinate with sections and/or units to determine their capabilities to support the demobilization effort.
☐	Establish a communications link with appropriate off-incident facilities.
☐	Prepare the demobilization plan, including following sections:
	■ General - Discussion of the demobilization procedure.

- Responsibilities - Specific implementation responsibility and activity.

- Release Priority - Take into account the assisting agency requirements and kinds and types of resources.

- Release Procedures - Detailed steps and processes to be followed.

- Travel Restrictions - Restrictions and instructions for travel.

COMPLETED/ NOT APPLICABLE	TASKS
❏	Prepare appropriate directories (e.g., maps, instructions, etc.) for inclusion in the demobilization plan.
❏	Obtain approval of the demobilization plan.
❏	Distribute the plan to each section and processing point (on-and off-incident).
❏	Ensure that all sections and/or units understand their responsibilities within the demobilization plan.
❏	Ensure that all personnel receive a critical incident stress debriefing.
❏	Coordinate and closely supervise the demobilization process.
❏	Brief the Planning Section Chief on the progress of demobilization.
❏	Complete all records prior to departure.
❏	Maintain the unit log (ICS Form 214 or local form).

OPERATIONS SECTION CHIEF CHECKLIST

RESPONSIBILITIES

The Operations Section Chief, a member of the general staff, is responsible for the management of all operations directly applicable to the primary mission. The Operations Section Chief activates and supervises operations, organizational elements, and staging areas in accordance with the incident action plan. The Operations Section Chief also assists in the formulation of the incident action plan and directs its execution. The Operations Section Chief also directs the formulation and execution of subordinate unit operational plans and requests or releases resources and recommends these to the incident commander. He or she also makes expedient changes to the incident action plan (as necessary) and reports such to the Incident Commander.

The Operations Section Chief reports directly to the Incident Commander. The Operations Section Chief may have a deputy. The deputy's responsibilities will be as delegated by the Operations Section Chief, and the deputy must serve in the same operational period.

Instructions: The checklist below presents the minimum requirements for Operations Section Chiefs. Note that some activities are one-time actions, while others are ongoing throughout the incident.

COMPLETED/ NOT APPLICABLE	TASKS
❏	Obtain a briefing from the Incident Commander.
❏	Develop the operations portion of the incident action plan.
❏	Brief and assign operations personnel in accordance with the incident action plan.
❏	Supervise operations.
❏	Establish staging areas.
❏	Determine need and request additional resources.
❏	Review the suggested list of resources to be released and initiate recommendations for the release of resources.
❏	Assemble and disassemble strike teams assigned to the Operations Section.
❏	Report information about activities, events, and occurrences to the Incident Commander.

The major responsibilities of the Operations Section Chief are stated below. Following each are tasks for implementing the activity.

RESPONSIBILITY	TASKS
Obtain Briefing from Incident Commander	❏ Receive briefing from Incident Commander and obtain: ■ Incident Briefing (ICS Form 201 or local form). ■ Summary of resources dispatched to the incident. ■ Initial instructions concerning work activities.

RESPONSIBILITY	TASKS
Develop Operations Portion of Incident Action Plan	❏ Discuss incident situation with immediate subordinates and obtain control actions planned for each operational period. ❏ Review control operations based on information provided by the Planning Section relating to: ■ Resource availability. ■ Situation status. ■ Fire behavior prediction. ■ Weather. ■ Communications capability. ■ Environmental impact and cost/resources use information. ❏ Develop planned control operations for each division/group. ❏ Make resource assignments for each division/group in conjunction with the Resources Unit.
Brief Operations Personnel on Incident Action Plan	❏ Contact the Resources Unit to identify branch directors and division/group supervisors who have been dispatched to the incident. ❏ Conduct a briefing meeting for branch directors and division/group supervisors on the incident action plan and attachments. ❏ Make sure subordinates have the incident action plan. ❏ Establish reporting requirements concerning execution of the operations portion of the incident action plan. ❏ Provide additional information as requested by subordinates.
Supervise Operations	❏ Receive information routinely or as requested about operations activities from Situation Unit field observers and operations personnel. ❏ Determine the adequacy of operations progress by:

- Approving changes to incident action plan as necessary.
- Providing information on the above changes to the Incident Commander and Planning Section Chief.
- ❑ Implement necessary changes in operations.
- ❑ Handle unresolved problems within the Operations Section.
- ❑ Provide for the general welfare and safety of operations personnel.

RESPONSIBILITY	TASKS
Establish and Maintain Staging Areas	❑ Identify appropriate location(s) for staging area(s).
	❑ Identify expected number and type of resources to be assembled in each area.
	❑ Identify anticipated duration for use of each area.
	❑ Determine if there is any need for temporary assignment of logistics service and support to staging areas.
	❑ Make arrangements for temporary logistics, if required, by notifying the Logistics Section Chief.
	❑ Assign a Staging Area Manager to each staging area, as appropriate.
Determine Need for Additional Resources	❑ Evaluate the progress of operations by obtaining the latest situation report, and the latest fire behavior prediction, and receive and evaluate reports form operations personnel.
	❑ Determine the reason(s) for inadequacies in operations, if they exist.
	❑ Request any additional required resources from the Resources Unit and provide the type and quantity, time and location of need, and supervisor and communications channel to use.

Review Suggested List of Resources to be Released	❑ Review the list of potential resources to be released provided by the Planning Section Chief. ❑ Evaluate the adequacy of operations by reviewing the latest situation status information, the latest fire behavior prediction information, and reports from field personnel. ❑ Estimate current and future resource requirements. ❑ Submit a list of resource requirements to the Resource Unit.

RESPONSIBILITY	TASKS
Assemble Strike Team from Resources Assigned to Operations Section	❑ Periodically review operations control activity to determine need for assembling strike teams from individual resources. ❑ Determine strike teams to be assembled by type, location, and strike team leader. ❑ Select specific resources to assign to each strike team. ❑ Notify strike team leaders and resources to assemble into assigned strike teams by preparing a list of assignments and submitting the list to Resources Unit which will assign strike team identification numbers and change the status of the assigned resources. Note: The specifications for each type of strike team must be followed, and all units within a strike team must have a common communications link.
Disassemble Strike Teams Assigned to Operations Section	Note: Strike teams are not disassembled unless there is a need for a specific resource or a need for fewer resources than in a strike team. When strike teams are disassembled at the incident, the individual units are identified and status maintained by the

Resources Unit. A strike team assembled at the incident may be disassembled for demobilization. Strike teams disassembled for purposes other than demobilization will reassemble at the earliest possible time.

- ❏ Review alternative control actions to determine anticipated need for types of resources other than strike teams such as single resources and task forces.
- ❏ Review resource status.
- ❏ Determine if there are an adequate number of single resources to fill needs at the incident.
- ❏ Determine strike teams to be disassembled.
- ❏ Reassign or release strike team leader.

RESPONSIBILITY	TASKS
	❏ Reassign resources to specific missions as applicable.
	❏ Notify strike team leaders of disassembly and reassignment of single resources within their strike teams by:
	■ Preparing a list of assignments.
	■ Submitting the list to the communications center for transmitting assignments.
	■ Submitting the list to the Resources Unit for changes to status of resources.
Initiate Recommendation for Release of Resources	❏ Designate resources recommended for release by type, quantity, location, and time.
	❏ Present recommendations to the Incident Commander with supporting information.
Report Special Incidents/Accidents	❏ Obtain information about special events, personal observations, and operations personnel from subordinates. This information should include the nature of the event, location, magnitude, personnel involved, initial action(s) taken, and appropriate subsequent action(s).

- ❏ Request needed assistance.
- ❏ Submit the report to the Incident Commander.
- ❏ Maintain the unit log (ICS Form 214 or local form) and give it to the Documentation Unit at the end of each operational period.

STAGING AREA MANAGER CHECKLIST

RESPONSIBILITIES

The Staging Area Manager is responsible for overseeing the staging area. The Staging Area Manager reports to the Operations Section Chief.

Instructions: The Staging Area Manager will accomplish the following checklist of activities. Note that some activities are one-time only actions, while others are ongoing or repetitive throughout the incident.

COMPLETED/ NOT APPLICABLE	TASKS
❏	Obtain a briefing from the Operations Section Chief.
❏	Proceed to a staging area.
❏	Establish a staging area layout.
❏	Determine any support needs for equipment, feeding, sanitation, and security.
❏	Establish check-in procedures, as appropriate.
❏	Determine required resource reserve levels from the Operations Section Chief or Incident Commander.
❏	Advise the Operations Section Chief or Incident Commander when reserve levels reach minimums.
❏	Post areas for identification and traffic control.
❏	Request maintenance service for equipment at staging areas, as appropriate.
❏	Respond to request for resource assignments.
❏	Obtain and issue receipts for radio equipment and other supplies distributed and received at the staging area.
❏	Report resource status changes as required.
❏	Maintain the staging area in orderly condition.
❏	Demobilize the staging area in accordance with the incident demobilization plan.
❏	Maintain the unit log (ICS Form 214 or local form).

BRANCH DIRECTOR (OPERATIONS SECTION) CHECKLIST
RESPONSIBILITIES

The Operations Branch Director is responsible for the implementation of the incident action plan within the branch. This includes the direction and execution of branch planning for the assignment of resources within the branch. Branch directors will be activated only when and as needed in accordance with incident characteristics, the availability of personnel, and the requirements of the Incident Commander and Operations Section Chief. The deputy, if activated, must serve in the same operational period as the director. The Operations Branch Director reports to the Operations Section Chief.

Instructions: The checklist below presents the minimum requirements for Operations Branch Directors. Note that some activities are one-time actions, while others are ongoing or repetitive throughout the incident.

COMPLETED/ NOT APPLICABLE	TASKS
❏	Obtain a briefing from the Operations Section Chief.
❏	Develop with subordinates alternatives for branch control operations.
❏	Interact with the Operations Section Chief and other Branch Directors to develop tactics to implement incident strategies.
❏	Attend planning meetings at the request of the Operations Section Chief.
❏	Review the Division/Group Assignment List (ICS Form 204 or local form) for divisions/groups within the branch. Modify lists based on the effectiveness of current operations.
❏	Assign specific work tasks to division/group supervisors.
❏	Resolve logistics problems reported by subordinates.
❏	Report to the Operations Section Chief when: ■ The incident action plan must be modified. ■ Additional resources are needed. ■ Surplus resources are available. ■ Hazardous situations or significant events occur.
❏	Approve accident and medical reports originating with the branch.
❏	Maintain the unit log (ICS Form 214 or local form).

DIVISION/GROUP SUPERVISOR (OPERATIONS SECTION) CHECKLIST
RESPONSIBILITIES

Divisions divide an incident into natural separations where resources can be effectively managed under span-of-control guidelines. Examples of divisions are floors of a building or segments of a line. Groups are functional and describe activity. Examples of groups are ventilation, salvage, or secondary line construction.

The Division or Group Supervisor is responsible for:

- The implementation of the assigned portion of the incident action plan.
- The assignment of resources within the division or group.
- Reporting on the progress of control operations.
- The status of resources within the division or group.

The Division/Group Supervisor reports to the Branch Director or, in the event that Branch Directors are not activated, to the Operations Section Chief.

Instructions: The checklist below presents the minimum requirements for Division or Group Supervisors. Note that some activities are one-time actions, while others are ongoing or repetitive throughout the incident.

COMPLETED/ NOT APPLICABLE	TASKS
❑	Obtain a briefing from a Branch Director or the Operations Section Chief.
❑	Implement the incident action plan for the division or group.
❑	Provide the incident action plan to Strike Team Leaders, when available.
❑	Identify the resources assigned to the division or group.
❑	Review the division or group assignments and incident activities with subordinates and assign tasks.
❑	Ensure that the Communications and/or Resources Unit are advised of all changes in status of resources assigned to the division or group.
❑	Coordinate activities with the adjacent division or group.
❑	Monitor and inspect progress and make changes as necessary.
❑	Determine the need for assistance on assigned tasks.
❑	Submit situation and resource status information to the Branch Director or Operations Section Chief.

- ☐ Report special occurrences or events (e.g., accidents, sickness, hazardous situations, etc.) to the immediate supervisor.
- ☐ Resolve logistics problems within the division or group.
- ☐ Ensure that assigned personnel and equipment get to and from their assignments in a timely and orderly manner.
- ☐ Participate in the development of branch plans for the next operational period.
- ☐ Maintain the unit log (ICS Form 214 or local form).

STRIKE TEAM/TASK FORCE LEADER CHECKLIST

RESPONSIBILITIES

The Strike Team or Task Force Leader is responsible for performing operations assigned to a strike team or task force. The leader reports work progress, resource status, and other important information to a division supervisor and maintains work records on assigned personnel. The Strike Team/Task Force Leader reports to a Division/Group Supervisor.

Instructions: The checklist below presents the minimum requirements for Strike Team or Task Force Leaders. Note that some activities are one-time actions, while others are ongoing and repetitive throughout the incident.

COMPLETED/ NOT APPLICABLE	TASKS
☐	Obtain a briefing from the division or group supervisor.
☐	Review strike team or task force assignments with subordinates and assign tasks.
☐	Travel to and from active assignment area with assigned resources.
☐	Monitor work progress and make changes when necessary.
☐	Determine the need for assistance on assigned tasks.
☐	Coordinate activities with adjacent strike teams or task forces and single resources.
☐	Submit situation and resource status information to the division or group supervisor.
☐	Retain control of assigned resources while in avail-

- ❏ able or out-or-service status.
- ❏ Report special events.
- ❏ Request service and/or support.
- ❏ Report status and location changes.
- ❏ Maintain the unit log (ICS Form 214 or local form).

LOGISTICS SECTION CHIEF CHECKLIST
RESPONSIBILITIES

The Logistics Section Chief, a member of the general staff, is responsible for providing facilities, services, and materials in support of the incident. The Logistics Section Chief participates in the development of the incident action plan and activates and supervises the branches and units within the Logistics Section.

Instructions: The checklist below presents the minimum requirements for Logistics Section Chiefs. Note that some items are one-time actions, while others are ongoing or repetitive throughout the incident.

COMPLETED/ NOT APPLICABLE	TASKS
❏	Obtain a briefing from the Incident Commander.
❏	Plan the organization of the Logistics Section.
❏	Assign work locations and preliminary work tasks to section personnel.
❏	Notify the Resources Unit of the Logistics Section units which have been activated, including the names and locations of assigned personnel.
❏	Assemble and brief unit leaders and branch directors.
❏	Participate in the preparation of the incident action plan.
❏	Identify the service and support requirements for planned and expected operations.
❏	Provide input to and review the communications, medical, and traffic plans.
❏	Coordinate and process requests for additional resources.
❏	Review the incident action plan, and estimate section needs for the next operational period.
❏	Ensure that the incident communications plan is prepared.

	Advise on current service and support capabilities.
❏	
❏	Prepare the service and support elements of the incident action plan.
❏	Estimate future service and support requirements.
❏	Receive the demobilization plan from the Planning Section.
❏	Recommend the release of unit resources in conformity with the demobilization plan.
❏	Ensure the general welfare and safety of Logistics Section personnel.
❏	Maintain the unit log (ICS Form 214 or local form).

The major responsibilities of the Logistics Section Chief are stated below. Following each are tasks for implementing the responsibility.

RESPONSIBILITY	TASKS
Obtain Briefing from Incident Commander	❏ Receive an incident briefing, summary of resources dispatched to the incident, and initial instructions concerning work activities.
	❏ Obtain a copy of the incident action plan, if available.
Activate Logistics Section Units	❏ Determine from the incident briefing what Logistics Section personnel have been ordered.
	❏ Confirm order of appropriate Logistics Section personnel.
	❏ Plan preliminary organization of the Logistics Section.
	❏ Compare the preliminary incident action plan with personnel ordered, as appropriate.
	❏ Identify additional personnel needed.
	❏ Request additional personnel.
	❏ Assign work locations and work tasks to logistics section personnel.
	❏ Notify the Resources Unit of Logistics Section units activated, including names and locations of assigned personnel.
Organize Logistics	

Section personnel.	❑ Confirm arrival of dispatched Logistics Section
	❑ Assemble and brief Logistics Section personnel.
	❑ Review initial operations of Logistics Section with section personnel.
	❑ Give instructions for initial operations to section personnel.
Assist in Preparation of the Incident Action Plan	
	❑ Attend planning meeting.
	❑ Review suggested strategy and operations for next operational period.
	❑ Advise on current service and support capabilities.

RESPONSIBILITY	**TASKS**
	❑ Estimate logistic capabilities with current capabilities.
	❑ Compare required capabilities with current capabilities.
	❑ Determine additional service and support requirements corresponding to the incident action plan.
	❑ Prepare service and support elements of the incident action plan.
	❑ Identify potential future control operations so as to anticipate logistics requirements.
Request Additional Incident Resources	Note: The Logistics Section Chief performs this function only if the Incident Commander has delegated the corresponding authority.
	❑ Receive requests for resources to be ordered from outside of the incident from members of the general staff or the Resources Unit.
	❑ Coordinate requests for additional resources so as to eliminate duplicate requests.
	❑ Submit the request through the communications center for additional resources from outside the incident. The request goes through normal channels and includes a confirmation/denial of request and ETAs.
Perform Operational	

RESPONSIBILITY	TASKS
Planning for Logistics Section	❏ Obtain the incident action plan from the Planning Section Chief and review with section personnel as appropriate.
	❏ Identify service and support requirements for planned and expected incident operations.
	❏ Plan organization of the Logistics Section.
	❏ Compare organization plan requirements with dispatched personnel.
	❏ Identify needed or surplus personnel.
	❏ Notify the Resources Unit of names of personnel available for assignment or reassignment.
	❏ Notify personnel being reassigned.
	❏ Request additional personnel needed.
	❏ Request additional support from the Incident Commander if personnel are not available from incident sources.
	❏ Notify the Resources Unit of resources assigned by Logistics Section for support and service needs.
	❏ Assign work locations and specific work tasks to section personnel.
Update Logistics Section Planning	❏ Review current situation status, resource status, and fire behavior prediction information.
	❏ Obtain information concerning future operations through discussions with incident personnel.
	❏ Estimate future service and support requirements.
	❏ Compare estimated future requirements with expected logistics capabilities.
	❏ Obtain changes to the incident action plan from the Planning Section Chief.
	❏ Obtain the demobilization plan from the Planning Section Chief.
	❏ Identify required modifications to Logistics Section planning. and modify planning as appropriate.
	❏ Inform Logistics Section branch directors, Planning Section Chief, Resources Unit, and oth-

FEDERAL EMERGENCY MANAGEMENT AGENCY (FEMA) 519

	ers as appropriate of planning modifications.
Direct Operations of Organizational Elements	❑ Receive reports of significant events.
	❑ Periodically check work progress on assigned tasks of support and service branches and units, as appropriate.
	❑ Coordinate and supervise activities of Logistics Section units.
	❑ Ensure general welfare and safety of logistics personnel.
	❑ Provide input to and review communications, medical, and traffic plans.

RESPONSIBILITY	TASKS
Recommend Release of Resources/Supplies	❑ List resources/supplies recommended for release by type, quantity, location, and time.
	❑ Present recommendations to the Planning Section Chief.
	❑ Coordinate with the Demobilization Unit on the demobilization plan.
Maintain Logs and Records	❑ Record Logistics Section activities on the unit log (ICS Form 214 or local form).
	❑ Maintain agency records and reports.
	❑ Provide unit logs to the Documentation Unit at the end of each operational period.

LOGISTICS SUPPORT BRANCH DIRECTOR CHECKLIST
RESPONSIBILITIES

The Support Branch Director is responsible for the management of all support activities at the incident.

The Support Branch Director position will be activated only as needed in accordance with incident characteristics, the availability of personnel, and the requirements of the Incident Commander and Logistics Section Chief. The Support Branch Director reports to the Logistics Section Chief.

Instructions: The checklist below presents the minimum requirements for

Support Branch Directors. Note that some items are one-time actions, while others are ongoing or repetitive throughout the incident.

COMPLETED/ NOT APPLICABLE	TASKS
❏	Obtain working materials from the logistics kit.
❏	Identify the Support Branch personnel dispatched to the incident.
❏	Determine initial support operations in coordination with the Logistics Section Chief and Service Branch Director.
❏	Prepare the initial organization and assignments for the initial support operations.
❏	Assemble and brief Support Branch personnel.
❏	Determine if assigned branch resources are sufficient.

COMPLETED/ NOT APPLICABLE	TASKS
❏	Monitor the work progress of units, and keep the Logistics Section Chief informed of activities.
❏	Resolve problems associated with requests from the Operations Section.
❏	Maintain the unit log (ICS Form 214 or local form).

GROUND SUPPORT UNIT LEADER CHECKLIST

RESPONSIBILITIES

The Ground Support Unit Leader is primarily responsible for:

- Providing for the transportation of personnel, supplies, food, and equipment.
- Providing for the fueling, service, maintenance, and repair of vehicles and other ground support equipment.
- Collecting and recording information about the use of rental equipment and services initiated and requested.
- Implementing the traffic plan for the incident.

Instructions: The checklist below presents the minimum requirements for Ground Support Unit Leaders. Note that some activities are one-time actions and others are ongoing or repetitive throughout the incident.

COMPLETED/

NOT APPLICABLE	TASKS
❏	Obtain a briefing from the Support Branch Director or Logistics Section Chief.
❏	Participate in Support Branch and/or Logistics Section planning activities.
❏	Implement the traffic plan developed by the Planning Section.
❏	Support out-of-service resources.
❏	Notify the Resources Unit of all status changes on support and transportation vehicles.
❏	Arrange for and activate the fueling, maintenance, and repair of ground resources.
❏	Maintain an inventory of support and transportation vehicles (ICS Form 218 or local form).
❏	Provide transportation services.

COMPLETED/ NOT APPLICABLE	TASKS
❏	Collect information on rented equipment.
❏	Requisition maintenance and repair supplies (e.g., fuel and spare parts).
❏	Maintain incident roads.
❏	Submit reports to the Support Branch Director as directed.
❏	Maintain the unit log (ICS Form 214 or local form).

FOOD UNIT LEADER CHECKLIST

RESPONSIBILITIES

The Food Unit Leader is responsible for determining feeding and cooking facility requirements at all incident facilities, menu planning, food preparation, serving, providing potable water, and general maintenance of the food service areas.

The Food Unit Leader reports to the Service Branch Director (if activated) or the Logistics Section Chief.

Instructions: The checklist below presents the minimum requirements for Food Unit Leaders. Note that some activities are one time actions, and others are ongoing or repetitive throughout the incident.

COMPLETED/ NOT APPLICABLE	TASKS
❏	Obtain a briefing from the Service Branch Director or

COMPLETED/ NOT APPLICABLE	TASKS
❏	Logistics Section Chief.
❏	Determine the location of the working assignment and the number of personnel assigned to the base and camps.
❏	Determine the method of feeding the best fits each situation.
❏	Obtain the necessary equipment and supplies to operate the food service facilities at the base and camps.
❏	Ensure that sufficient potable water is available to meet all incident needs.
❏	Set up food unit equipment.
❏	Prepare menus to ensure incident personnel of well-balanced meals.
❏	Ensure that all appropriate health and safety measures are taken.

COMPLETED/ NOT APPLICABLE	TASKS
❏	Supervise cooks and other Food Unit personnel.
❏	Keep an inventory of food on hand, and check in food orders.
❏	Provide the Supply Unit Leader with food supply orders.
❏	Demobilize the Food Unit in accordance with the incident demobilization plan.
❏	Maintain the unit log (ICS Form 214 or local form).

COMMUNICATIONS UNIT LEADER CHECKLIST
RESPONSIBILITIES

The Communications Unit Leader, under the direction of the Service Branch Director or Logistics Section Chief, is responsible for developing plans for the effective use of incident communications equipment and facilities. These include:

- Installing and testing of communications equipment.
- Supervision of the incident communications center.
- Distribution of communications equipment to incident personnel.
- Maintenance and repair of communications equipment.

Instructions: The checklist below presents the minimum requirements for

Communications Unit Leaders. Note that some activities are one-time actions, and others are ongoing or repetitive throughout the incident.

COMPLETED/ NOT APPLICABLE	TASKS
❏	Obtain a briefing from the Service Branch Director or Logistics Section Chief.
❏	Determine the Communications Unit personnel needs.
❏	Advise on the communications capabilities and/or limitations during preparation of the incident action plan.
❏	Prepare and implement the incident radio communications plan (ICS Form 205).
❏	Ensure that the incident communications center and message center is established.
❏	Set up the telephone and public address system.

COMPLETED/ NOT APPLICABLE	TASKS
❏	Establish appropriate communications distribution and/or maintenance locations within the base and camp(s).
❏	Ensure that communications systems are installed and tested.
❏	Ensure that an equipment accountability system is established.
❏	Ensure that personal portable radio equipment from cache(s) is distributed per radio plan.
❏	Provide technical information as required on: ■ Adequacy of communications systems currently in operation. ■ Geographic limitations on communications systems. ■ Equipment capabilities. ■ Amount and types of equipment available. ■ Anticipated problems in the use of communications equipment.
❏	Supervise Communications Unit activities.

	Maintain records on all communications equipment as appropriate.
☐	Ensure that all equipment is tested and repaired.
☐	Recover equipment from relieved or released units.
☐	Maintain the unit log (ICS Form 214 or local form).

LOGISTICS SERVICE BRANCH DIRECTOR CHECKLIST

RESPONSIBILITIES

The Service Branch Director is responsible for the management of all service activities at the incident. The Service Branch Director position will be activated only as needed in accordance with incident characteristics, the availability of personnel, and the requirements of the Incident Commander and Logistics Section Chief. The Service Branch Director reports to the Logistics Section Chief.

Instructions: The checklist below presents the minimum requirements for Service Branch Directors. Note that some items are one-time actions, and others are ongoing or repetitive throughout the incident.

COMPLETED/ NOT APPLICABLE	TASKS
☐	Obtain working materials from the logistics kit.
☐	Determine the level of service required to support operations.
☐	Confirm the dispatch of branch personnel.
☐	Participate in the planning meetings of Logistics Section personnel.
☐	Review the incident action plan.
☐	Organize and prepare assignments for Service Branch personnel.
☐	Coordinate the activities of branch units.
☐	Inform the Logistics Section Chief of branch activities.
☐	Resolve Service Branch problems.
☐	Maintain the unit log (ICS Form 214 or local form).

MEDICAL UNIT LEADER CHECKLIST

RESPONSIBILITIES

The Medical Unit Leader is primarily responsible for the development of the medical emergency plan, obtaining medical aid and transportation for injured and ill incident personnel, and preparation of reports and records. The Medical Unit may also assist operations in supplying medical care and assistance to civilian casualties at the incident. The Medical Unit Leader reports to the Service Branch Director. The Medical Unit Leader may require the services of a Welfare Officer to assist in resolving personal matters or to support the general well-being of personnel assigned to the incident.

Instructions: The checklist below presents the minimum requirements for Medical Unit Leaders. Note that some activities are one-time actions, and others are ongoing or repetitive throughout the incident.

COMPLETED/ NOT APPLICABLE	TASKS
❏	Obtain a briefing from the Service Branch Director or Logistics Section Chief.
❏	Participate in Logistics Section and/or Service Branch planning activities.
❏	Determine the level of emergency medical activities performed prior to activation of Medical Unit.

COMPLETED/ NOT APPLICABLE	TASKS
❏	Activate the Medical Unit.
❏	Prepare the medical emergency plan (ICS Form 206 or local form).
❏	Prepare procedures for a major medical emergency.
❏	Declare a major medical emergency, as appropriate.
❏	Respond to requests for medical aid.
❏	Respond to requests for medical transportation.
❏	Respond to requests for medical supplies.
❏	Prepare medical reports.
❏	Submit the reports, as directed.
❏	Maintain the unit log (ICS Form 214 or local form).

FACILITIES UNIT LEADER CHECKLIST
RESPONSIBILITIES

The Facilities Unit Leader is primarily responsible for the activation of incident facilities (i.e., the base, camp(s), and Incident Command Post). The unit provides sleeping and sanitation facilities for incident personnel, and manages base and camp operations. Each facility is assigned a manager who reports to the Facilities Unit Leader and is responsible for managing the operation of the facility. The basic functions or activities of the base and camp manager are to provide security service and facility maintenance. The Facilities Unit Leader reports to the Support Branch Director. Close liaison must be maintained with the Food Unit Leader, who is responsible for providing food for all incident facilities.

Instructions: The checklist below presents the minimum requirements for Facilities Unit Leaders. Note that some of the activities are one-time actions, while others are ongoing or repetitive throughout the incident.

COMPLETED/ NOT APPLICABLE	TASKS
❏	Receive the incident action plan.
❏	Participate in Logistics Section and/or Support Branch planning activities.
❏	Determine the requirement for each facility to be established.
❏	Prepare layouts of incident facilities.
❏	Notify unit leaders of facility layouts.

COMPLETED/ NOT APPLICABLE	TASKS
❏	Activate incident facilities.
❏	Obtain personnel to operate facilities.
❏	Provide sleeping facilities.
❏	Provide security services.
❏	Provide facility maintenance services (sanitation, lighting, clean up, etc.).
❏	Supervise out-of-service resources and unassigned personnel.
❏	Demobilize base and camp facilities.
❏	Maintain the Facilities Unit records.
❏	Maintain the unit log (ICS Form 214 of local form).

SUPPLY UNIT LEADER CHECKLIST
RESPONSIBILITIES

The Supply Unit Leader is primarily responsible for:

- Ordering personnel.
- Ordering, receiving, and storing all supplies for the incident.

- Maintaining an inventory of supplies.
- Servicing non-expendable supplies and equipment.

The major functions of the unit are grouped into the ordering of equipment and supplies and the receiving and/or distribution of equipment, other than primary supplies. The Supply Unit Leader reports to the Support Branch Director.

Instructions: The checklist below presents the minimum requirements for Supply Unit Leaders. Note that some activities are one-time actions, and others are ongoing or repetitive throughout the incident.

COMPLETED/ NOT APPLICABLE	TASKS
❑	Obtain a briefing from the Support Branch Director or Logistics Section Chief.
❑	Participate in the Logistics Section and/or Support Branch planning activities.
❑	Provide kits to Planning, Logistics, and Finance/Administration Sections.
❑	Determine the type and amount of supplies en-route.

COMPLETED/ NOT APPLICABLE	TASKS
❑	Arrange for receiving ordered supplies.
❑	Review the incident action plan for information on operations of the Supply Unit.
❑	Develop and implement safety and security requirements.
❑	Order, receive, distribute, and store supplies and equipment.
❑	Receive and respond to requests for personnel, supplies, and equipment.
❑	Maintain an inventory of supplies and equipment.
❑	Service reusable equipment.
❑	Demobilize the Supply Unit.
❑	Submit reports to the Support Branch Director.

| | Maintain the unit log (ICS Form 214 or local form). |

FINANCE/ADMINISTRATION SECTION CHIEF CHECKLIST
RESPONSIBILITIES

The Finance/Administration Section Chief, a member of the general staff, is responsible to organize and operate the Finance/Administration Section within the guidelines, policy, and constraints established by the Incident Commander and the responsible agency. The Finance/Administration Section Chief participates in the development of the incident action plan and activates and supervises the units within the section.

The finance/administration function within the Incident Command System is heavily tied to agency-specific policies and procedures. The Finance/Administration Section Chief will normally be assigned from the agency with incident jurisdictional responsibility. The organization and operation of the finance/administration function will require extensive use of agency-provided forms. The Finance/Administration Section Chief reports directly to the Incident Commander.

Instructions: The checklist below presents the minimum requirements for Finance/Administration Section Chiefs. Note that some activities are one-time actions, and others are ongoing or repetitive throughout the incident.

COMPLETED/ NOT APPLICABLE	TASKS
❏	Obtain a briefing from the Incident Commander.
❏	Manage all financial aspects of an incident.
❏	Provide financial and cost analysis information as requested.
❏	Attend a briefing with the responsible agency to gather information.
❏	Attend a planning meeting to gather information on overall strategy.
❏	Identify and order supply and support needs for the Finance/Administration Section.
❏	Develop an operations plan for the finance/administration function at the incident.
❏	Prepare work objectives for subordinates, brief staff, make assignments, and evaluate performance.

❏ Determine the need for a commissary operation.

❏ Inform the Incident Commander and general staff when the section is fully operational.

❏ Meet with assisting and cooperating agency representatives, as required.

❏ Provide input in all planning sessions on finance matters.

❏ Ensure that all personnel time records are transmitted to home agencies according to policy.

❏ Participate in all demobilization planning.

❏ Ensure that all obligation documents initiated at the incident are properly prepared and completed.

❏ Brief agency administration personnel on all incident-related business management issues needing attention and follow-up prior to leaving the incident.

❏ Maintain the unit log (ICS Form 214 or local form).

The major responsibilities of the Finance/Administration Section Chief are stated below. Following each are tasks for implementing the responsibility.

RESPONSIBILITY	TASKS
Obtain Briefing from Incident Commander	❏ Obtain an Incident Briefing and a copy of the incident action plan, if available.
Attend Briefing With Responsible Agency to Gather Information	Note: This briefing may by held at an off-incident location prior to arrival at the incident. The purpose of the briefing is to obtain financial information and administrative guidelines and constraints.
Attend Planning Meeting	❏ Gather information on overall strategy and resource use planning.
Identify and Order Supply and Support Needs for Finance Section	❏ Arrange for personnel to support Finance Section's unit-level operations. ❏ Arrange for equipment facilities and supplies nec-

	essary to support finance operation.
Develop an Operating Plan for Finance Function	❑ Consider the size and complexity of incident.
	❑ Consider the role of the Finance Section in serving/assisting other agencies on incident.
	❑ Consider guidelines and policy established by agency.
	❑ Consider personnel assignments, work loads, and welfare.
Meet with Assisting and Cooperating Agency Representatives as Required	❑ Establish contact with the Liaison Officer.
	❑ Obtain list of assisting and cooperating agencies supporting incident.
	❑ Ensure that the Liaison Officer is advised as to the Finance Section operation.

RESPONSIBILITY	TASKS
Provide Input in All Planning Sessions on Finance Matters	❑ Provide cost analysis data on control operations as required.
	❑ Provide financial summary information as required.
Participate in All Demobilization Planning	❑ Provide input to demobilization planning.
	❑ Ensure that all required documentation is available at time of demobilization.
Ensure that All Documents are Prepared and Completed	❑ Maintain required agency records and reports.
	❑ Transfer fiscal documents from incident to responsible agency.

TIME UNIT LEADER CHECKLIST

RESPONSIBILITIES

The Time Unit is responsible for establishing files, collecting employee time reports, and providing a commissary operation to meet incident needs. The Time Unit Leader reports directly to the Finance/Administration Section Chief.

Instructions: The checklist below presents the minimum requirements for Time Unit Leaders. Note that some of the activities are one-time actions, while others are ongoing or repetitive throughout the incident.

COMPLETED/ NOT APPLICABLE	TASKS
❑	Obtain a briefing from the Finance/Administration Section Chief.
❑	Determine the incident requirements for the time-recording function.
❑	Establish contact with appropriate agency personnel and/or representatives.
❑	Organize and establish the Time Unit.
❑	Establish unit objectives, make assignments, and evaluate performance.

COMPLETED/ NOT APPLICABLE	TASKS
❑	Ensure that daily personnel time recording documents are prepared and compliance to time policy is met.
❑	Establish a commissary operation as required.
❑	Submit cost-estimate data forms to the Cost Unit, as required.
❑	Provide for record security.
❑	Ensure that all records are current or complete prior to demobilization.
❑	Release time reports from assisting agencies to the respective agency representatives prior to demobilization.
❑	Brief the Finance/Administration Section Chief on current problems, recommendations, outstanding issues and follow-up requirements.
❑	Maintain the unit log (ICS Form 214 or local form).

COST UNIT LEADER CHECKLIST

RESPONSIBILITIES

The Cost Unit Leader is responsible to prepare summaries of actual and estimated incident costs. The unit also prepares information on costs of resource use and provides cost effectiveness recommendations. The Cost Unit Leader reports to the Finance/Administration Section Chief.

Instructions: The checklist below presents the minimum requirements for Cost Unit Leaders. Note that some of the activities are one-time actions, while others are ongoing or repetitive throughout the incident.

COMPLETED/ NOT APPLICABLE	TASKS
❏	Obtain a briefing from the Finance/Administration Section Chief.
❏	Coordinate with agency headquarters on cost-reporting procedures.
❏	Ensure that all equipment and/or personnel requiring payment are identified.
❏	Obtain and record all cost data.
❏	Prepare incident cost summaries.

COMPLETED/ NOT APPLICABLE	TASKS
❏	Prepare resource-use cost estimates for planning.
❏	Make recommendations for cost savings to the Finance/Administration Section Chief.
❏	Maintain cumulative incident cost records.
❏	Ensure that all cost documents are accurately prepared.
❏	Complete all records prior to demobilization.
❏	Provide reports to the Finance/Administration Section Chief.
❏	Maintain the unit log (ICS Form 214 or local form).

COMPENSATION/CLAIMS UNIT LEADER CHECKLIST

RESPONSIBILITIES

Compensation for injury and claims are handled together within one unit in ICS.

The Compensation/Claims Unit is responsible for:

- The prompt preparation and processing of all forms required in the event of injury or death to any person.

- Gathering evidence and preparing claims documentation for any event involving damage to public or private properties which could result in a claim against the agency. The Compensation/Claims Unit Leader must have firsthand knowledge of all required agency procedures on claims handling.

The unit leader and assigned specialist must work in close coordination with the Medical Unit, Safety Officer, and Agency Representatives. The Compensation/Claims Unit Leader reports to the Finance/Administration Section Chief.

Instructions: The checklist below presents the minimum requirements for Compensation/Claims Unit Leaders. Note that some activities may be one-time actions, and others are ongoing or repetitive for the duration of an incident.

COMPLETED/ NOT APPLICABLE	TASKS
❑	Obtain a briefing from the Finance/Administration Section Chief.
❑	Establish contact with the Safety Officer and Liaison Officer or Agency Representatives if no Liaison Officer is assigned.

COMPLETED/ NOT APPLICABLE	TASKS
❑	Determine the need for injury and claims specialists and order personnel if needed.
❑	Determine with the Medical Unit if the injury took place in a work area, whenever feasible.
❑	Obtain a copy of the incident medical plan.
❑	Ensure that injury and claims specialists have adequate work space and supplies.
❑	Brief compensation or claims specialists on incident activity.
❑	Coordinate with the Procurement Unit on procedures for handling claims.
❑	Periodically review all logs and forms produced by compensation/claims specialists to ensure that:

- Work is complete.
- Entries are accurate and timely.

☐	■ Work is in compliance with agency requirements and policies.
☐	Keep the Finance/Administration Section Chief briefed on unit status and activity.
☐	Obtain the demobilization plan and ensure that injury and claims specialists are adequately briefed on the demobilization plan.
☐	Ensure that all injury and claims logs and forms are up to date and routed to the proper agency for post-incident processing prior to demobilization.
☐	Demobilize the unit in accordance with the demobilization plan.
☐	Maintain the unit log (ICS Form 214).

PROCUREMENT UNIT LEADER CHECKLIST

RESPONSIBILITIES

The Procurement Unit Leader is responsible to develop a procurement plan for the incident and to perform equipment time recording. The Procurement Unit Leader will ensure that goods and services are procured to meet the needs of the incident within his or her authority and the constraints of the Finance/Administration Section and the jurisdictional agency.

The Procurement Unit will work closely with the Supply Unit, which will implement the procurement plan and perform all incident ordering. The Procurement Unit Leader reports to the Finance/Administration Section Chief.

Instructions: The checklist below presents the minimum requirements for Procurement Unit Leaders. Note that some activities are one-time actions, and others are ongoing or repetitive throughout the incident.

COMPLETED/ NOT APPLICABLE	TASKS
☐	Obtain a briefing from the Finance/Administration Section Chief.
☐	Contact the appropriate unit leaders about incident needs and any special procedures.
☐	Coordinate with the local jurisdiction on plans and supply sources.
☐	Obtain the incident procurement plan.
☐	Prepare and sign contracts and land use agreements, as needed.
☐	Draft memorandums of understanding.
☐	Establish contracts with supply vendors, as required.

❑	Provide for coordination between the Ordering Manager, agency dispatch, and all other procurement organizations supporting the incident.
❑	Ensure that a system is in place which meets agency property management requirements. Ensure proper accounting for all new property.
❑	Interpret contracts and/or agreements and resolve claims or disputes within delegated authority limits.
❑	Coordinate with the Compensation/Claims Unit on procedures for handling claims.
❑	Finalize all agreements and contracts.
❑	Coordinate the use of funds, as required.
❑	Organize and direct the equipment time-recording function.
❑	Complete final processing and send documents for payment.
❑	Coordinate the cost data in contracts with the Cost Unit Leader.
❑	Maintain the unit log (ICS Form 214 or local form).

Endnotes

[1] http://www.fema.gov/txt/onp/toolkit_unit_11.txt

C

DISASTER PREPAREDNESS GUIDE FOR INCIDENTS INVOLVING CHEMICAL, BIOLOGICAL, RADIOLOGICAL AND ENVIRONMENTAL (CBRE) AGENTS

Navy Environmental Health Center
620 John Paul Jones Cir Ste 1100
Portsmouth VA 23708-2103
Phone: (757) 953-0700
After Hours: (757) 621-1967
DSN: 377

Preface

This manual has been developed to assist military emergency and medical planners in preparing for the possibility of an intentional or unintentional CBRE incident occurring at their installations or in their adjacent communities. It is meant to supplement current disaster planning efforts for immediate, local response. The various states and the federal government have elaborate plans and many resources which can be brought to bear to mitigate the effects of disasters, including those caused by CBRE agents. Unfortunately, activation of and response by these various state and federal agencies and their enormous resources takes time (12-18 hours for state resources, 24-48 hours for federal services), and, as is true with most disasters, the most devastating events tend to occur during the initial hours. Disaster response, regardless of the cause, is and will remain primarily a local issue.

Disasters involving CBRE agents are different in several respects from other disasters. First, most disasters are due to natural causes (hurricanes, tornadoes, earthquakes, etc.) whereas CBRE disasters are exclusively man-made, whether intentional or not intentional. Secondly, there is usually little if any warning, especially for intentional CBRE incidents. Thirdly, CBRE incidents may have a high probability of affecting a significant part of the community population.

Table of Contents

Preface

Table of Contents

  Medical Aspects of Disaster Planning

    General Concepts

    Planning Team

    Medical Services and Mass Care Requirements

    Availability of Medical Resources

    Medical Services/Mass Care Annexes and CBRE Appendices

      Medical Annex

    Mass Care Annex

      CBRE Appendices - Medical considerations
      Chemical Warfare Agents
      Biological Warfare Agents
      Radiological Agents
      Environmental (Hazardous Industrial) Materials

    Plan Evaluation
    Corrective Actions

Appendices:

1. Disaster Planning Process Flow Sheet
2. Disaster Preparedness Planning Team Membership
3. Threats and Vulnerabilities
4. Medical Services Resources
5. Sample Medical Services Function Annex
6. Sample Generic CBRE Appendices
7. References

Medical Aspects of Disaster Planning

General Concepts

Medical planning for incident response is a 6-step process (See Appendix 1). This process should be followed whether developing initial, generic plans, or when reviewing and/or modifying existing plans for specific potential disasters, including CBRE incidents. The process entails:

a. Establishing a planning team;

b. Reviewing existing disaster plans;

c. Estimating medical services requirements:

(1) Risk and vulnerability assessment

(2) Population at risk

(3) Critical modifiers

(4) Realistic parameters

d. Identifying available resources:

(1) Installation resources;

(2) Local/regional resources available for augmentation;

e. Developing the medical annex and CBRE appendices the disaster plan;

f. Testing the plan and identifying deficiencies and taking corrective action:

(1) Programming, planning and budgeting to improvement response posture;

(2) Advising installation commanders and higher echelons;

Planning for disasters due to CBRE incidents is similar to the planning process for any disaster situation. This manual will not discuss disaster planning in general, except to the extent that general planning applies specifically to the CBRE environment. It is important to not attempt to start from scratch - for efficiency of planning, and subsequent operations, it is advisable to build on existing plans, rather than to create a new, separate plan for these situations.

Disaster plans may take a variety of formats. However, per 29 CFR 1910, immediate response to disasters involving hazardous materials must follow the organization structure of the Incident Command System (ICS) first developed by the Firefighting Resources of Southern California Organized for Potential Emergencies (FIRESCOPE) in the early 1970s after a series of devastating wildland fires. Nearly all Federal Fire Departments, and a vast majority of civilian local and regional disaster response organizations, utilize the ICS system for incident on-scene organization. It is therefore imperative that medical planners become aware of this system.

Federal agencies will be involved in the crisis management of terrorist attacks involving CBRE agents, and most likely will also be required at some point during either the mitigation of or recovery from such incidents. Disaster plans developed along parallel functional capabilities as these plans will provide for improved interface and transition from unassisted local response to state and federally assisted services.

Resources for plan development include: the Federal Response Plan (FRP), the Federal Radiological Emergency Preparedness Plan (FREPP), the Hazardous Materials Emergency Planning Guide, and the Guide for All-Hazard Emergency Operations Planning.

Planning Process

1. Establish a Planning Team

A team approach should be taken in developing a CBRE Incident Annex to the existing disaster plan. This team should provide representation from key offices or functions involved with disaster response. Both internal organizations (facilities management, emergency department) and external organizations which might either support the medical services function, or have significant interface with the medical services should be represented. Just as it is important for all affected organizations to have representation during general ("core") planning, the capabilities and limitations of these other organizations and agencies may have significant impact on ultimate medical services functions and plan design. Potential team membership is included in Appendix (2).

2. Review Existing Disaster Plans and Other Policies and Legislation

Prior to developing a CBRE Response Annex, existing plans should be reviewed. Many of the logistical, supply, mutual assistance, and communications issues will generically apply to all disasters, to one extent or another, and duplication might not be necessary. Military medical treatment facility, installation, and community plans should all be perused. All existing Memoranda of Understanding/Agreement and Interservice Support Agreements should be reviewed. For large disasters, many regions have assigned one medical treatment facility to serve as the overall coordinating body for all hospitals. Other potential treatment locations (doctor's offices, urgent care centers) have frequently been identified, and procedures may already be in place to activate these centers. If this is the case, this must be factored into disaster planning.

Existing service and Defense Department instructions and directives, as well as applicable state or federal statutes and rules and regulations, should be reviewed. Many of these are included in the reference section of this manual (Appendix (9)).

3. Estimating Medical Services Requirements

Medical Services requirements include both immediate response and extended/mass care needs. Initial estimates are developed based on determination of the population at risk, an identification of actual risks, and application of critical modifiers that might increase or reduce medical requirements:

Risk evaluations estimate both potential threats and vulnerabilities. Where possible, actions are taken to reduce of eliminate these. Threats and vulnerabilities should include those in the proximate local community that might adversely impact installation operations or that would affect the ability of local agencies to assist with installation disaster mitigation. Additional threats might exist due to heightened terrorist activities or movements, which is the responsibility of intelligence communities to track and report. Vulnerabilities might include unprotected waterfronts, high visibility targets (national shrines),or an open-gate policy at the installation. Appendix (3) includes a general approach to hazardous materials analysis.

Population at risk would include not only active duty service members assigned to the installation, but would include those resident dependents, civil servants, contractors, and guests at the installation at the time of the incident. Since installation populations may vary by more than 400% throughout the day, potential medical system demands might be extreme.

Critical modifiers increase or decrease medial service requirements. Certain assets, including assigned personnel, are designated as Critical/Key Assets, which are considered vital to national security and might require additional services. The availability and proximity of state and federal level resources will affect the time that installation resources must function with only local assistance, and in certain locations, these local resources may not exist. Finally, guidance from higher authority may be provided to increase or decrease readiness posture.

Once needs assessments have been made, **realistic parameters** must be set. In most locations, regional and state assistance should be available within 12 hours, and federal assistance through the Stafford Act should be on-scene within 24 hours. However, the potential exists for a worst case scenario, especially if biological weapons are used: A man portable 8-liter aerosol sprayer could release a biological agent in less than 5 minutes, and could contaminate a 2200 km^2 area. Metropolitan emergency managers plan for between 10,000 and 10% of the population affected by biological agents, and up to 1,000 for chemical, environmental, and non-thermonuclear radiological agents. However, for certain aspects of planning, 100% of the population could be affected. Examples would include both post-exposure prophylaxis, which must be started within 24 hours of exposure, shelters, and Critical Incident Stress Debriefing (CISD) mental health services. A variety of medical planning guides exist for estimating casualty numbers.

Finally, an estimate is made of the various medical resources (personnel, equipment and supplies, facilities, and transportation services) which would be required to provide immediate response and mitigation during the initial phase of the CBRE disaster.

4. Identifying Available Medical Resources

Required resources must be matched with those available. Since it is unlikely that installation resources will be sufficient to manage CBRE incidents, this would include those resources available locally and regionally to augment installation medical services. Consider resources from a functional aspect. In addition to prehospital and acute care services, provisions for mass care for extended periods must be made.

Prehospital care services require personnel, equipment, supplies, and vehicles. Personnel must be trained and experienced in triage, treatment, and transportation functions. They must also be trained in hazardous materials to the OSHA Operations level, and in the pre hospital treatment of specific medical consequences of exposure to CBRE agents. Personnel must be identified to provide sustained operations for extended periods of time. By doctrine, decontamination is not a medical services function; however, history has shown that decontaminated patients do access pre hospital care services—EMS personnel must be trained in field decontamination procedures. Additional equipment would include such items as OSHA-approved Individual Protective Equipment, Nerve Agent Antidote Kits, and contamination detection supplies (M8/9 paper).

Vehicles available for transport of ambulatory victims, such as buses and trucks, should be identified as well.

Ideally, all critical patients should be transported to the nearest appropriate medical treatment facility. In a disaster situation, this is not always possible. Approximately half of all disaster victims will arrive at a treatment facility (which might not be a hospital) by methods other than traditional prehospital care (EMS) services. Between 15-40% of potential victims will arrive at the nearest recognized medical facility, and will do so within the first hour of the disaster. The installation medical treatment facility might be within the contaminated area, requiring redeployment of facility personnel to other locations, possibly off installation, to provide medical services. In addition to identification of hospital capacity (bed capacity, average daily occupancy, specialized capabilities - ICU, isolation, and ventilator beds, and emergency department/urgent care capacity, knowledge of caches of critical stores, such as antibiotics and antidotes, is a requirement. Decontamination capabilities (primarily for large numbers of ambulatory personnel) must be identified. Again, sufficient personnel trained and experienced in decontamination, triage, treatment and personal protection must be identified to provide sustained operations.

Mass and extended care require special attention. Have facilities been designated as mass shelters, and are there sufficient shelters outside potentially affected areas? Not all individuals at these sites will be healthy and might need medical care prior to restoration of normal society locally. Approximately 80% of individuals affected by disasters seek shelter with family or friends. However, it might be necessary to quarantine all personnel within a given distance of the attack until the biological agent is identified. With mass gatherings under rather austere conditions, diseases which are considered relatively innocuous or rare can have devastating effects - in third world countries, for example, measles is a leading cause of death from infectious diseases. Potable water, sanitation, vector control, and animal control might become necessary, depending on the circumstances and length of time for restoration.

5. Develop Medical Services and Mass Care Annexes and CBRE Appendices

It is presumed that a working disaster preparedness plan is in existence. National consensus panels reviewing CBRE response initiatives have almost uniformly recommended that all measures taken in response to CBRE incidents be incorporated into existing structures and response plans as the most efficient and cost-effective means of dealing with this threat.

When developing these annexes and appendices, it is important to consider all aspects of disaster preparedness: Routine operations, pre-incident planning considerations, immediate post-incident response operations, and longer term recovery operations.

Routine operations would include those actions, which are performed as a medical defense against CBRE events. Such items would include environmental surveillance operations. Additionally, because covert biological attacks must be detected as early as possible in order for post-exposure prophylaxis to be effective, it might be necessary to establish a health monitoring program which provides daily reports on symptom based medical problems (e.g. diarrheal disease, upper respiratory track infection symptoms). Also included in routine operations would be those training requirements of personnel who would be involved in the various phases of disaster mitigation - EMS person, hospital providers, and environmental health officers, to name a few.

Pre-incident planning would include those actions that should be undertaken in the event of prior notification of a terrorist attack or CBRE incident. For terrorist attacks, this might be best accomplished by coupling those actions with the installation threat condition.

Post-incident planning encompasses all actions to be taken in immediate response (first 24-48 hours) after an incident. Many of these items will have already been accomplished given prior notification.

Special attention should be taken concerning the stages of operations directed against biological incidents, since these are most apt to be detected and initially responded to outside the usual emergency (9-1-1) system. A proposed template for operations may be found in the appendices.

Although federal assistance should be available within 24 hours, in the case of mitigation, the federal agencies (as coordinated through the Federal Emergency Management Agency) are there to primarily assist local and state agencies; therefore, continued active participation by local emergency management personnel will still be required. Medical aspects of mass care and infrastructure restoration (both physical and functional) are accomplished during this phase. Prior to complete functional restoration, abnormal operations will continue to be in place.

Many planning tools are available through the Federal Emergency Management Agency and other organizations involved with disaster preparedness, and some of these are available through the Internet.

6. Testing Plans, Identifying Deficiencies and Taking Corrective Action

As with all disaster plans, the need to validate material resources, human resources and training, and policies, procedures and protocols requires that the plan be tested and revised periodically. There are two primary methods of testing these plans, which are educational processes in themselves: tabletop exercises involving key members from all functional organizations, and full-scale field exercises.

By testing CBRE incident disaster plans in a variety of plausible scenarios, deficiencies in those plans may be readily apparent. Once a deficiency is identified, there are several options:

   a. Modifications in policies, procedures, protocols or the disaster plan might eliminate or alleviate the deficiencies.

   b. Additional resources may be obtained through programming, planning and budgeting.

c. Installation commanders or higher authority may take steps to reduce vulnerabilities or threats, thereby reducing medical services requirements and eliminating the deficiency.

   d. The deficiency might be considered an acceptable risk.

Potential Disaster Planning Team Members

Security

Fire and Emergency Services

Public Works

Social Services Agencies

American Red Cross

Public Affairs Officials

Mortuary Affairs

Facilities Management

Installation Operations

Installation Safety

Installation Media Services

Legal Services

Veterinary Services

Community Emergency Medical Services Organizations

Community Hospitals

Community Disaster and Emergency Response Organizations

Regional and State Offices, as appropriate

Threats and Vulnerabilities

Potential Threats

   1. Hazardous Materials Identification

      a. Types and quantities of hazardous materials located in or transported through a community.

      b. Locations of hazardous materials facilities and routes

      c. Nature of the effects most likely to ensue as the result of a release of these substances

   2. Specific sites with a high probability of hazardous materials

      a. Chemical plants

      b. Biological research centers

c. Refineries

d. Major industrial facilities

e. Petroleum or natural gas farms or depots

f. Storage facilities

g. Trucking terminals

h. Railroad Yards

i. Hospitals and educational institutions

j. Waste disposal and treatment plants

k. Port authorities

l. Airports

m. Nuclear reactors and nuclear processing plants

n. Major transportation corridors

3. Intelligence community information

   a. Location of domestic or international terrorist cells

   b. Other extreme cultural or religious sects

   c. Current intelligence on most probable chemical or biological warfare agents

Potential Vulnerabilities

1. Public Access

   a. Open gate policy

   b. Shared military/civilian facilities or properties

   c. Unguarded or unmonitored perimeter fencing

   d. Unprotected waterfronts

   e. Dual use airfields

   f. Uncontrolled airspace

   g. Major thoroughfares or railways which pass through installations

2. Proximate civilian targets

   a. Government buildings

   b. Educational institutions

   c. Stadiums or arenas

   d. Nuclear power plants

   e. Chemical production plants

  f. Chemical storage facilities

  g. Abortion clinics

  h. National shrines

  i. Animal research centers

Critical Modifiers

 1. Isolated locations

  a. Island installations

  b. Rural or wilderness areas

  c. Overseas locations

 2. Special sites

  a. Critical Asset Assurance Program sites

  b. Key Asset Assurance program sites

  c. Major logistics or supply centers

  d. Major shipyards

  e. Low frequency communications facilities

  f. Other major C4I locations

  g. Central commands/type commands

 3. Locations designated by higher authority as requiring additional mitigation efforts

Medical Services Capabilities

Function

Installation

Local/Regional Resources

Prehospital Care

Transport Vehicles

Air

Fixed Wing

Rotorcraft

Ground

Ambulances

Patient Transport Vans

Patient Buses

Other
Personnel
EMTs
EMT-Basic
EMT-Intermediate
Paramedics
Specialty Response Teams
Equipment & Supplies
Individual Protective Equipment
Decontamination Equipment
Contaminant Detection Equipment
Personal Antidotes
Personal Prophylaxis

Medical Treatment Facilities

Receiving Areas
Contaminated ambulance areas
Helicopter landing zones
Mass decontamination areas
Holding Areas
Medevac staging areas
Mortuary services
Ambulatory patient areas
Bed Capacity
Emergency Department
In-patient Beds
Isolation Units
Intensive Care Units
Supplies & Equipment
Antidotes
Post-exposure prophylaxis
Individual Protection Equipment

Decontamination supplies

Contaminant detection equipment

Mass Care Capabilities

Shelters

Arenas

Gymnasiums

Campgrounds

Auditoriums

Special Needs Locations for Nursing Home patients, children, ill or injured not qualifying for hospitalization

Specialized Medical Services

American Red Cross

Social Services

Environmental health Officers

Public Health Officers

Veterinary Services Personnel

Food Services Officials

Sanitation Specialists

Mass Care Medical Facilities

Prophylaxis Distribution Centers

Medical Services Function Annex (SAMPLE)

I. Purpose: {This section describes in general terms the purpose of the annex, e.g., to provide emergency medical services (EMS), treatment facility, public and environmental health services, mental health services and mortuary services in the event of a declared disaster situation. It also describes the activities related to those services.}

II. Situation and Assumptions: {This section outlines general assessment and overview of existing capabilities. It further describes limitation that might degrade health care operations. Specific situations and assumptions involving CBRE incidents which should be considered include:}

- CBRE incidents would be primarily large-scale events that would rapidly overwhelm local medical resources
- Local health care resources might be degraded due to location in a contaminated areas

- Exclusively existing local resources will most likely provide health care services during the first 12-24 hours (longer overseas)
- Medical care resources might be forced to relocate operations under austere conditions in temporary structures
- Volunteers are less likely to come foreword in situations with unknown CBRE agents than in most disasters

III. Concept of Operations: This section describes how operations will be conducted. It further describes interaction with other jurisdictions and State or Federal agencies.

  A. General: Areas to be described include:

- Routine operations directed against CBRE incidents:
- Environmental surveys
- Health surveillance
- Training and exercises involving:
  - EMS personnel
  - Hospital providers and ancillary staff
  - Hospital emergency operations personnel
  - Specialized personnel or medical teams
- Equipment and supply inventory, preventive maintenance
- Antidote and post-exposure prophylaxis cache maintenance and exchange
- Actions to be taken in the event of heightened threat conditions
- Actions to be taken in the event of a suspected or actual incident
- Methods of determination that an event has occurred
- Conditions and provisions for mobilization of medical care services
- Separate algorithms for dealing with potential threats, actual threats, unannounced releases, or probable hoaxes should be addressed
- Medical services command post(s)
- Overall structure and coordination of medical services
- Treatment considerations
- Triage
- Decontamination procedures, especially for large populations
- Methods of identification of hazardous materials
- Specific actions to be taken for long term recovery operations in the event of a CBRE incident

B. Relationships with other Jurisdictions
- Mutual aid agreements
- Other arrangements with other agencies, such as USAMRICD, CDC, etc

IV. Organization and Assignment of Responsibilities

A. Command Suite

B. Medical Control Officer

C. Emergency Medical Services {Specific discussion of:
- Response to a scene
- Command and control at the scene
- Treatment of victims in contaminated areas
- Triage
- Command and control of assisting agencies
- Evacuation and contaminated vehicles.}

D. Medical Treatment Facilities {CBRE issues to be discussed include;
- Conditions and procedures for activation of hospital disaster plan
- Mass decontamination - 40-50% of CBRE victims arrive at MTFs through other methods than EMS
- Communications nets
- Isolation of infected patients
- Handling of large number of cadavers, some of whom are still contaminated or are infectious
- Handling of large quantities of contaminated clothing.}

E. Public Health Services
- Public health and epidemiological services are vitally important in both early detection of biological agent release, and in following the course of the attack
- Coordination with specialty laboratories for identification of agents
- Coordination of immunizations or prophylaxis and distribution centers
- Monitoring of food handling and mass feeding and sanitation

F. Environmental Health Services {Coordination for such services as:
- Vector control
- Potable water availability
- Structural inspections for habitability}

G. Mental Health Services {In the event of a CBRE incident, with or without large numbers of casualties, a large percentage of the population will require early and extended mental health services. services will also be needed for hospital personnel and emergency responders.}

H. Mortuary Services {Anticipate very large numbers of fatalities, who might be contaminated or infectious. Temporary morgues might be needed.}

I. American Red Cross {Additional ARC volunteers will most likely be needed for assistance in all social services functions.}

J. Social Services

K. Animal Control

L. Security {Up to 45% of victims, or those who feel they have been exposed to these agents, will go to the nearest medical treatment facilities, including urgent care centers or individual physicians' offices, and will arrive within hours of the exposure. Crowd control is therefore advisable.}

V. Administration and Logistics

A. Administration {This section outlines administrative procedures to be followed.}

B. Logistics

VI. Plan Development and Maintenance

VII. Authority and References

CBRE Appendix to Disaster Plan
(SAMPLE/GENERIC)

I. Hazard Defined

II. Risk Areas {Identify high risk areas, as outlined under vulnerability assessment. For industrial chemicals, radiological stores, identify specific storage areas.}

III. Estimates of Vulnerability Zones

- For industrial hazardous materials, estimate the approximate area affected from an airborne release, which is worst case scenario.
- Identify also most probable release scenarios.
- For terrorist activities, a reasonable initial zone would be 2000 meters in all directions (unknown wind condition) since EOD sets initial contaminated zones as 2000 meters downwind and 450 meters in all other directions

IV. Determine Vulnerability

- Identify populations within vulnerability zones
- Identify other significant structures in those zones (e.g. barracks, hospitals, etc.)

V. Risk Assessment

- Estimate risk using Operational Risk Management guidelines

VI. Direction and Control

- 29 CFR 1910 requires that an Incident Command System be used for on-scene management of response activities involving industrial hazardous materials.

    a. Response Action

- Notification procedures for local response organizations
- Determination that release has occurred
- Estimation and accurate determination of areas and populations affected
- Identification and designation of special technical experts (e.g. chemists, toxicologists, etc.) to augment response organizations
- Identification and response of special teams
- Methods of identifying specific hazards

    b. On-scene Response Actions

- EMS personnel have appropriate individual protective equipment, including antidotes and personal prophylaxis
- EMS personnel and equipment are staged upwind and outside hot zone of area
- EMS personnel set up a contaminated treatment area, holding areas (deceased/alive)
- Identification of contaminated patients, equipment and supplies
- Disposal areas for contaminated clothing, etc

VII. Public Information

- Determination of appropriate release of public health information
- Provision of personal protective instructions through the media
- Diversion of scheduled patients from medical treatment facilities

VIII. Evacuation Procedures

IX. Mass Care

- Upwind, out of range facilities
- Provisions for prophylaxis distribution points
- Provisions for mass care medical services, including for special needs patients

X. Medical Services

- Designation of medical treatment facilities with capabilities for treatment
- Designation of medical treatment facilities with capabilities for decontamination
- Monitoring air, soil and water for hazardous materials
- Provisions for continued medical surveillance

XI. Resource Management

- Antidote, prophylaxis, antibiotic stockpiles
- Replacement protective equipment for medical decontamination and EMS personnel
- Provisions for medical hazardous material detection or sampling devices and handling thereof
- Points of contact and procedures for obtaining stockpiles

References

DODD 3025.1 Military Support to Civil Authorities (MSCA), 15 January 1993.

OPNAVINST 3440.16C Navy Civil emergency Management Program, 10 March 1995.

BUMEDINST 3400.1 Operational Concept for Medical Support and Casualty Management in Chemical and Biological Warfare Environments, 28 February 1994.

NAVMEDCOMINST 3440.4 Activity Disaster preparedness Plans and Material for Disaster Preparedness Teams, 28 march 1989.

OPNAVINST 3400.10F Chemical, Biological and Radiological (CBR) Defense Requirements Supporting Operational Fleet Readiness, 22 May 1998.

National Response Team (NRT-1): Hazardous Materials Emergency Planning Guide, G-WER/12, 2100 2nd Street SW, Washington, DC 20593

State and Local Guide (SLG 101): Guide for All-Hazard Emergency Operations Planning, Federal Emergency Management Agency, September 1996. Available on the internet at: http://www.fema.gov/pte/gaheop.htm

Lewis CP and RV Aghababian, *Disaster Planning, Part I. Overview of hospital and emergency department planning for internal and external disasters.* Emerg Med Clin North Am 1996 May; 14(2)

Auf der Hiede, Erick. *Disaster Planning, Part II. Disaster Problems, Issues and Challenges Identified in the Research Literature.* Emerg Med Clin North Am 1996 May; 14(2)

Biological Warfare Improved Response Program: Final Report. 1998 Summary Report on BW Response Template and Response Improvements, Mar 1999.

National Institute of Medicine, National Research Council, *Improving Civilian Medical Response to Chemical or Biological Terrorist Incidents,* National Academy Press, Washington, DC 1998.

National Institute of Medicine, National Research Council, *Chemical and Biological Terrorism: Research and Development to Improve Civilian Medical Response,* National Academy Press, Washington, DC 1999.

Innovative Emergency Management, Inc. *Restoration of Operations (RestOps) of Fixed Sites: Literature Review and Analysis (DRAFT),* Baton Rouge, LA 1998.

NATO (Unclassified) Study Draft3: Medical Planning Guide for the Estimation of NBC Battle Casualties, Vol. II: Biological, 1998.

American College of Emergency Physicians, *Community Medical Disaster Planning and Evaluation Guide,* ACEP, Dallas, TX 1998.

Auf der Heide, Erick, *Disaster Response: Principles of Preparation and Coordination,* CV Mosby, St. Louis: 1989.

D

CATASTROPHIC EVENTS: EQUIPMENT REQUIREMENTS

Rapid Assessment and Initial Detection Equipment Requirements

Cat	Nomenclature	NSN	Quantity	Cost	Total FY 99
C	10 Bank PAPR Battery Charger	COTS	1	597	597
C	1-C United Charger (Radio Charger)	COTS	1	517	517
C	Cellular Phone	COTS	9	200	1,800
C	Cellular Phone/Fax (STU III)	COTS	2	7,500	15,000
C	Commo/Computer Maintenance Contracts	COTS	1	2,000	2,000
C	INMARSAT	COTS	1	12,000	12,000
C	Lap Top Computer	COTS	9	3,000	27,000
C	Lap Top Computer Printer	COTS	9	300	2,700
C	"Pagers, Skytel w/ Message Screen"	COTS	24	100	2,400
C	Saber Radios (Secure)	COTS	15	3,000	45,000
C	"Water Test kit, M272"	6665-01-134-0885	1	178	178
D	Air Inflatable Tents	COTS	1	7,140	7,140
D	Collapsible Pool	COTS	1	200	200
D	High Test Hypochlorite (50 lbs)	COTS	2	81	162
D	M-295 Individual Decon Kit	6850-01-357-8456	24	27	656
D	Plastic lined Drums	COTS	5	201	1,005

D	Portable Shower Kits	COTS	1	768	768
D	Roll Sheet 6mm (Box)	COTS	1	26	26
D	Sodium Carbonate (10 kg)	COTS	2	214	428
D	Sodium Hydroxide (10 Kg)	COTS	2	182	364
D	Sodium Hypochlorite (15 lb Bucket)	6810-00-598-7316	2	58	116
D	Spill Containment pillows	COTS	6	174	1,044
D	Waste Water Containment Device	COTS	1	2,584	2,584
D	Water Bladder, Decon Shower Waste Collection	COTS	1	1,762	1,762
M	"Convulsant Antidote, Nerve Agent (CANA)"	6505-01-274-0951	22	10	222
M	Cover, Chem Protective Patient Wrap	8415-01-311-7711	2	75	150
M	Cyanide Antidote Kit	COTS	22	15	330
M	Med Equip Set - Chem Agent Decon Kit	6545-01-176-4612	1	4,791	4,791
M	Med Equip Set - Chem Agent Treatment Kit	6545-01-141-9469	1	12,736	12,736
M	Nerve Agent Antidote Kit (Mark 1) - 3 Autoinjectors	6505-01-174-9919	22	12	257
M	Nerve Agent Antidote Kit (Mark 1) – Trainer		2	12	23
M	Nerve Agent pretreatment - PB Tablets	6505-01-178-7903	44	18	775
P	Apron, Toxicological Agent protective	8415-00-281-7815	6	57	341
P	Best "N" Dex Nitrile Gloves (25 Pr/Box)	COTS	2	15	29
P	Boots, Toxicological Agent protective	8430-00-820-6300	44	100	4,402
P	Cotton Blankets	COTS	12	24	288
P	Cotton Clothing, Coveralls	COTS	40	50	1,980

P	Cotton Drawers	COTS	44	2	88
P	Cotton Socks	COTS	44	2	88
P	Cotton Towels	COTS	44	3	132
P	Cotton T-Shirts	COTS	44	3	132
P	Coveralls, Toxicological Agent protective	8415-00-099-6970	6	791	4,746
P	Drawers, Underwear, CP	8415-01-363-8689	66	91	6,013
P	Filter Canister, C2	4240-00-165-5026	66	9	574
P	Gloves, Toxicological Agent protective	8415-00-753-6553	40	12	484
P	HAZMAX 16"" Kneeboots"	COTS	6	43	260
P	Hood, Toxicological Agent protective	8415-00-261-6690	6		0
P	ILC Dover Cool Vest	COTS	4	189	756
P	Interspiro SCBA	COTS	6	3,000	18,000
P	Level A Encapsulated Suits with Boots/Gloves (Kappler or equal)	COTS	12	685	8,220
P	M30 Modified Hoods		44		0
P	Mask M40	4240-01-143-2017	22	290	6,380
P	Replacement Ice Pack for cool Vest	COTS	12	75	900
P	Safety Glasses	COTS	22	3	74
P	SCBA Spare Cylinders	COTS	6	1,400	8,400
P	TOPPS Nomex IIIA Coveralls	COTS	8	83	664
P	TYVEK-F Decon Suits	COTS	66	81	5,346
P	"Undershirt, Underwear, CP"	8415-01-363-8693	66	99	6,501
R	44-6 Side Window GM Probe (Beta/Gamma)	COTS	1	145	145
R	44-9 Pancake G-M Detector (Alpha)	COTS	1	175	175
R	4-Gas Portable Monitor	COTS	1	1,700	1,700
R	Battery, CAM US	6665-99-760-9742	4	18	71

R	BIOS 600 Air Sampling Air Pumps/ DAAMS Tubes	COTS	1	3,675	3,675
R	CAMSIM	6910-01-275-4833	1	5,100	5,100
R	"Chemical Detector, M90"		2	16,500	33,000
R	"Detector Kit, Chemical, M18A2"	6665-01-903-4767	1	294	294
R	"Detector Kit, M256A1"	6665-01-133-4964	4	40	160
R	Dosimeter Charger	COTS	1	95	95
R	Draeger Quantimeter	COTS	1	2,200	2,200
R	Draeger Multiwarn II	COTS	2	2,800	5,600
R	Draeger Kit	COTS	1	853	853
R	Extra Draeger Tubes	COTS	1	368	368
R	Hand Held Reader for Immunoassay Tickets	COTS	1		0
R	ICAD Mini Chemical Agent Detector	COTS	1	2,798	2,798
R	Immunoassay Tickets	COTS	400	3	1,000
R	Improved Chemical Agent Monitor (ICAM) W/alarm & tips	665-01-199-4153	3	5,500	16,500
R	Ludlum Digital Scaler Ratemeter 2241-2	COTS	1	725	725
R	M256 Training Kits	6665-01-112-1644	2	100	199
R	M8 Detection Paper	6665-00-050-8529	52	1	41
R	M9 Detection Paper	6665-01-226-5589	13	4	56
R	Monitox Phosgene Detector	COTS	4	200	800
R	Photoionization Detector	COTS	1	1,325	1,325
R	Pocket Radiac (AN/UDR-13)		16	627	10,032
R	"RADIAC Monitoring System, Alpha (AN/PDR-77)"	6665-01-347-6100	2	4,800	9,600

R	"RADIAC Monitoring System, Beta- Gamma (AN/VDR-2)"	6665-01-222-1425	2	2,026	4,052
R	Sample Collection Equipment Kits (Trelborg or equal)	COTS	1	500	500
R	Sample Transfer Container/ Small Black Infectious Substance Container	COTS	4	150	600
R	SAW MiniCAD	COTS	2	5,495	10,990
R	Sensidyne Detector Tubes	COTS		650	0
R	"Soil Sampler, M34"	6665-00-776-8817	1		0
R	Spray Simulant for Tng CAM		1	550	550
R	TEU Sampling Kit		4	300	1,200
X	Digital Camera	COTS	1	500	500
X	EMS Book	COTS	16	10	160
X	HAZMAT Gear bag	COTS	24	40	960
X	Med Mgt of Bio Casualty Book	COTS	24	2	50
X	Med Mgt of Chemical Casualty Book	COTS	24	2	50
X	Nicad batteries (Extra)	COTS	4	109	437
X	REAC/TS Trans. Of Radiological materials - Q&A About Incident Response	COTS	1	0	0
X	SCBA Maintenance Contract	COTS	1		0
X	Video Camera	COTS	1	800	800
				Total:	336,888
			54 Teams Total:		18,191,953

Decontamination Element Equipment Requirements

Cat	Nomenclature	NSN	Quantity	Cost	Total FY 99
D	Air Inflatable Tents	COTS	3	7,140.00	21,420
D	Collapsible Pool	COTS	1	200.00	200
D	High Test Hypochlorite (50 lbs)	COTS	2	81.16	162
D	M-295 Individual Decon Kit	6850-01-357-8456	0	27.35	0
D	Plastic lined Drums	COTS	20	200.95	4,019
D	Portable Shower Kits	COTS	5	768.00	3,840
D	Roll Sheet 6mm (Box)	COTS	10	26.00	260
D	Sodium Carbonate (10 kg)	COTS	2	214.03	428
D	Sodium Hydroxide (10 Kg)	COTS	2	182.03	364
D	Sodium Hypochlorite (15 lb Bucket)	6810-00-598-7316	2	58.12	116
D	Spill Containment pillows	COTS	6	174.00	1,044
D	Waste Water Containment Device	COTS	1	2,584.00	2,584
D	Water Bladder, Decon Shower Waste Collection	COTS	1	1,762.00	1,762
M	Convulsant Antidote, Nerve Agent (CANA)	6505-01-274-0951	20	10.08	202
M	Cyanide Antidote kit	COTS	20	15.00	300
M	Med Equip Set - Chem Agent Decon Kit	6545-01-176-4612	1	4,790.77	4,791
M	Nerve Agent Antidote Kit (Mark 1) - 3 Autoinjectors	6505-01-174-9919	20	11.66	233
M	Nerve Agent Antidote Kit (Mark 1) – Trainer		2	11.66	23
M	Nerve Agent pretreatment - PB Tablets	6505-01-178-7903	40	17.61	704
P	Apron, Toxicological Agent protective	8415-00-281-7815	20	56.90	1,138
P	Best "N" Dex Nitrile Gloves (25 Pr/Box)	COTS	2	14.50	29
P	Boots, Toxicological Agent Protective	8430-00-820-6300	40	100.05	4,002
P	Cotton Blankets	COTS	60	24.00	1,440

P	Cotton Clothing, Coveralls	COTS	40	49.50	1,980
P	Cotton Drawers	COTS	40	2.00	80
P	Cotton Socks (Size 10)	COTS	40	2.00	80
P	Cotton Towels	COTS	60	3.00	180
P	Cotton T-Shirts	COTS	60	3.00	180
P	Drawers, Underwear, CP"	8415-01-363-8689	60	91.10	5,466
P	Filter Elements, M13A2"	4240-00-165-5026	60	18.83	1,130
P	Gloves, Toxicological Agent protective"	8415-00-753-6553	60	12.10	726
P	Level B Chemical Suits with boots/Gloves	COTS	20	126.00	2,520
P	TYVEK-F Decon Suits	COTS	60	81.00	4,860
P	Undershirt, Underwear, CP"	8415-01-363-8693	60	98.50	5,910
R	44-6 Side Window GM Probe (Beta/Gamma)	COTS	1	145.00	145
R	44-9 Pancake G-M Detector (Alpha)	COTS	1	175.00	175
R	Battery, CAM US	6665-99-760-9742	4	17.81	71
R	CAMSIM	6910-01-275-4833	1	5,100.00	5,100
R	Chemical Detector, M90		2	16,500.00	33,000
R	Detector Kit, M256A1	6665-01-133-4964	4	39.93	160
R	ICAD Mini Chemical Agent Detector	COTS	1	2,798.00	2,798
R	Improved Chemical Agent Monitor (ICAM) W/alarm & tips	665-01-199-4153	3	5,500.00	16,500
R	M256 Training Kits	6665-01-112-1644	2	99.55	199
R	M8 Detection Paper	6665-00-050-8529	52	0.78	41
R	M9 Detection Paper	6665-01-226-5589	13	4.27	56
R	Pocket Radiac (AN/UDR-13)		5	627.00	3,135
R	"RADIAC Monitoring System, Alpha (AN/PDR-77)"	6665-01-347-6100	1	4,800.00	4,800
R	"RADIAC Monitoring System, Beta-Gamma (AN/VDR-2)"	6665-01-222-1425	1	2,026.00	2,026

CATASTROPHIC EVENTS

R	SAW MiniCAD	COTS	2	5,495.00	10,990
R	Spray Simulant for Tng CAM		1	550.00	550
X	HAZMAT Gear bag	COTS	20	39.99	800
			Per Team		152,719
			Total:	127 Teams	19,395,275

Reconnaissance Element Equipment Requirements

Cat	Nomenclature	NSN	Quantity	Cost	Total FY 99
C	Water Test kit, M272	6665-01-134-0885	1	178.00	178
D	Collapsible Pool	COTS	1	200.00	200
D	M-295 Individual Decon Kit	6850-01-357-8456	20	27.35	547
M	Convulsant Antidote, Nerve Agent (CANA)	6505-01-274-0951	20	10.08	202
M	Cyanide Antidote kit	COTS	20	15.00	300
M	Nerve Agent Antidote Kit (Mark 1) - 3 Autoinjectors	6505-01-174-9919	20	11.66	233
M	Nerve Agent Antidote Kit (Mark 1) – Trainer		2	11.66	23
M	Nerve Agent pretreatment - PB Tablets	6505-01-178-7903	40	17.61	704
P	Best "N" Dex Nitrile Gloves (25 Pr/Box)	COTS	2	14.50	29
P	Cotton Clothing, Coveralls	COTS	60	49.50	2,970
P	Cotton Drawers	COTS	60	2.00	120
P	Cotton Socks (Size 10)	COTS	60	2.00	120
P	Cotton Towels	COTS	60	3.00	180
P	Cotton T-Shirts	COTS	60	3.00	180
P	Drawers, Underwear, CP	8415-01-363-8689	60	91.10	5,466
P	Filter Elements, M13A2	4240-00-165-5026	60	18.83	1,130
P	ILC Dover Cool Vest	COTS	4	189.00	756
P	Interspiro SCBA	COTS	6	3,000.00	18,000
P	Level A Encapsulated Suits with Boots/Gloves (Kappler or equal)	COTS	12	685.00	8,220

CATASTROPHIC EVENTS: EQUIPMENT REQUIREMENTS

P	Replacement Ice Pack for cool Vest	COTS	12	75.00	900
P	SCBA Spare Cylinders	COTS	6	1,400.00	8,400
P	TYVEK-F Decon Suits	COTS	60	81.00	4,860
P	Undershirt, Underwear, CP	8415-01-363-8693	60	98.50	5,910
R	44-6 Side Window GM Probe (Beta/Gamma)	COTS	2	145.00	290
R	44-9 Pancake G-M Detector (Alpha)	COTS	2	175.00	350
R	4-Gas Portable Monitor	COTS	1	1,700.00	1,700
R	Battery, CAM US	6665-99-760-9742	4	17.81	71
R	BIOS 600 Air Sampling Air Pumps/ DAAMS Tubes	COTS	1	3,675.00	3,675
R	CAMSIM	6910-01-275-4833	1	5,100.00	5,100
R	Chemical Detector, M90		2	16,500.00	33,000
R	Deluxe Pump Kit for Draeger kit	COTS	1	374.50	375
R	Detector Kit, Chemical, M18A2	6665-01-903-4767	1	294.00	294
R	Detector Kit, M256A1	6665-01-133-4964	4	39.93	160
R	Draeger Kit	COTS	1	852.75	853
R	Extra Draeger Tubes	COTS	2	367.55	735
R	Hand Held Reader for Immunoassay Tickets	COTS	1		0
R	ICAD Mini Chemical Agent Detector	COTS	2	2,798.00	5,596
R	Immunoassay Tickets	COTS	400	2.50	1,000
R	Improved Chemical Agent Monitor (ICAM) W/alarm & tips	665-01-199-4153	3	5,500.00	16,500
R	Ludlum Digital Scaler Ratemeter 2241-2	COTS	1	725.00	725
R	M256 Training Kits	6665-01-112-1644	4	99.55	398
R	M8 Detection Paper	6665-00-050-8529	40	0.78	31
R	M9 Detection Paper	6665-01-226-5589	20	4.27	85
R	Monitox Phosgene Detector	COTS	4	200.00	800

R	Photoionization Detector	COTS	1	1,325.00	1,325
R	Pocket Radiac (AN/UDR-13)		20	627.00	12,540
R	"RADIAC Monitoring System, Alpha (AN/PDR-77)"	6665-01-347-6100	2	4,800.00	9,600
R	"RADIAC Monitoring System, Beta- Gamma (AN/VDR-2)"	6665-01-222-1425	2	2,026.00	4,052
R	Sample Collection Equipment Kits (Trelborg or equal)	COTS	1	500.00	500
R	Sample Transfer Container/ Small Black Infectious Substance Container	COTS	4	450.00	1,800
R	SAW MiniCAD	COTS	2	5,495.00	10,990
R	Sensidyne Detector Tubes	COTS	0	650.00	0
R	Soil Sampler, M34	6665-00-776-8817	1		0
R	Spray Simulant for Tng CAM		1	550.00	550
R	TEU Sampling Kit		4	300.00	1,200
X	EMS Book	COTS	16	10.00	160
X	HAZMAT Gear bag	COTS	20	39.99	800
X	REAC/TS Trans. Of Radiological materials - Q&A About Incident Response	COTS	1	0.00	0
X	SCBA Maintenance Contract	COTS	1		0
				Total:	174,883
			Total:	43 Teams	7,519,979

Triage Medical Response Element Equipment Requirements

Description
Trelborg Level B Suit
Trelborg Training Suit
Rebreather Masks
Advantage 1000
Filters
Tyvek Saranex 23-P Suit
Disposable Surgical Mask

Disposable Gloves
Triage Kit
Blankets
Laptop Computer w/Modem
Printer
Field Medical Tag
CAM
M8 Paper
M9 Paper
Traffic Cones
Marking Pens
Flashlights
I-V Fluids /Basic Dressings
Communications equipment
Muti-passenger vans

Trauma Medical Response Team Equipment Requirements

Description
Trelborg Level B Suit
Trelborg Training Suit
Rebreather Masks
Advantage 1000
Filters
Tyvek Saranex23-P Suit
M8 Paper
M9 Paper

Stress Management Element Equipment Requirements

Trelborg Level B Suit
Kappler Training Suit
Trelborg Training Suit
Rebreather Masks
Advantage 1000
Filters
Tyvek Saranex23-P Suit
M8 Paper
M9 Paper
Neuropsych patient treatment bag
Medical aid bag with basic physical and neurological exam inst.
ID Vests

NBC Medical Response Element Equipment Requirements

Trelborg Level B Suit
Kappler Training Suit
Trelborg Training Suit
Rebreather Masks
Advantage 1000
Filters
Tyvek Saranex23-P Suit
Disposable Surgical Mask
Disposable Gloves

Sample Collection Kits (water, entomology, industrial hygiene, occupational health)
Radiac meters
CAM
M8 Paper
M9 Paper
M272 Chemical agent water testing kit
Medical specimen collection and transport kits
Commo equipment
Muti-pax vans
ID Vests

Crisis/Stress Management Element Equipment Requirements

Ruggedized computer w/printer, FAX, and PC MCIA LAN
Communication Equipment
Hand held radios
Multi-pax vans
Expand-o Van (Cmd Ctr)
Expand-o Van (Commo Ctr)
Fields Tables
Traffic Cones
Barrier Material
Bull Horn
Engineer Tape
Light Sets
30k Generators
Porta-Johns
Refrigeration Trailer

Information and Planning Element Equipment Requirements

Copiers
FAX machines
Printers
Calculators
Clocks displaying local and Zulu time
Laptop or notebook-type personal computers
Computer work stations with file servers
Fax boards on file servers
Laser printers
Optical scanners
Display projector
Surge protectors
Backup power sources
Portable televisions with built-in Video Cassette Recorder (VCR)
Portable, battery-powered AM/FM radios
Flashlights with extra batteries
Wall charts/display boards
Maps/overlays
Supply of preformatted computer diskettes, ribbons and VCR tapes

Mortuary Affairs Elements Equipment Requirements

Personal Protective Equipment
Body Bags
Refrigerated trucks: A mass fatality situation may require additional refrigeration capacity to store remains until final disposition. Military units with this capability need to be identified and included in the DoD Resources Database. Operators need both awareness training and training in operating the equipment in a contaminated environment.

Communications Element Equipment Requirements

Mobile or transportable telecommunications equipment
Multichannel radio systems
Base station and hand held portables
Mobile or transportable microwave systems
Mobile or transportable switchboards and station equipment
Aircraft suitable as platforms for airborne radio repeaters
Trained installation and operations personnel available for deployment to the field
Naval ship(s) as appropriate to act as relay platforms

Endnotes

[1] http://www.defenselink.mil/pubs/wmdresponse/annex_f.html

E

HOMELAND SECURITY PROGRAM DEVELOPMENT TEAM

Project Director
Daniel Byram, MA

DANIEL BYRAM brings over twenty years of law enforcement experience to his role as educator. He has been involved for more than thirty years in program management, tactical security operations, intelligence operations, and law enforcement.

He has been widely involved in training and program development for law enforcement agencies and business, including the creation of a covert operations training program for law enforcement special operations personnel, the development of hostage rescue and crisis survival responses for covert operations personnel, and the designing of a "blueprint" for corporate terrorism response for the Insurance Education Association.

Daniel, with Dr. Julie Brown, designed the model for the homeland security degree programs offered in the Corinthian College network, and then he brought together experts from across the United States and Canada to develop the program materials.

Daniel holds a Masters Degree in Human Behavior and has provided over fifteen years of leadership in the post secondary educational experience. Mr. Byram is currently the National Director of Security, Justice and Legal Programs for Corinthian Colleges, Inc.

Team Leader
Jeff Hynes

JEFF HYNES has twenty-three years of experience with the Phoenix Police Department and currently serves as Commander. He has managed the advanced training for the Department in defensive tactics, firearms, driving, and physical fitness and wellness. He also coordinated and facilitated the yearly forty hour in-service advanced training tactical module for 2700 police officers, handles curriculum review, records keeping, and facilitates the yearly Citizen's Forum. Mr. Hynes has served as a liaison between the Phoenix Police Department and other federal, state and local agencies' proficiency related training at the Arizona Law Enforcement Training Academy.

Commander Hynes received his B.S. in Police Science and holds a Masters Degree in Educational Leadership. He is currently finishing his Doctorate in Education from Northern Arizona University and is an adjunct faculty member of several state and community colleges.

Jeff has also received numerous national, state and Phoenix Police Department Excellence Awards and nominations for Outstanding Community Based Policing Initiatives.

Team Leader
Jean Goodall

DR. JEAN GOODALL'S diverse career has been distinguished by the broad range of experiences she has had in her community that have brought her first-hand knowledge of the link between the criminal justice system and government operations. Throughout her efforts in this field, she has interlaced her strong background in public administration and teaching.

Following her study of criminology and public administration, she completed an MA in Criminal Justice. Jean Goodall has worked for over thirty years with legal issues in the criminal justice environment. She has also combined an MA in Management and a Doctorate in Public Administration with knowledge gleaned from numerous FBI, CBI, OEM and weapons seminars, camps and schools. She has received specialized FEMA training regarding Emergency Management and Homeland Security. This in-depth combination provides Dr. Goodall with a unique perspective on the role of the Civil and Criminal Justice System in Homeland Security issues.

Dr. Goodall holds a teaching certificate from the State of Colorado and has twelve years' experience in adult education at the university level. She recently served as the chairman of the Criminal Justice Department at Blair College, Colorado Springs, Colorado.

Team Leader
Julie Brown MD

If ready response to disasters is the mark of an exemplary security specialist, DR. JULIE BROWN'S career sets an excellent example. Whether she's been the physician assisting with refugees or a member of the crime-prevention posse for the sheriff's department of Maricopa County, Arizona, Julie has actively engaged in disaster management throughout her career.

An experienced professional in forensic pathology, law enforcement and disaster medicine, Julie is a registered nurse and became a physician in 1988. Dr. Brown serves on several federal disaster teams that respond to a variety of disasters.

Her areas of study include nursing, biology, psychology, human behavior, medicine and general business. She has a teaching certificate in biology, psychology and the medical sciences and has extensive experience as an educator at the college level. She received an MBA in General Business from the University of Phoenix. She formerly served as the program manager for business and accounting for Corinthian College.

Subject Matter Expert
Richard Wilmot, General (retired), United States Army

GENERAL RICHARD WILMOT comes from a varied and unusual career in both the government and private sectors. Before retiring from the Army, General Wilmot, a Vietnam War veteran, held several key positions in the defense sector as the Commanding General of the US Army Intelligence Center and School, the Director of Intelligence Systems in the Pentagon, and as a former Special Forces soldier.

Having been in 106 countries, his life story is replete with unusual true adventures. In Afghanistan he was an advisor to the Afghan rebels when they were fighting the Soviet Union in the mid-1980s. This foray alone has resulted in many interesting anecdotes, realizations, and a broad understanding of events that are tied to the international terrorist situation we now face.

Today, General Wilmot is a successful entrepreneur and an international businessman. He practices leadership in tense areas of the world where leading in crisis situations leaves no room for error. He is a motivator, a strategic planner, and the stories of his adventures provide international insight. He is a graduate of Michigan State University, the US Army War College, the Industrial College of the Armed Forces, the Command and General Staff College and he studied at Oxford University.

Subject Matter Expert
Jane Chung—Examiner, CSC LA Joint Drug Intelligence Group

The analysis of intelligence data is JANE CHUNG'S expertise. As an examiner for the Los Angeles Joint Drug Intelligence Group (LA JDIG), she provides analytical support for narcotics cases for the LA JDIG's Southwest Border Team. She researches commercial and law enforcement databases, evaluates and analyzes the data extracted and presents the findings to the case agent and other analysts.

Jane's experience with data analysis includes work in international intelligence. In Kosovo, she conducted over 800 personnel interviews to determine the threat level against US Forces. She has also developed intelligence threat and damage products focusing on foreign intelligence services, terrorist, paramilitary, law enforcement, political and criminal organizations. Jane managed the classified segment of the Migrated Defense Intelligence Threat Data System (MDITDS) database for the US Army Europe Analytical Control Element.

Jane graduated in 1998 from the University of California, Irvine with her B.A. in Criminology, Law and Society.

Subject Matter Expert
Stewart Kellock, Ost J CD Detective 897, Toronto Police Service

With twenty-six years of policing for the Toronto Police Service and serving with the military, STEWART KELLOCK offers a hands-on perspective to anti-terrorist intelligence. He has a strong and varied investigative background that includes work with a major crime unit, leading multi-unit investigations, investigating several major political incidents, plain-clothes experience, and work with Provincial Weapons Enforcement and Intelligence and is currently attached to the anti-terrorist unit of Intelligence Support.

Stewart has extensive military experience. He was commissioned in 1981 and currently holds the rank of Captain. Most recently he was the Leadership Company Commander at 32 CBG Battle School. Previously he was an Intelligence officer at LandForce Central Area Headquarters, which involved him in numerous intelligence operations both domestic and international. His most recent international mission was commanding the Regional Crime Squad as part of the contingent of the United Nation's Mission in Kosovo.

Stewart majored in International Terrorism at Humbar College and has completed numerous courses in military training and policing. He currently serves as the Unit Training Officer for the 53 Division Canadian Regional Unit.

Subject Matter Expert
James McShane MPA—Deputy Chief (retired)—Executive Officer, Narcotics Division New York City Police Department

JAMES MCSHANE was second in command of the 2,300 person Narcotics Division of the New York City Police Department where he was responsible for all narcotics enforcement activities in New York City, including all "Buy and Bust" operations, as well as the investigation of all narcotics complaints. He directed all major narcotic investigations in the City of New York.

James has a law degree and is a member of the Bar of the State of New York and has been admitted to practice in the U.S. Supreme Court and the Federal Courts of New York. He also holds permanent certification as a New York Secondary School Teacher and a New York City teacher's license. James taught Math and served as the Dean of a South Bronx High School where he taught, counseled students and adjudicated conflicts.

James received his BA in Communications from Fordham and was a Fulbright scholar and lecturer at the Police College of Finland in Helsinki. He attended the Police Management Institute at Columbia University School of Business, received a Master of Public Administration from Harvard University's Kennedy School of Government.

Subject Matter Expert
Bruce Tefft—Senior Associate, Orion Scientific Systems

BRUCE TEFFT is a well-seasoned intelligence investigator with 30 years of service in foreign affairs and intelligence operations as Headquarter's Branch Chief and field Chief of Station in the Central Intelligence Agency's Directorate of Operations. He served in several African countries, Europe, South Asia, and the Middle East. His multiple responsibilities and activities with the CIA varied from developing and teaching intelligence collection and analysis courses, and running intelligence collection and counter-terrorist operations against Islamic fundamentalists, to developing and implementing logistics and training programs for over several thousand U.S. and foreign personnel. He planned and organized the first joint CIA-FBI-US Military operation, successfully capturing a foreign terrorist.

Bruce has managed liaison relationships and operations with major Allied nation-intelligence organizations and the U.S. government departments such as State and Defense, the Federal Bureau of Investigation, the Drug Enforcement Agency, Defense Intelligence Agency, Defense HUMINT Service, and the U.S. Marine Corps.

He is currently a successful executive as the Senior Associate, Community Research Associates, and is the counter-terrorism advisor to the New York City Police Department. He has a B.A and a Master's degree in History, and in 1974 received his law degree from the University of Denver.

Subject Matter Expert
Lieutenant Colonel Xavier Stewart

Throughout the last twenty years, LTC STEWART has been widely recognized for distinguished service in his military career and in the field of healthcare. He joined the Army National Guard following the Marine Corps and is the Commander of a Weapons of Mass Destruction Civilian Support Team with the Pennsylvania National Guard. His military duties have included assignments within the Military Police, Military Intelligence, Physical Security, Military Academy and Medical arenas bringing him several prestigious honors including three Meritorious Service Metals and the Guarde Nationale Trophy for Outstanding Service. He is currently a member of the Executive Advisory Board for Homeland Security and has been recognized by Congress with a Congress Special Award.

With a Doctorate in Public Health, LTC Stewart has held numerous faculty positions in respiratory therapy programs, rehabilitation services, biology, physician assistant and nurse practitioner programs.

He has been recognized as one of the Top Ten Respiratory Care Practitioners by the AARC Journal and is listed in Who's Who Among College Professors, Who's Who in America and Who's Who in the World. LTC Stewart is a board certified Forensic Examiner and is Board Certified in Forensic Medicine. He is currently a first responder as a nationally registered EMT, HAZMAT firefighter and former deputy sheriff.

He is a graduate of Command and General Staff College and earned a Master's in Education with a Concentration in Health Services.

Subject Matter Expert
Christopher J. Wren

CHRISTOPHER WREN is a Security Specialist whose talents have repeatedly been tested in the field. During the Atlanta Olympic games, where he was responsible for athlete and venue security, he assisted in the design and evaluation of security plans for three high profile locations. While the Director of Security for a large downtown Phoenix hotel, he was responsible for completely overhauling the hotel security monitoring system to include state-of-the-art camera, motion sensors and sound monitors. Christopher's training in dignitary/VIP protection brought him a commendation from the White House Security Detail for his assistance in protection of the President and Vice-President of the United States.

As a commissioned law enforcement officer, Mr. Wren coordinated security efforts with Federal, state and local law enforcement agencies, and he has over four years' experience in the planning, set-up and supervision of threat assessment teams. He currently serves on the Homeland Defense Planning and Advisory Team for the City of Phoenix, Arizona.

Christopher served in the U.S. Marine Corps, receiving the rank of Meritorious Sergeant, and was awarded the Navy/Marine Corps Achievement Medal for Excellence. He has studied criminal justice and vocational education and has special training in advanced detective work, media relations, negotiation techniques for first responders and technical aspects of covert operations.

Subject Matter Expert
Dr. James Stanger

DR. JAMES STANGER, a prolific author and PhD, is the Director of Certification and Product Development at ProsoftTraining. His credentials include Symantec Technology Architect, Convergence Technology Professional, CIW Master Administrator, Linux+, A+ and he has led certification development efforts in these proficiencies for various organizations including ProsoftTraining, Symantec, and Linux Professional Institute. Dr. Stanger's specialties include network auditing, risk management, business continuity planning, intrusion detection and firewall configuration. He has coordinated audits for various clients, which have recently included Brigham Young University, Fuelzone.com and the William Blake Archive.

As an author, James has created titles for Symantec Education Services, designed executive training seminars concerning firewall and Virtual Private Network (VPN) management, and written other titles concerning security, Cisco routing and system administration for many companies.

In addition to his development work, James finds time to serve on several certification boards and advisory councils where, among his many responsibilities, he works to ensure that exams remain relevant and to protect certification exam intellectual property.

Subject Matter Expert
Matt Pope, CPP

MATT POPE is the founder of The Security, Integrity and Perception Standard, a private consultancy, which advises on the impact of global security and business integrity on government and economic stability. He also specializes in identifying and creating cutting-edge marketing trends, services and technologies to improve homeland security and public trust.

Matt has sixteen years' experience in business, public safety and military force protection. He is certified in professional security management, and holds a degree in political science. Most recently Matt has developed a specific expertise with contemporary issues of public security and integrity, privacy legislation, and security ethics. He has worked with some of the world's leading corporations on a broad range of security and emergency planning projects. Matt also serves as an adjunct instructor of homeland security and has written extensively on topics relating to security, law, contingency planning, ethics, privacy and legislation.

Subject Matter Expert
Lieutenant Kevin Kazmaier

With over 20 years of experience in law enforcement, LIEUTENANT KEVIN KAZMAIER has an extensive background in explosive devices, SWAT and special operations. A member of the International Association of Bomb Technicians and Investigators since his certification as a bomb technician in 1987, he has taught many courses in firearms and advanced explosive technique.

Lieutenant Kazmaier is a graduate of Protective Operations courses from the Association of Chiefs of Police, Secret Service Debriefings, the United States Army Military Police School and the Phoenix Police Department. He has worked as a consultant for America West Airlines and has provided security at such events as the Super Bowl, the Senior PGA Tour and the World Series.

Lieutenant Kazmaier holds an A.A.S. in Law Enforcement and a B.S. in Social Justice Professions.

Subject Matter Expert
Harold M. Spangler, MD

Dr. Spangler is Chief Resident of Emergency Medicine at North Carolina Baptist Hospital and Bowman-Gray School of Medicine. He is board certified in Emergency Medicine and holds licenses as an Advanced Cardiac Life Support Instructor, an Advanced Trauma Life Support Provider and as a Basic Trauma Life Support Instructor.

After receiving his B.S. degree in Biology, Dr. Spangler went on to receive honors throughout medical school graduating from Jefferson Medical College, Thomas Jefferson University in Philadelphia, Pennsylvania. He is a member of the American Medical Association, the American College of Emergency Physicians, the North Carolina Chapter of ACEP and the National Association of EMS Physicians.

Subject Matter Expert
Alan Pruitt CPP

Mr. Pruitt's extensive expertise has made him an integral part of the development of the Homeland Security Specialist Program at Bryman College in San Jose, California, where he currently serves as the Homeland Security Program Chair. His experiences in the field of intelligence investigation and gathering are diverse. They include his duties as a Marine Corps intelligence officer, his work in corporate security and his service as a licensed private investigator.

In the military, Alan served as a Counterintelligence Agent with the US Army National Guard and as a Counterintelligence Specialist with the U.S. Marine Corps. His military education introduced him to the skill of tactical intelligence photography and he completed courses in counterintelligence and qualified as an anti-terrorist instructor. He was awarded the Navy Commendation Medal from the Secretary of the Navy for superior achievements in security management.

A member of the California Association of Licensed Investigators and the Association of Certified Fraud Examiners, Alan has over 18,000 hours of investigative experience. His many corporate clients have included Paramount Pictures, the Department of Justice and the U.S. Customs Service.

He holds a B.S. in Business Management and is a member of the American Society of Law Enforcement Trainers and the American Society of Industrial Security.

HOMELAND SECURITY PROGRAM DEVELOPMENT TEAM

Subject Matter Expert
Steve Martin—Security Operations CEO

MR. MARTIN has over twenty-one years of international experience in the government and private sectors. His career highlights are wide-ranging in the fields of international security, communications, management, business and paralegal.

As a Special Agent with the Defense Department's National Security Agency, Steve faced the complexities of providing physical security for NSA/CSS personnel and facilities. His overall mission was to create and maintain security activities that detected and protected against acts of espionage, sabotage and terrorism. His accomplishments include authoring complex government policies and procedures, team leading for the NSA Strategic Planning Sessions, and coordinating actions and policies on counter-terrorism and counter-intelligence measures. He has trained security officers, managing many of the duties of a 500-man NSA police force. He lectures at home and abroad on security and advanced technology.

When Mr. Martin was with the NSA he maintained a liaison with his counterparts in other government agencies and private industry. In 1993, Mr. Martin founded a private, multi-division company of which he is part owner. From this vantage he is able to provide a very informed analysis of the critical need in homeland security for entrepreneurial enterprise.

Subject Matter Expert
Master Sergeant Rocky Dunlap

MASTER SERGEANT ROCKY DUNLAP retired as Program Manager and Inspector General Team Member for the Air Force Space Command Explosive Ordnance Disposal at Peterson Air Force Base. In this capacity, he oversaw command objectives relating to anti-terrorism issues, homeland defense initiatives, conventional, nuclear and biological improvised devices, weapons of mass destruction and reducing the vulnerability of Air Force Space Command installations within the U.S. and abroad. He was responsible for six major installations and nine remote sites worldwide. Master Sergeant Dunlop established the first regional post-9-11 bomb squad in the Department of Defense to combat weapons of mass destruction. He led NASA's pyrotechnic recovery operations of the Columbia Space shuttle disaster.

He is a graduate of many military courses that deal with weaponry, terrorism and disaster control. He completed courses in anti-terrorism, weapons and ordinance disposal, chemical and biological school, counter insurgency, nuclear, and HAZMAT first responders and has studied to be a post-blast investigator.

Master Sergeant Dunlop has completed a BS in Workforce Education and Development, an AS in Explosive Technology and currently holds the highest certification possible for a Department of Defense bomb Technician.

Subject Matter Expert
Mr. William Oberholtzer

WILLIAM OBERHOLTZER, a military consultant with Vector Incorporated, provides expertise on weapons of mass destruction to a variety of security and defense teams. As Chief of Weapons of Mass Destruction Counter Technology Integration, Mr. Oberholtzer provides the National Guard, civilian emergency response agencies and those of the first responder community with recommendations on organization, training and equipment alterations best suited to meet unit requirements.

Credited with making major contributions toward the establishment of the Nation's premier Weapons of Mass Destruction Civil Support Team, he also developed the organization and curriculum requirements for the National Weapons of Mass Destruction and Counter Terrorism Training and Simulation Center in response to initiatives pertaining to homeland defense.

William has a B.S. in Education Administration and Master Degrees in Management and Human Relations. He graduated from Defense Systems Management College and has many FEMA courses and certifications to his credit, including: Emergency Program Manager and Emergency Preparedness, Radiological Emergency Management and Response, Hazardous Material, and the Role of the Emergency Operations Center.

Subject Matter Expert
Dr. Victor Herbert

DR. VICTOR HERBERT has dedicated a lifetime to teaching and administration. He is currently Dean of Instruction for the New York City Fire Department (FDNY) where he coordinates all training for FDNY personnel and directs FDNYC, AmeriCorp, and Fire Safety Education. He came to this position following his work in public and higher education.

With a master's degree in English Education and a Doctorate in Educational Leadership, he taught English and Spanish and went on to win the Fund for the City of New York's award as Educator of the Year in 1983. After obtaining a number of professional certificates in educational administration and Spanish, he received a Fulbright Award to study relationships between Mexico and the United States. He has traveled extensively throughout Latin America.

Besides his teaching positions, Dr. Herbert has assumed many roles in public school administration working as department chair, principal and school superintendent in school districts in New York,

Arizona and Connecticut. Dr. Herbert has been an associate faculty member at Chapman University, Arizona State University, Norwalk Community College and St. Joseph's College in New York. His expertise has enabled him to teach both technical and academic subjects. He also has directed several programs for emergency responders in the acquisition of Spanish.

ENDNOTES

Chapter 1—Planning as a Front-End Concept

1. EM-DAT: The OFDA/CRED International Disaster Database. http://www.em-dat.net, UCL-Brussels, Belgium. www.unisdr.org.
2. FEMA, "2005 Federal Disasters Declared," Declared Disasters Archives, 2005. http://www.fema.gov/news/disasters.fema FEMA 500 C Street, SW Washington, D.C. 20472 Phone: (202) 566-1600
3. FEMA, Virtual Library & Electronic Reading Room, News Media, Top Ten Natural Disasters, March 2005. http://www.fema.gov/library/df_8.shtm. FEMA 500 C Street, SW Washington, D.C. 20472 Phone: (202) 566-1600
4. United Nations Development Programme.
5. UNDP/World Bank/FAO, "Joint Tsunami Disaster Assessment Mission," 4–8 January 2005, Livelihood Recovery & Environmental Rehabilitation: Thailand, 10 January 2005. www.undp.or.th/tsunami/documents/MissionReport-UNDP-WB-FAO4-8Jan05_001.doc
6. FEMA, "MITIGATION—An INVESTMENT for the FUTURE: Quantifying the Post-Disaster Benefits of Three Elevated Homes in Baldwin County, Alabama," Prepared by FEMA Region IV Mitigation Division in cooperation with Alabama Emergency Management Agency and Baldwin County Emergency Management Agency, 10 February 1999. FEMA 500 C Street, SW Washington, D.C. 20472 Phone: (202) 566-1600
7. www.faqs.fema.gov—FEMA 500 C Street, SW Washington, D.C. 20472 Phone: (202) 566-1600
8. FEMA, Education & Training, Multi-Hazards, HAZUS-MH Flood Loss Estimation Models, Loss Estimation Software, http://www.fema.gov/hazus/fl_main.shtm. FEMA 500 C Street, SW Washington, D.C. 20472 Phone: (202) 566-1600.
9. READYBusiness, U.S. Department of Homeland Security website, "Plan to Stay in Business: Continuity of Operations Planning." http://www.ready.gov/business/st1-planning.html
10. READYBusiness, U.S. Department of Homeland Security website, "Plan to Stay in Business: Emergency Planning for Employees." http://www.ready.gov/business/st1-empwellbeing.html

11. READYBusiness, U.S. Department of Homeland Security website, "Plan to Stay in Business: Emergency Supplies." http://www.ready.gov/business/st1-emersupply.html
12. READYBusiness, U.S. Department of Homeland Security website, "Protect Your Investment: Review Insurance Coverage." http://www.ready.gov/business/st3-reviewins.html
13. READYBusiness, U.S. Department of Homeland Security website, "Protect Your Investment: Prepare for Utility Disruptions." http://www.ready.gov/business/st3-prepareutility.html
14. READYBusiness, U.S. Department of Homeland Security website, "Protect Your Investment: Secure Facilities, Buildings and Plants." http://www.ready.gov/business/st3-securefac.html
15. READYBusiness, U.S. Department of Homeland Security website, "Protect Your Investment: Secure Your Equipment." http://www.ready.gov/business/st3-secureequip.html
16. READYBusiness, U.S. Department of Homeland Security website, "Protect Your Investment: Assess Building Air Protection." http://www.ready.gov/business/st3-assessair.html
17. READYBusiness, U.S. Department of Homeland Security website, "Protect Your Investment: Improve Cyber Security." http://www.ready.gov/business/st3-improvecyber.html

Chapter 2—Key Elements in Response Protocol

1. TEXAS FOREST SERVICE, PROJECT DIRECTOR, UNITED STATES FOREST SERVICE, REGION 8 (CO-SPONSOR), FEDERAL EMERGENCY MANAGEMENT AGENCY, REGION VI (CO-SPONSOR), "INCIDENT MANAGEMENT TEAM ALL-RISK OPERATIONS AND MANAGEMENT STUDY," August, 2003, Prepared by Dr. Amy K. Donahue, Principal Investigator, Center for Policy Analysis and Management, Institute of Public Affairs, University of Connecticut, 421 Whitney Road Unit 1106, Storrs CT 06269-1106, (860) 486-4519. http://www.myfirecommunity.net/documents/IMT_Shuttle_Response_FINAL.pdf
2. National Commission on Terrorist Attacks Upon the United States, Chapter 9, "Heroism and Horror," http://www.9-11commission.gov/report/911Report_Ch9.htm; this website was frozen on September 20, 2004, at 12:00am EDT. It is now a Federal record managed by the National Archives and Records Administration, pp. 22-28.
3. Federation of American Scientists, "Federal Radiological Emergency Response Plan (FRERP)-Operational Plan," Published May 1, 1996. This plan, signed by FEMA Director James L. Witt on May 1, 1996, was originally published in the Federal Register as a Notice, dated May 8, 1996, Part III, pp. 20944-20970. http://www.fas.org/nuke/guide/usa/doctrine/national/frerp.htm#_1_1

Chapter 3—Planning with Other Agencies

1. City of Los Angeles, Emergency Preparedness Department, Emergency Operations Master Plan and Procedures, Part V, "Multi-Agency Coordination," Revised 9/96. http://www.lacity.org/epd/pdf/eompp/part5.pdf
2. Los Angeles County, Chief Administrative Office, "What Is Sems*?" (*California Government Code Section 8607, effective January 1, 1993). http://www.lacoa.org/sems.shtml
3. Kern County, California, Kern County Office of Emergency Services, Executive Overview of SEMS http://www.co.kern.ca.us/fire/oes/overview1.htm
4. New York State, State Emergency Management Office. http://www.nysemo.state.ny.us/TRAINING/ICS/explain.htm
5. Northern Virginia EMS Council, "VIRGINIA MASS CASUALTY INCIDENT PLAN," July 1, 2004-June 30, 2005, http://www.northern.vaems.org/Documents/No%20Va%20MCI%20 Plan.pdf
6. U.S. Environmental Protection Agency, Emergency Response Program, "Inside the Emergency Response Program," http://www.northern.vaems.org/Documents/No%20Va%20MCI%20Plan.pdf
7. CDC, NIOSH (National Institute for Occupational Safety and Health, Protecting Emergency Responders, Vol. 3, "Safety Management in Disaster and Terrorism Response, NIOSH Publication No. 2004-144, http://www.cdc.gov/niosh/docs/2004-144/chap1.html
8. U.S. Department of Homeland Security, Fact Sheet, NATIONAL RESPONSE PLAN (NRP), January 6, 2005. http://www.dhs.gov/interweb/assetlibrary/NRP_FactSheet_2005.pdf

Chapter 4—Lookout, Communication, Escape, and Safety (LCES)

1. City of Boulder, Fire-Rescue, 2004 Wildland Fire Management Standard Operations Plan, Personnel and Equipment Mobilization Guide, prepared by Marc R. Mullenix, Division Chief, Wildland Fire. http://www.ci.boulder.co.us/fire/wfire/MOB.htm
2. "LCES and Other Thoughts," by Paul Gleason, June, 1991. http://www.wildlandfire.com/docs/gleason/lces.htm
3. National Wildfire Coordinating Group, Fireline Handbook, March, 2004, NWCG Handbook 3, PMS 410-1, NFES 0065, http://www.nwcg.gov/pms/pubs/410-1/410-1.pdf

Chapter 5—Logistics

1. FEMA, Tool Kit for Managing the Emergency Consequences of Terrorist Incidents, Interim Planning Guide for State and Local Governments, Unit XI, "Roles and Responsibilities in a Terrorist Incident." July 2002. http://www.fema.gov/txt/onp/toolkit_unit_11.txt

2. New York City, Office of Emergency Management, "OEM Staff Profiles: Field Operations and Logistics," NYC.gov website: http://www.nyc.gov/html/oem/html/other/sub_news_pages/911/profiles_oplog.html
3. National Incident Management System-Virginia, Components of the ICS, "Predesignated Incident Facilities," http://www.vdfp.state.va.us/components.htm#facilities
4. DMORT, Disaster Mortuary Operational Response Teams, "Incident Morgue Requirements," http://www.dmort.org/DNPages/DMORTDPMU.htm.
5. Federal Bureau of Investigation website, "Disaster Squad Marks a Milestone." http://www.fbi.gov/hq/lab/disaster/disaster.htm
6. GFN (Genesee FreeNet-nonprofit organization providing Internet service for the Flint and Genesee County communities, "Triage and Disaster Planning." http://www.gfn.org/emsregb/shiawasee/TRIAGE.htm
7. El Paso County, Colorado, Sheriff, "Preparing for the Media Mega-Event," by Ken Hilte, http://shr.elpasoco.com/media_event.asp
8. Statement of the American Hospital Association before the National Committee on Vital and Health Statistics Panel on National Preparedness and a National Health Information Infrastructure, February 26, 2002, http://www.ncvhs.hhs.gov/020226p1.htm
9. U.S. Department of Homeland Security, National Disaster Medical System, "What is a Disaster Medical Assistance Team (DMAT)?," http://ndms.dhhs.gov/dmat.html
10. http://egov.cityofchicago.org/city/webportal/
11. Washington State Department of Health, Office of Emergency Medical Services and Trauma System, Emergency Medical Services Communications,"Concepts of EMS Communications," http://www.doh.wa.gov/hsqa/emstrauma/communic.htm
12. Council on Foreign Relations, "Emergency Responders: Drastically Underfunded, Dangerously Unprepared," by Richard A. Clarke, et al., Publication #6086, June 29, 2003. http://www1.cfr.org/pub6086/press_release/publication.php?id=6085
13. City of Sunnyvale, California, Emergency Preparation, "Project ARK," http://sunnyvale.ca.gov/Departments/Public+Safety/Emergency+Preparedness/Volunteer+Opportunities/Project+ARK.htm
14. City of Los Angeles, Emergency Preparedness Department, Emergency Operations Master Plan and Procedures, Part IV, "Primary Emergency Operations Center," Revised 9/96. http://www.lacity.org/epd/pdf/eompp/part4.pdf
15. FEMA, Emergency Personnel, Response and Recovery, "Mobile Operations Capability Guide for Emergency Managers and Planners." FEMA, 500 C Street, SW Washington, DC., 20472, http://www.fema.gov/rrr/mers01.shtm

Chapter 6—First Responder Responsibilities

1. New York State, Department of Health, "Job Description: Certified First Responder," http://www.health.state.ny.us/nysdoh/ems/pdf/srgcfr.pdf
2. http://www.mfbb.vic.gov.au/default.asp?casid=154
3. Town of Upton, Massachusetts, US., Inter-Agency, Incident Command System, Chapter 8, "Roles of the other Emergency Services." http://www.upton.ma.us/html/inter-agency.html

Chapter 7—Emergency Operations Planning (EOP) vs. Incident Management System.

1. FEMA, Emergency Personnel, "National Incident Management Systems [NIMS]," at http://www.fema.gov/nims/
2. Definitions of terms and acronyms used in this document are given in Tabs F and G, respectively.
3. Table 5 provides an overview of events likely to occur in a WMD incident. It is designed to help planners better understand the interface that State and local response will likely have with Federal response organizations. The table includes both crisis management and consequences management activities that would be operating in parallel and is intended to illustrate the complex constellation of responses that would be involved in a WMD incident.

* For facilities or materials regulated by the Nuclear Regulatory Commission (NRC), or by an NRC Agreement State, the technical response is led by NRC as the LFA (in accordance with the Federal Radiological Emergency Response Plan) and supported by DOE as needed.

Chapter 8—Mass Casualty/Fatality Planning

1. The Alberta College of Paramedics, "Multiple Casualty Incidents," 2001, http://www.collegeofparamedics.org/con_ed/2001/mci.htm
2. U.S. Department of Homeland Security, Disaster Mortuary Operational Response Team, August 2004 DMORT News, http://www.dmort.org/news/August2004.htm
3. *Basic Trauma Life Support for Paramedics and Other Advanced Providers*, John E. Campbell, 4th ed., Prentice Hall, 2000, ISBN 0130845841.

Chapter 9—Federal Response Planning (FRP)

1. U.S. Department of Homeland Security, National Response Plan, Press Release, "Fact Sheet: National Response Plan," http://www.dhs.gov/dhspublic/interapp/press_release/press_release_0581.xml
2. FEMA, Response and Recovery, "Concept of Operations," http://www.fema.gov/rrr/conplan/conpln4.shtm

Chapter 10—On-Scene Complication Planning

1. http://www.lacity.org/epd/pdf/eompp/
2. City of Upton, Massachusetts, Incident Command System, "Psychology of Command," http://www.upton.ma.us/html/_psychology_of_command.html
3. City of Upton, Massachusetts, Incident Command System, "Legal Considerations," Appendix A2, http://www.upton.ma.us/html/legal_considerations.html
4. City of Upton, Massachusetts, Incident Command System, Chapter 4 "Incident Risk Management," Chapter 4, http://www.upton.ma.us/html/risk_management.html
5. Disaster Management, "Lost Innocents," Mike Byrd, Miami-Dade Police Department, Crime Scene Investigations, http://www.crime-scene-investigator.net/disaster.html
6. SARS and Public Health, *Renewal of Public Health in Canada*, Chapter 9, "Some Legal and Ethical Issues Raised by SARS and Infectious Diseases in Canada."

APPENDICES

A.

Greater Metropolitan Washington Area Mutual Aid Operation Plan (MAOP) *Northern Virginia Mass Casualty Incident Plan*, July 2, 2004–June 30, 2005. http://www.northern.vaems.org/Documents/No%20Va%20MCI%20Plan.pdf

B.

FEMA, Tool Kit for Managing the Emergency Consequences of Terrorist Incidents, July 2002, pp. XI-74. http://www.fema.gov/txt/onp/toolkit_toc.txt

C.

Navy Environmental Health Center, Disaster Preparedness Guide for Incidents involving Chemical, Biological, Radiological and Environmental (CBRE) Agents. Navy Environmental Health Center, 620 John Paul Jones Cir Ste 1100, Portsmouth, VA 23708-2103, Tel. 757 953 0700. http://www-nehc.med.navy.mil/plansops/Disaster_Preparedness.htm

D.

United States Department of Defense, Department of Defense Plan for Integrating National Guard and Reserve Component Support for Response to Attacks Using Weapons of Mass Destruction, "Annex F: Equipment Requirements," prepared by Department of Defense, January 1998. http://www.defenselink.mil/pubs/wmdresponse/annex_f.html